Flame and Combustion

Flame and Combustion

Third edition

J. F. Griffiths
Reader in Physical Chemistry,
University of Leeds

and

J. A. Barnard
Sometime Newton Drew Professor of
Chemical Engineering and Fuel Technology,
University of Sheffield

CRC Press is an imprint of the
Taylor & Francis Group, an **informa** business

CRC Press
Taylor & Francis Group
6000 Broken Sound Parkway NW, Suite 300
Boca Raton, FL 33487-2742

Visit the Taylor & Francis Web site at
http://www.taylorandfrancis.com

and the CRC Press Web site at
http://www.crcpress.com

Preface

The first edition of this text, entitled *Flame and Combustion Phenomena*, was published by Professor John Bradley in 1969. Subsequent to John Bradley's untimely death, the second edition, *Flame and Combustion*, was published in 1985 by Professor John Barnard. The intention of my predecessors was that the book should be suitable for final year undergraduates and as an introductory book about combustion phenomena for those involved in research and development in a wide range of disciplines. It is my hope that the same is true of this third edition, with particular attention paid to chemical aspects.

The potential market for an introductory text has changed considerably since the appearance of John Bradley's monograph. There has been a considerable growth in concern for efficiency, safety and minimisation of the environmental impact of combustion, whereas the development of rocket fuels, explosives and propellants do not command the same intensity of effort as formerly. Thus, it seems prudent to shift the emphasis from some parts of the earlier texts, and to expand others that are more in line with current combustion activities.

By its very nature, combustion is a multidisciplinary subject, embracing chemistry, engineering, mathematics and physics. Communication amongst the disciplines has always been excellent, but it is extremely important to acquire one's own foundation of understanding as a basis for fostering that communication. It is hoped that the revisions within this third edition not only contribute to the wider knowledge base that is now required by combustion scientists but also help to cultivate an interest in the subject and offer the foundation for more peripheral needs amongst environmental scientists, safety engineers or other specialist groups.

It is my privilege to have been given the opportunity to revise the text of *Flame and Combustion*, since I have enjoyed a close and long-standing friendship with both of my predecessors. One connection began when I was an undergraduate at Liverpool University, during which time I was taught by John Bradley. The other started at the end of my PhD studentship; John Barnard was my external examiner. Our respective paths became intertwined and friendships blossomed through the common links in combustion research during the years prior to John

Bradley's death and to John Barnard's retirement from Sheffield University.

My own contribution to this book would not have been possible without the the inestimable guidance and encouragement from the late Dr Charles Tipper and Dr Geoffrey Skirrow in the early years, and from Professor Peter Gray throughout most of the years thereafter. Concerning this edition, Des O'Hara prepared most of the diagrams. Advice was sought from and most willingly given by Professors Derek Bradley, Peter Gray and Chris Sheppard, and Drs Allan Hayhurst, Andy McIntosh, Steve Scott and David Smith. I am most grateful to each of them, and I bear full responsibility for any misrepresentation or misunderstanding of their advice. I hope that the end-product does not disappoint these mentors, John Barnard, or you, the reader.

J.F.G.
November 1994

Glossary

List of Symbols

A	Pre-exponential factor in rate constant	
A	Area	
AIT	Minimum autoignition temperature	(Chapter 12)
A_B	Burner tube area	(Chapter 3)
A_F	Orifice area for gas inlet in burner tube	(Chapter 3)
a	Sound speed	(Chapter 5)
a	Thermal diffusivity	
a	Coefficient of thermal expansion	(Chapter 8)
a	Radius of droplet or particle	(Chapter 11)
B	Dimensionless exothermicity	(Chapter 5)
B	Dimensionless adiabatic temperature excess	
B'	Dimensionless maximum temperature excess at a surface	(Chapter 10)
$B^{\bullet}$	Dimensionless maximum temperature in a non-adiabatic CSTR	(Chapter 10)
b	Biot number	(Chapter 8)
b.d.c.	Bottom dead centre	(Chapter 13)
CR	Compression ratio	(Chapter 13)
CCR	Critical compression ratio	(Chapter 13)
C_p	Molar heat capacity at constant pressure	
C_v	Molar heat capacity at constant volume	
C	Heat capacity	
c	Concentration	
$\bar{c}$	Mean velocity	
$\boldsymbol{D}$	Detonation velocity	(Chapter 5)
Da	Damkohler number	(Chapter 4)
D	Diffusion coefficient	
d_T	Flame quenching diameter in tube	(Chapter 3)
d_Q	Flame quenching distance between parallel plates	(Chapter 3)
d	Diameter of tube	
E	Activation energy	
E_i	Ionisation energy	(Chapter 6)

e	Internal energy	
exp	2.71828	
G	Gibbs function (molar free energy)	
g	Gravitational acceleration	(Chapter 8)
g_F	Critical gradient at flash-back limit	(Chapter 3)
g_B	Critical gradient at blow-off limit	(Chapter 3)
H	Enthalpy	
ΔH_{vap}	Latent heat of vaporisation	
h	Enthalpy per unit mass	(Chapter 5)
h	Heat transfer coefficient	
$\boldsymbol{h}$	Planck constant	
K	Equilibrium constant	
K	Karlovitz stretch factor	(Chapter 4)
Ka	Dimensionless Karlovitz stretch factor	
$\boldsymbol{k}$	Boltzmann constant	
k	Rate constant	
k_b	Burning rate constant	(Chapter 11)
L	Integral length scale	
$\boldsymbol{L}$	Rate of heat loss	
Le	Lewis number	
l	Length	
l	Heat transfer per unit volume	(Chapter 9)
l_k	Kolmogorov length scale of turbulence	(Chapter 4)
$\dot{M}$	Molar flux	(Chapter 11)
M_a	Mach number	(Chapter 5)
M_r	Relative molar mass	
MON	Motor octane number	(Chapter 13)
m	Mass	
m	Coefficient in chain branching reaction	(Chapter 9)
$\dot{m}$	Mass flow rate	
N	Molar density	
n	Number density of species	
p	Pressure	
ppm	Parts per million	
q	Exothermicity	
R	Gas constant ($=8.314\ \mathrm{J\,mol^{-1}\,K^{-1}}$)	
$\boldsymbol{R}$	Rate of heat release	
R'	Gas constant/unit mass	(Chapter 5)
Ra	Rayleigh number	(Chapter 8)
Re	Reynolds number	
RON	Research octane number	(Chapter 13)
r	Radius of sphere or characteristic dimension of solid body	
r	Rate of wave propagation	(Chapter 4)
r_f	Flame radius	

S	Molar entropy	
S	Surface area	
S/V	Surface/volume ratio	
S_b	Burning velocity with respect to burned gas	
S_u	Burning velocity with respect to unburned gas	
S_l	Laminar burning velocity	
S_t	Turbulent burning velocity	
T	Temperature	
t.d.c.	Top dead centre	(Chapter 13)
t	Time	
t_{th}	Characteristic convective heat transport time from a hot surface	(Chapter 10)
t_{tr}	Characteristic convective mass transport time from a hot surface	(Chapter 10)
t_{res}	Mean residence time in a CSTR	
t_N	Newtonian cooling time	
t_{ch}	Characteristic chemical time	
$t_{1/2}$	Half-life	
U	Internal energy	
u	Flow velocity	
u_{mf}	Minimum fluidisation velocity	(Chapter 11)
u'	rms turbulence velocity	(Chapter 4)
u'_k	Effective rms turbulence velocity	(Chapters 4 and 13)
u_s	Shock velocity	(Chapter 5)
v	Reaction rate	
V_i	Initiation reaction rate	(Chapter 10)
V	Volume or specific volume	
$\dot{w}$	Particle velocity	(Chapter 5)
w	mol kg^{-1} of material	(Chapter 10)
w	Amount	
x	Linear dimension	
x	Mole fraction	
x, y, z	Cartesian co-ordinates	
y, z	Stoichiometric coefficients	(Chapter 9)
∇^2	Laplacian operator	
[X]	molar concentration of species X	

Greek Letters

α	Fractional extent of reaction	(Chapter 10)
α	Area ratio ($=A_F/A_B$)	(Chapter 3)
α	Coefficient of expansion	(Chapter 8)
β	Surface mass transfer coefficient	(Chapter 10)
β	Relative efficiency of third body	(Chapter 9)
γ	Ratio of principal heat capacities (C_p/C_v)	
Γ	Specific volume ratio	(Chapter 5)

δ	Flame thickness	(Chapter 4)
δ	Frank-Kamenetskii parameter	(Chapter 8)
ε	Dimensionless activation energy ($= RT_a/E$)	
ε	Eddy dissipation term	(Chapter 4)
η	Dynamic viscosity	
η	Engine efficiency	(Chapter 13)
θ	Dimensionless temperature excess ($=(E(T-T_a)/RT_a^2)$)	
Θ	Dimensionless temperature excess ($=(E(T-T_b)/RT_b^2)$)	(Chapter 4)
κ	Thermal conductivity	
λ	Mean free path	(Chapter 2)
λ	Taylor microscale of turbulence	(Chapter 4)
λ	Wavelength of electromagnatic radiation	
λ	Root of characteristic equation	(Chapter 10)
ν	Frequency of electromagnetic radiation	
π	Pressure ratio	(Chapter 5)
Π	Spectroscopic term symbol	(Chapter 6)
ρ	Molecular diameter	
σ	Molar density or density	
Σ	Spectroscopic term symbol	(Chapter 6)
τ	Dimensionless time	(Chapter 8)
τ	Period of oscillatory ignition	(Chapter 9)
υ	Kinematic viscosity	
ν	Stoichiometric coefficient	(Chapter 2)
ϕ	Fuel to air ratio (Stoichiometric ratio, $\phi = 1$)	
ϕ	Net branching factor	(Chapter 9)
ψ	Semenov parameter	

Subscripts

a	Ambient conditions	
ad	Adiabatic	
b	Backward	
b	Burned gas	
c	Critical condition	
ch	Chemical	
ext	Extinction transition	
e	Electron	
f	Flame surface	(Chapters 4 and 11)
f	Forward	
g	Spectroscopic term symbol	
F	Fuel	
ign	Ignition transition	
l	Laminar	
M	Mixture	

N	Newtonian cooling
o	Initial condition
ox	Oxidant
P	Product
res	Mean residence time in vessel
S	Surface
s	Stable stationary state
t	Turbulence
th	Heat transfer
tr	Mass transfer
u	Unstable stationary state
u	Unburned gas
p, s, t	Primary, secondary, tertiary (C–H bonds in hydrocarbons)
i, b, p, t	Initiation, branching, propagation, termination (rate constants in chain reactions)
1	Region of undisturbed gas
2	Region behind shock or detonation front
∞	Infinity
+	Positive ion

Superscripts

n	Temperature coefficient in reaction rate constant
$\ominus$	Standard state
+	Positive ion
−	Negative ion

A time derivative is noted by an overdot, e.g. $\dot{T} = \mathrm{d}T/\mathrm{d}t$

100 kPa = 1 bar = 0.987 atm

Contents

Introduction 1

1.1 Introduction and background

Combustion provided early man with his first practical source of energy; it gave him warmth and light, it extended the range of foodstuffs which he could digest, and it enabled him to 'work' metals. Throughout the world today, combustion still provides more than 95% of the energy consumed and, despite the continuing search for alternative energy sources, there is little doubt that combustion will remain important for a very considerable time to come. The use of combustion fuels remains especially attractive where convenient 'energy storage' is required, as in transport applications, for example.

It is becoming increasingly important to ensure that combustion processes are utilised in the most efficient manner. Foremost are the need to minimise waste of energy, to avoid unnecessary emissions of carbon dioxide as a contributor to the 'greenhouse effect', and to minimise adverse effects on the environment, which may arise through pollutant emissions. Reserves of fossil fuels are also becoming depleted to varying extents, so that fuels themselves are becoming less accessible, and more expensive to recover. Thus, the study of combustion processes is an important area of scientific endeavour, and it is destined to remain so having regard to the requirements of society and to the demands of national and international legislation which control the environmental impact.

Combustion phenomena arise from the interaction of chemical and physical processes. The heat release originates in chemical reactions, but its exploitation in combustion involves heat transport processes and fluid motion. Thus, the theoretical interpretation draws heavily on physics, fluid mechanics and applied mathematics. Numerical analysis often forms a vital bridgehead between experiment and theory.

The aim of this book is to introduce many of the different aspects of combustion and its environmental impact, and to address the interdisciplinary nature of combustion at a sufficiently elementary level that an understanding may be gained by readers from a range of academic and technical backgrounds. In order to concentrate the information in a manageable way, much detailed material has been omitted and experimental evidence in support of scientific arguments has been excluded. Generalisations sometimes have to be made to maintain clarity. The reader is asked to bear these points in mind, in the knowledge that more specialised texts are available which address each topic in considerably greater depth. There is an extremely rich 'combustion literature' available which comprises learned journals and conference proceedings.

The most comprehensive coverage from conferences is to be found in the proceedings of the biennial 'International Combustion Symposia of The Combustion Institute'. However, more selective material can be drawn from the proceedings of international conferences organised regularly by such bodies as The Institution of Mechanical Engineers (IMechE), The Society of Automotive Engineers (SAE) and the Society for Industrial Applied Mathematics (SIAM). The combustion orientated journals which cover the subject widely are led by *Combustion and Flame*, *Combustion Science and Technology*, and *Combustion, Explosion and Shock Waves*. Combustion topics are reviewed in *Progress in Energy and Combustion Science*. More specialised areas are addressed in other journals, such as *Fuel*, *Fire Safety Engineering*, *Fire and Materials* or the *Journal of Loss Prevention*. There is a wealth of information about combustion to be drawn from many chemistry, physics, engineering or mathematics series. The intention of the references and suggestions for further reading given here is to provide a route into the literature related to particular topics.

1.2 Survey of combustion phenomena

Combustion begins in chemistry with a self-supported, exothermic reaction. The physical processes involved are principally those which pertain to transport of material and energy. The conduction of heat, the diffusion of chemical species and the bulk flow of gas may all develop from the release of chemical energy in an exothermic reaction as a consequence of the thermal and concentration gradients that are set up in the vicinity of the reaction zone. It is the interaction of these various processes that leads to the phenomenon that is observed. Other effects, such as light emission, depend on specific chemical processes which may have only a negligible bearing on the main phenomenon.

The chemical reaction usually involves two components, one of which is termed the *fuel* and the other the *oxidant* (normally air), because of the part each plays in the reaction. The simplest circumstances for combustion to take place are when the two gaseous, premixed components, are introduced to a container maintained at a uniformly controlled temperature. If the vessel is hot enough, measurable exothermic *oxidation* of the fuel will occur. If the heat produced by the reaction is transported sufficiently rapidly to the container walls by conduction and convection, a steady (or *stationary state*) reaction is maintained. This balance of the heat release and loss rates, such that the reaction proceeds smoothly to completion, is usually referred to as *slow reaction* or *slow oxidation*, although there is no absolute criterion of reaction rate alone from which we could answer the question: 'How slow is slow?'.

Above a certain temperature of the container, which depends on the physical properties of the reactants and the size and shape of the container, the rate of energy release from the chemical reaction may exceed the rate at which it can be transported to the vessel walls by the various heat transfer processes. The temperature within the system then increases, and the rate of reaction (and therefore the rate of heat release) also increases. This acceleration of reaction rate leads to a further increase in temperature, and the combined effects culminate in an *explosion*. The term 'explosion' refers to the violent increase in pressure which must accompany the rapid self-acceleration of reaction, usually manifest physically by its damaging consequences.

This type of explosion is driven solely by the rate of energy release through *thermal feedback*. The state of self-acceleration is termed *ignition* and the phenomenon described here is, therefore, called a *thermal ignition* or *thermal explosion*. Thermal ignition will be discussed in more quantitative detail in Chapter 8. The present description illustrates, perhaps in the simplest way of all, how an interaction between the 'physics' (i.e. the heat transport processes) and the 'chemistry' (determined here solely by the rate of heat of release and how it is affected by the temperature) governs whether or not explosion will occur. The underlying properties that make such an event possible are that reaction rates normally increase exponentially with temperature, whereas the heat transfer often depends almost linearly on temperature.

In a propagating combustion wave, called a *deflagration* or *flame*, reaction is initiated by a spark or other energy stimulus. Reaction is then induced in the layer of reactant mixture ahead of the *flame front* by two possible mechanisms, that is by *heat conduction* or by *diffusion* of reactive species from the hot burned gas or *reaction zone* behind the flame front. Thus, the thermal and chemical reaction properties of the combustion system may still drive the reaction but now there is a spatial structure, and both the heat and mass transport processes have to be described within that framework.

If the premixed reactants are forced to flow towards the flame front, and their velocity is equal to the rate at which the flame would propagate into stagnant gas, i.e. the *burning velocity*, the flame itself would come to a standstill. This is put into practice in combustion applications involving burners, the design of the appliance being aimed at holding the flame in one position and render it stable towards small disturbances.

An alternative to the *premixed flame* is the *diffusion flame*, in which the separate streams of fuel and oxidant are brought together and reaction takes place at their interface. The candle flame must surely be the commonplace example, although the description of it is complicated by the role of heat from the flame acting in a supplementary context of causing melting and evaporation of the wax which then burns in the gas-phase. The supply of air to the reaction zone is sustained by the convection currents set up by the flame itself. This flow also provides cooling to the sides of the cup of the melted fuel [1]. Flame processes are introduced in Chapter 3 and combustion in turbulent flames is discussed in Chapter 4.

The velocities of premixed flames are limited by transport processes, for example, heat conduction and species diffusion. The velocities cannot exceed the speed of sound in the reactant gas. However, it is often found that a propagating combustion wave undergoes a transition to a quite different type of wave, a *detonation wave*, which travels at a velocity much higher than the speed of sound. In this type of wave the chemical reaction is initiated by a supersonic compression, or *shock wave*, travelling through the reactants. The chemical energy that is released in the hot, compressed gases behind the shock front provides the driving force for the shock wave. It is necessary to consider the chemistry of the system only to the extent that it provides a source of energy at a rate which is governed by the prevailing temperature, pressure and reactant concentrations. These phenomena are discussed in more detail in Chapter 5.

1.3 Summary of principles and overview of applications

The phenomena described so far are not necessarily restricted to gaseous media. Most of them occur also in liquids and solids, and in dispersions of one phase within another (e.g. in droplet mists or dust clouds). Combustion can also occur at the interface between bulk phases (*heterogeneous combustion*), as discussed in later chapters.

Illustrated in this introductory summary are three broad subject areas within combustion, namely (i) chemical kinetics and spontaneous processes in an essentially homogeneous reactant mixture, (ii) flame propagation, and (iii) detonation and shock. There are many ramifications which may either subdivide or extend the scope of each, or which may be relevant to more than one area. Discussion of the applications may make these divisions still more diffuse. Nevertheless, from these various examples the *self-sustaining* characteristics inherent to combustion systems (a feature that sets the discipline of combustion apart from most other aspects of chemistry) should be emerging more clearly. Heat and free radicals originate in the chemistry itself, whereas the physical processes, heat and mass transport in particular, enable these sources to be used to promote (i.e. sustain) the reaction without any other external agency being applied.

The energy released in combustion may turned directly into mechanical energy and used as a source of motive power, as in spark ignition engines, diesel engines or gas turbines, or it may be used indirectly to generate mechanical energy, as is the case when steam is generated to drive the turbines in an electricity power station. Combustion processes are also involved in heating applications, ranging from domestic gas, oil or solid-fuel central heating units to large industrial furnaces. In all of these applications the oxidant is atmospheric air, which may impose constraints on efficiency and which may contribute to pollutant emissions. The efficiency of combustion processes may be enhanced by raising the pressure above atmospheric, as in turbocharging of spark ignition or diesel engines.

In specialist applications, especially where extremely high energy liberation and also high density of the fuel are essential, the oxidant may not be air, it may not even be oxygen, it may not be in the gaseous phase, or it may not be molecularly separated from the fuel. The class of substances that fall within these categories are ***high explosives*** for commercial or military use, or *low-order explosives* that are used in propellants for armaments and rocketry. In high explosives the maximum amount of energy is liberated in a minimum time, as is characteristic of detonations. Typical materials are the nitrate esters, such as trinitrotoluene (TNT), nitroglycerine (NG), or nitrocellulose (NC). The oxidisers are the fuel-bound nitrogen and oxygen in these compounds. Nitrogen and oxygen also feature in a common commercial, blasting explosive, ANFO. This comprises a mixture of ammonium nitrate, the oxidiser, and fuel oil. Subject to certain criteria being satisfied, the practical qualities of ANFO include the inherent safety of being able to mix the ingredients 'on site' and to be able to pump the slurry into bore holes.

The smoother and more prolonged energy release and the production of considerable volumes of gaseous products required from propellants relies on the deflagration process. The design of the appliance may govern whether or not deflagration or detonation occurs, thus NC and NG also find favour as military propellants. By virtue of their high power/density performance, liquid hydrogen and oxygen serve as the primary fuels for launching craft into space, but hydrazine (N_2H_4) or its derivatives may be exploited as fuel for the more subtle control of craft in outer space or for maneouverability, as in moon landings. Although some of the principles of the combustion processes involved in these types of systems are discussed in later sections, the main emphasis of this book is directed towards the more common domestic, commercial or industrial applications of combustion, especially in which hydrocarbon and fossil fuels are used. Aspects of the combustion of explosives and propellants were discussed in the second edition by Barnard and Bradley.

1.4 The nature of hydrocarbon and other fuels

Arguably the ozone decomposition flame, given by the overall stoichiometry

$$2O_3 = 3O_2 \qquad \Delta H^{\ominus}_{298} = -142.5\,\mathrm{kJ\,mol^{-1}} \tag{1.1}$$

is the simplest of all gaseous combustion systems, but it is of rather specialist interest. The oxidation of hydrogen, given by the stoichiometric equation

$$2H_2 + O_2 = 2H_2O \qquad \Delta H^{\ominus}_{298} = -242.2\,\text{kJ mol}^{-1} \tag{1.2}$$

is the simplest gaseous combustion system that is to be addressed here. The understanding of the kinetics and mechanism of this reaction over an extremely wide temperature range (700–3000 K) is central to the understanding of the mechanisms involved in the combustion of more complex fuels. Equation (1.2) represents the overall process, and the convention, reactants = products, will be retained throughout to signify a 'global' representation of the chemistry. $\Delta H^{\ominus}_{298}$ is the standard enthalpy change (or 'minus the heat release', as explained in Chapter 2) associated with the overall conversion. The kinetic detail which describes the complete behaviour of hydrogen oxidation, and which we should have to incorporate in a model if we were to predict the behaviour by numerical methods comprises nearly 100 reversible *elementary reactions* [2]. Hydrogen oxidation also provides a kinetic framework which may be used to explore the intellectual discipline of branching chain reaction theory and combustion wave propagation, and much can be established by making simplifying assumptions in the kinetic scheme, or by deriving an abbreviated scheme by formal mathematical methods [3, 4].

Hardly less important than hydrogen oxidation is that of carbon monoxide:

$$2CO + O_2 = 2CO_2 \qquad \Delta H^{\ominus}_{298} = -281.5\,\text{kJ mol}^{-1} \tag{1.3}$$

Carbon monoxide arises as a major pollutant from combustion applications because, in hydrocarbon oxidation, virtually the only route to carbon dioxide, as *the* final carbon-containing product, is via carbon monoxide. Its formation cannot be avoided, and the efficiency and cleanliness of the combustion system is governed by the ability to ensure the complete oxidation of CO to CO_2. As with hydrogen oxidation, the process does not occur by a fortuitous association of two molecules of CO and one of O_2 to yield two molecules of CO_2, but by a sequence of consecutive and competitive elementary reactions, closely connected with those which are involved in hydrogen oxidation, as discussed in Chapters 6 and 9.

Methane (CH_4) is the simplest of the hydrocarbons and is the major constituent of natural gas, typically comprising 93–96% by volume. Although there are small variations in composition according to the source, the minor components present are mainly ethane (C_2H_6), propane (C_3H_8) and butane (C_4H_{10}). These compounds are not strongly influential in the energy content, but the presence of these components can affect spontaneous ignition, which presents a combustion hazard.

Although there is some common ground, the chemistry of methane combustion is not strictly typical of all alkanes throughout the entire temperature range of general interest (500–2500 K, say), and to understand more about alkanes in general it is necessary to begin investigations with butane, taking into account the distinctions between its two isomeric structures, normal butane, $CH_3(CH_2)_2CH_3$, and isobutane, $(CH_3)_3CH$. The chemical consequences of the structural distinctions in combustion will be addressed in more detail in Chapter 7, and some of the repurcussions on combustion processes are discussed in Chapters 12 and 13.

Liquid hydrocarbon fuels comprise enormous numbers of components embracing the major classifications, alkanes, alkenes, alkynes and aromatic compounds. A chemical analysis of the fuel does not yield sufficient information to be able to predict combustion characteristics or performance, because the physicochemical interactions involved in combustion are not derived in a way that is proportionate to the mass or volume fraction of each component.

Blends of petrol (or gasoline) are drawn mainly from the fraction of crude oil in the normal boiling point range approximately 320–450 K.

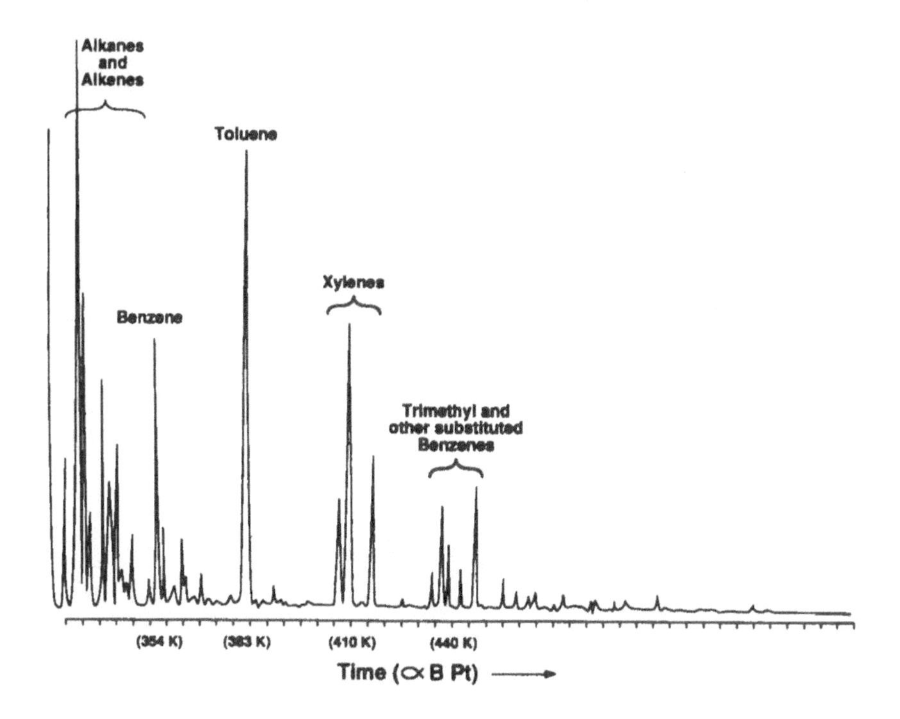

Figure 1.1 A gas chromatographic analysis of aromatic hydrocarbons in a typical unleaded gasoline. The time of elution from the column is proportional to the boiling point of the component (as marked in representative cases) and the peak height is proportional to the amount (analysis courtesy of B. Frere).

This can include components of carbon number distributed between C_4 and C_{10}, with the main constituents being around C_8. The blend itself is governed by the source of the crude oil, and the purpose for which it has been refined, such as winter or summer use, or for hot or cold climatic conditions. Blends for lead-free or leaded '4-star' petrol also differ. A chromatogram of a typical, unleaded petrol is shown in Fig. 1.1, which highlights the aromatic fractions, of which toluene and the xylene isomers are particularly prominent. The aromatics may be up to 40% of this type of fuel, by volume, with alkanes probably constituting 30–40% and the remainder being unsaturated hydrocarbons, the alkenes. The relationship between composition and combustion efficiency in engines is addressed in Chapter 13.

Commercial jet, aircraft fuel (kerosene) is taken from a slightly lower volatility fraction than petrol (normal boiling range 430–550 K) and comprises hydrocarbons in the range C_{10}–C_{16}. Typical diesel fuels are taken from a still higher boiling fraction (550–750 K) and they are distinguished by the predominance of the straight chain (i.e. normal) alkanes, n-$C_{11}H_{24}$ to n-$C_{24}H_{50}$ (Fig. 1.2). Diesel fuels are not free of aromatic compounds though, comprising a few percent by volume in total as substituted naphthalenes, fluorenes and phenanthrenes. Sulphur-containing and nitrogen-containing compounds are also present in small proportions, but sufficient to be of concern in relation to pollutant

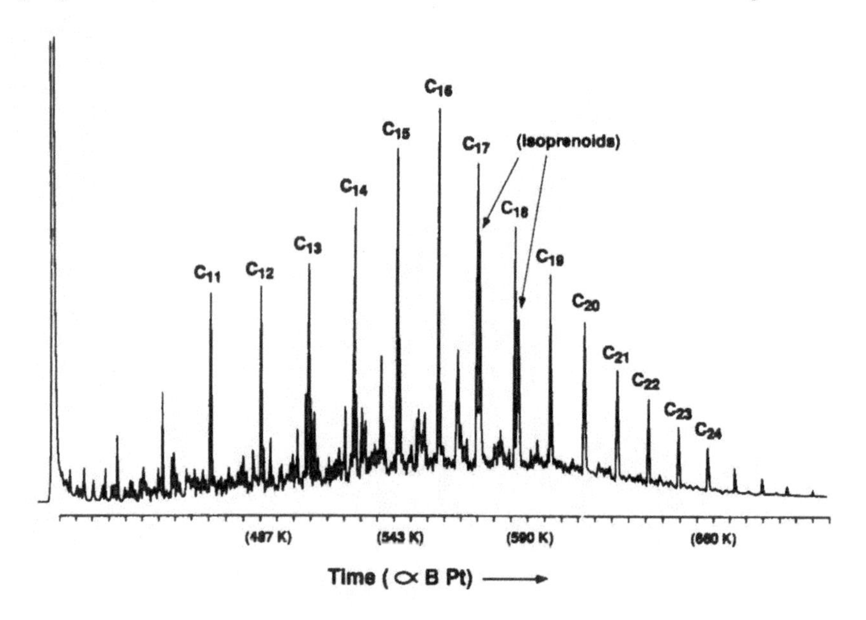

Figure 1.2 A gas chromatographic analysis of alkanes in a typical diesel. The principal components are the series of straight-chain alkanes, as marked. The isoprenoids are naturally occurring alkenes which belong to the 'essential oil' family of compounds. Other details as Fig. 1.1 (analysis courtesy of B. Frere).

formation. Heavy fuel oils are drawn from still higher molecular weight residues of crude oil.

Further reading

Fordham, S. (1980). *High Explosives and Propellants* (*2nd edn.*) Pergamon, Oxford, UK.

2 Physicochemical principles

2.1 Introduction

The principles of chemical thermodynamics and reaction kinetics are discussed in this chapter. Heat and mass transport processes, some aspects of the chemistry–fluids interaction and the non-dimensionalisation of parameters as they apply to combustion processes are also introduced.

2.2 Thermodynamics of combustion

2.2.1 Energy of reaction and overall heat release

Combustion involves the liberation of energy as the chemical reaction proceeds. The interpretation of the overall amount of energy released from and the state of equilibrium attained in the combustion process is part of the subject of thermodynamics. The principles of thermodynamics that are relevant to combustion are reviewed briefly here. More comprehensive treatments, including chemical equilibrium, are found in virtually all general physical chemistry or chemical engineering books and other more specialised texts.

How much energy can be liberated from a given chemical reaction is determined from the energies of the individual reactants and products. The precise products, and hence the overall stoichiometry of the reaction, must always be established by chemical analysis. From the examples given in Chapter 1 it might be presumed that the outcome can be guessed, but it is not necessarily the case. For example, methane may react with oxygen to produce either carbon monoxide and water:

$$CH_4 + 3/2O_2 = CO + 2H_2O \tag{2.1}$$

or carbon dioxide and water:

$$CH_4 + 2O_2 = 2H_2O + CO_2 \tag{2.2}$$

Whereas the second reaction corresponds to complete combustion, and hence to the maximum release of energy, in many circumstances the first reaction provides a better representation of what actually happens when methane is burned. Other possibilities arise from the combustion of methane at temperatures lower than those normally encountered in flames.

The energy of an individual chemical species may be given either in terms of its internal energy, U, or its enthalpy, H, where, by definition

$$H = U + pV \tag{2.3}$$

The product pV relates to the mechanical work done on the system. Since neither internal energy nor enthalpy can be measured absolutely, it is necessary to choose a reference state to which all other energies may be related. This standard state is generally taken as the stable state of the pure substance at atmospheric pressure and at a specified temperature (normally 298 K). The enthalpy of each element is arbitrarily assigned the value zero in the standard state. The enthalpy change involved at some temperature, T, in forming one mole of the chemical species in its standard state from its elements in their standard states is known as the standard enthalpy of formation, $\Delta H_f^\ominus$, of the compound (Table 2.1). There are extensive tabulations of these standard enthalpies [5,6], which are also available on computer disk [7].

From Hess's Law (which is a special case of the first law of thermodynamics), the enthalpy change in the reaction taking place under standard state conditions $\Delta H_T^\ominus$ is equal to the difference between the sums of the

Table 2.1 Some standard enthalpies of formation (all species are gaseous, unless otherwise stated)

Species	Formula	$\Delta H_{f298}^\ominus$ (kJ mol^{-1})
Hydrogen atom	H	217.99
Oxygen atom	O	249.19
Hydroxyl radical	OH	39.46
Hydroperoxy radical	HO_2	20.92
Water (liquid)	H_2O (l)	−285.83
Water (vapour)	H_2O	−241.81
Hydrogen peroxide (liquid)	H_2O_2 (l)	−136.10
Carbon monoxide	CO	−110.52
Carbon dioxide	CO_2	−393.51
Methyl radical	CH_3	145.70
Methane	CH_4	−74.87
Ethyne (acetylene)	C_2H_2	226.70
Ethene	C_2H_6	52.47
Ethane	C_2H_6	−84.67
Propane	C_3H_8	−103.85
Benzene	C_6H_6	82.93
Methanol vapour	CH_3OH	−201.20
Nitrogen atom	N	472.7
Amine radical	NH_2	167.7
Ammonia	NH_3	−45.94
Nitric oxide	NO	90.29
Sulphur dioxide	SO_2	−296.81

standard enthalpies of formation of the products and the reactants. For eqn (2.1), when the water is produced as vapour,

$$\Delta H^{\ominus}_{298} = \Delta H_f^{\ominus}(CO) + 2\Delta H_f^{\ominus}(H_2O, g) - \Delta H_f^{\ominus}(CH_4)$$

$$= (-110.52) + 2(-241.81) - (-74.80) = -519.34\,\text{kJ mol}^{-1} \quad (2.4)$$

The negative sign indicates that energy is released, and the reaction is said to be *exothermic*. Although combustion processes are exothermic overall, not all of the elementary reactions involved in the overall process are necessarily exothermic. A reaction for which ΔH is positive and in which, therefore, energy is absorbed by the system from the surroundings, is said to be *endothermic*.

For the general reaction

$$\nu_A A + \nu_B B + \ldots = \nu_P P + \nu_Q Q + \ldots \quad (2.5)$$

$$\Delta H^{\ominus}_{298} = \sum \nu_X \Delta H_f^{\ominus}{}_{298}(X) \quad (2.6)$$

where the summation is performed with the stoichiometric factors, ν_X, as positive for the products and negative for the reactants.

ΔH for a chemical reaction is quoted unambiguously only if the stoichiometric equation is also given. The units of ΔH are J mol^{-1}, that is Joules for an extent of reaction of one mole of the reactant. The stoichiometric equations given so far imply complete conversion of the reactant to the specified products. They convey nothing about the scale on which the reaction is carried out, whether or not conversion is complete or the absolute amount of energy released.

For some engineering purposes it is preferable to use the enthalpy released per unit mass of reactants. For eqn (2.1) this is

$$519.27/(16 + 3/2 \times 32) = 8.11\,\text{kJ g}^{-1} \text{ (of the reactant mixture)}$$

since 0.016 and 0.032 kg are the molar masses of methane and oxygen respectively. As atmospheric air is usually regarded as 'free', another useful quantity is the amount of energy available from unit mass of fuel. For methane burning to carbon monoxide, the energy available is

$$519.27/16 = 32.45\,\text{kJ g}^{-1}$$

These calculations relate the enthalpy changes taking place at constant pressure. The corresponding energy released when reaction takes place at constant volume is given by the internal energy change, ΔU. For reactions involving ideal gases there is a simple relationship between ΔH and ΔU, namely,

$$\Delta H = \Delta U + (\Delta \nu)\,RT \quad (2.7)$$

where Δv is the change in the number of moles of gaseous reactants to products and ideal gas behaviour is assumed. When only liquids or solids are involved, $\Delta H \approx \Delta U$.

Combustion reactions do not usually take place at room temperature, but the enthalpy change at other temperatures is easily obtained from

$$\Delta H_T = \Delta H_{298} + \int_{298}^{T} \Delta C_p \mathrm{d}T \tag{2.8}$$

where ΔC_p refers to the difference between the heat capacity of the products and that of the reactants, that is,

$$\Delta C_p = \sum v_X C_p(X) \tag{2.9}$$

where the summation is performed, as before, treating the stoichiometric factor v as positive for the products and negative for the reactants. $C_p(X)$ represents the molar heat capacity of reactant X. In practice, the enthalpy changes are not very sensitive to temperature because the heat capacities on each side of the chemical equation are approximately equal. For eqn (2.1), ΔH changes from $-519\,\text{kJ mol}^{-1}$ at 298 K to $-530\,\text{kJ mol}^{-1}$ at 2000 K and $-542\,\text{kJ mol}^{-1}$ at 3000 K.

In real systems, the energy is released over a range of temperatures, but since both internal energy and enthalpy are state properties, this means that their values depend only on the present state of the system and not the path by which it was reached. It follows that, as long as the initial and final states of the system are fixed, the change in internal energy or enthalpy associated with the process will be independent of the route by which the change takes place.

Thus, we can find the enthalpy change for a system which begins with reactants at an initial temperature T_0 and finishes with products at a final temperature T_f by first finding the enthalpy change at T_0 and then using heat capacity data to calculate the enthalpy change involved in heating the products from T_0 to T_f.

2.2.2 Adiabatic temperature rise and evaluation of heat capacities

The energy released during reaction in adiabatic conditions goes into raising the temperature of the system. For a constant volume process, such as a closed vessel explosion

$$\Delta U_{T_f} - \Delta U_{T_0} = \int_{T_0}^{T_f} C_v \text{ (products) d}T \tag{2.10}$$

and similarly for a constant pressure process, typified by a premixed flame,

$$\Delta H_{T_f} - \Delta H_{T_0} = \int_{T_0}^{T_f} C_p \text{ (products) } dT \tag{2.11}$$

Under adiabatic conditions,

$$\Delta H_{T_f} = \Delta U_{T_f} = 0 \tag{2.12}$$

Heat capacities are to be found in compilations of thermodynamic data either tabulated against temperature [8] or, more conveniently, they may be derived as polynomials in temperature of the following form [9]:

$$C = a + bT + cT^2 + dT^3 \tag{2.13}$$

or

$$C = a + b/T + c/T^2 \tag{2.14}$$

The heat capacities of some species involved in combustion are given in Table 2.2.

For approximate calculations it is necessary to know only that the maximum value of C_v for gaseous molecules is given by the sum of the classical contributions to molecular energy, which on a molar basis are given by

Translational motion:	3/2 R
Rotational motion:	$2R$ for a non-linear molecule
	R for a linear molecule
Vibrational motion:	R for each vibrational mode

Table 2.2 Molar heat capacities of some gases ($C_p(\text{J mol}^{-1}\text{ K}^{-1}) = a + b(T/K) + c(T/K)^2 + d(T/K)^3$)

Species	a	$10^2 b$	$10^5 c$	$10^9 d$	Temperature range (K)
H_2	29.09	−0.1916	0.400	−0.870	273–1800
O_2	25.46	1.519	−0.715	1.311	273–1800
N_2	27.32	0.6226	−0.095	—	273–3800
CO	28.14	0.1674	0.537	−2.220	273–1800
CO_2	22.24	5.977	−3.499	7.464	273–1800
H_2O	32.22	0.1920	1.054	−3.594	273–1800
CH_4	19.87	5.021	1.286	−11.00	273–1500
C_2H_2	21.80	9.208	−6.523	18.20	273–1500
C_2H_4	3.95	15.63	−8.339	17.66	273–1500
C_2H_6	6.89	17.25	6.402	7.280	273–1500
C_3H_8	−4.04	30.46	−15.71	31.71	273–1500
C_6H_6	−39.19	48.44	−31.55	77.57	273–1500
CH_3OH	19.04	9.146	−1.218	−8.033	273–1000
NH_3	27.55	2.563	0.990	−6.686	273–1500
NO	27.03	0.9866	0.322	0.365	273–3800
SO_2	25.76	5.791	−3.809	8.606	273–1800

There are $3n$–5 vibrational modes for a linear molecule and $3n$–6 modes for a non-linear molecule comprising n atoms. A mean heat capacity, suitable for approximate temperatures, is obtained by assuming that 50% of the classical vibrational modes are active, and therefore contribute to the heat capacity. The classical limits of translation and rotation contributions are established at all normal, or higher temperatures. The heat capacity, C_v, which is calculated in this way is assumed independent of temperature. Values of C_p may be then obtained from the following relationship:

$$C_p = C_v + R \tag{2.15}$$

in which R is the ideal gas constant.

2.2.3 *Extent of reaction and determination of chemical equilibrium*

Chemical reactions may not go to completion and so the final composition is not necessarily that shown simply by the stoichiometric equation, as in eqns (2.1) or (2.2), for example. The second law of thermodynamics provides criteria for the position of equilibrium, and in the case of a system at constant temperature and pressure, this may be represented by a minimum in the Gibbs function (or free energy change) for the reaction, ΔG, where

$$\Delta G = \Delta H - T\Delta S \tag{2.16}$$

If the reactants and products behave as ideal gases (as in most combustion systems at moderate pressures), then the change in the standard Gibbs function accompanying reaction at a temperature T is related to the equilibrium constant $K^{\ominus}$ by the expression

$$\Delta G_T^{\ominus} = -RT \ln K^{\ominus} \tag{2.17}$$

For the generalised reaction shown in eqn (2.5):

$$K^{\ominus} = \frac{(p_P/p^{\ominus})^{\nu_P}\,(p_Q/p^{\ominus})^{\nu_Q}}{(p_A/p^{\ominus})^{\nu_A}\,(p_B/p^{\ominus})^{\nu_B}} \ldots \tag{2.18}$$

where p_i is the partial pressure of component i. The equilibrium constant is dimensionless when the partial pressures are related to the standard pressure, $p^{\ominus}$. If $K_p^{\ominus}$ is defined in terms of the absolute pressures then it has dimensions of $[p]^{\Delta\nu}$, where $\Delta\nu$ represents the sum of the stoichiometric coefficients of the gaseous species. Just as there are tabulations of standard enthalpies of formation, there are lists of standard Gibbs functions of formation for a very large number of compounds [7]. $\Delta G_T^{\ominus}$

for a particular reaction may be obtained by combining the tabulated $\Delta G_f^\ominus$ values in a way analogous to the combination of $\Delta H_f^\ominus$ in eqn (2.6).

ΔG must change with temperature, even if both ΔH and ΔS for the reaction are constant over a certain temperature range (eqn (2.16)), and the equilibrium constant is a function of temperature, given by

$$\ln(K_2^\ominus/K_1^\ominus) = \frac{-\Delta H}{R}\left(\frac{1}{T_2} - \frac{1}{T_1}\right) \quad (2.19)$$

If the second term on the right of eqn (2.16) could be neglected, then the condition for equilibrium would correspond to a maximum in $(-\Delta H)$, the amount of heat evolved by the process. That is, the reaction would always go to completion. However, the presence of the second term means that a small reduction in $(-\Delta H)$ may be compensated by a corresponding change in $T\Delta S$. The entropy, S, is a measure of disorder in the system, and the entropy is increased by an increase in the number of different chemical species present as a result of the chemical reaction. Since mass must be conserved, an increase in entropy is favoured when the gaseous products comprise molecular species that are 'simpler' than the reactants; that is, they contain fewer atoms. In an adiabatic methane

Table 2.3 Mole fractions of reactants and products calculated at thermodynamic equilibrium for a stoichiometric propane + air mixture[a]

Species	Constant pressure[a]			Constant volume
	(a)	(b)	(c)	
CO_2	0.1004	0.1003	0.1111	0.0914
H_2O	0.1423	0.1439	0.1481	0.1374
N_2	0.7341	0.7347	0.7407	0.7276
CO	0.0099	0.0100	—	0.0182
H_2	0.0032	0.0033	—	0.0053
O_2	0.0048	0.0055	—	0.0075
CH_2O	$< 10^{-5}$	—	—	$< 10^{-5}$
CH_2H_4	$< 10^{-5}$	—	—	$< 10^{-5}$
C_3H_6	$< 10^{-5}$	—	—	$< 10^{-5}$
CHO	$< 10^{-5}$	—	—	$< 10^{-5}$
CH_3	$< 10^{-5}$	—	—	$< 10^{-5}$
C_2H_5	$< 10^{-5}$	—	—	$< 10^{-5}$
i-C_3H_7	$< 10^{-5}$	—	—	$< 10^{-5}$
H_2	$< 10^{-5}$	—	—	$\ll 10^{-5}$
H	0.0035	—	—	0.0009
O	0.0020	—	—	0.0008
OH	0.0027	—	—	0.0058
N	$< 10^{-5}$	—	—	$< 10^{-5}$
NO	0.0020	0.0022	—	0.0052
N_2O	$< 10^{-5}$	—	—	$< 10^{-5}$
Final temperature (K)	2219	2232	2324	2587

[a] The data in column (a) refer to calculations based on a comprehensive set of equilibria. Those in column (b) refer to calculations using a more limited set of equilibria. The data in column (c) were obtained by considering only the formation of CO_2 and H_2O.

flame, in addition to carbon dioxide and water as the final products of combustion, there may be traces of residual reactants (CH_4 and O_2), other molecular products (e.g. CO and H_2) and free radical intermediates (e.g. H, O, OH and CH_3).

In general, when any reaction has reached equilibrium there will be varying amounts of chemical species other than the expected final products. This is relatively unimportant in most combustion processes at low final temperatures, but the presence of the temperature multiplier in the $T\Delta S$ term means that the trace materials become more important at elevated temperatures. The effect is fairly small up to about 2000 K (Table 2.3) but is very marked at 3000 K (Table 2.4), as is appropriate to propane combustion in oxygen. A considerably lower temperature is predicted to be associated with the more complex equilibrium composition, by contrast to the final temperature (6340 K) associated with the final products of combustion, CO_2 and H_2O (Table 2.4).

An alternative approach to computation, which is viable if only a few species are present, is to consider the individual chemical equilibria involved. Although a condition of equilibrium is that all of the possible equilibrium relationships are satisfied, in practice many of them will be interdependent and it is therefore necessary to consider only an appropriate subset, as discussed by Gaydon and Wolfhard [10]. Manual calculations of the equilibrium composition of the products of combustion [10] are possible but computer methods are usually employed [11].

Table 2.4 Mole fractions of reactants and products calculated at thermodynamic equilibrium for a stoichiometric propane + oxygen mixture[a]

Species	Constant pressure			Constant volume
	(a)	(b)	(c)	
CO_2	0.1396	0.0896	0.4286	0.1293
H_2O	0.3000	0.3182	0.5714	0.2908
CO	0.1951	0.2544	—	0.2035
H_2	0.0697	0.1404	—	0.0693
O_2	0.0984	0.1974	—	0.0954
CH_2O	$< 10^{-5}$	—	—	$< 10^{-5}$
C_2H_4	$< 10^{-5}$	—	—	$< 10^{-5}$
C_3H_6	$< 10^{-5}$	—	—	$< 10^{-5}$
CHO	$< 10^{-5}$	—	—	$< 10^{-5}$
CH_3	$< 10^{-5}$	—	—	$< 10^{-5}$
C_2H_5	$< 10^{-5}$	—	—	$< 10^{-5}$
i-C_3H_7	$< 10^{-5}$	—	—	$< 10^{-5}$
H	0.0526	—	—	0.0460
O	0.0441	—	—	0.0443
OH	0.1005	—	—	0.1212
HO_2	3×10^{-5}	—	—	0.0001
Final temperature (K)	3081	3420	6340	3630

[a] Conditions as in Table 2.3.

However, for underlying kinetic reasons, a state of chemical equilibrium may not necessarily have been established in the combustion system.

2.3 Rates of chemical change in combustion

While all combustion processes depend on the total amount of energy released by chemical reaction, not all depend on the rate of reaction, provided that it exceeds some minimum value. Thus the overall behaviour in detonation waves, diffusion flames, burning droplets and liquid propellant rocket engines is virtually independent of chemical kinetics, whilst premixed flames, fires and internal combustion engines are sensitive to the detailed kinetics involved. The fundamentals of reaction kinetics are described in this section. Supplementary details of specific reactions and reaction mechanisms are considered in later chapters.

2.3.1 Reaction rate, kinetic rate laws and order of reaction

The quantitative behaviour of a chemical reaction is described by a *rate law* which specifies the rate of change of the concentration of chemical species in terms of the product of concentration terms and a rate constant (or rate coefficient) which is independent of concentration but, usually, not of temperature. For the reaction represented by the following stoichiometric equation:

$$\nu_A A + \nu_B B + \ldots = \nu_P P + \nu_Q Q + \ldots \qquad (2.5)$$

the rate law with respect to these four components takes the form

$$\frac{-1}{\nu_A}\frac{d[A]}{dt} = \frac{-1}{\nu_B}\frac{d[B]}{dt} = \frac{1}{\nu_P}\frac{d[P]}{dt} = \frac{1}{\nu_Q}\frac{d[Q]}{dt} = k[A]^a[B]^b \qquad (2.20)$$

where k is the rate constant. The powers a and b are known as the orders of reaction with respect to the concentrations of the reactants A and B, respectively, and the overall order is given by $(a+b)$. The individual orders of reaction quantify the dependence of the reaction rate on the species concentrations. The terms on the left-hand side of eqn (2.20) rationalise the reaction rate in terms of changes of any of the molecular reactants or product concentrations through the respective stoichiometric coefficient. Note that the rate of change of concentration is regarded to be a positive quantity, and so the convention of negative stoichiometric coefficients representing the reactant consumption is carried over to the rate expression. The square brackets signify either a molecular or a molar concentration of a species and, therefore, the reaction rate is expressed as molecules (or moles) per unit volume per unit time.

2.3.2 *Concentration dependences in global and elementary reaction rate expressions*

A distinction must be made between a rate expression which represents an *overall reaction*, often referred to as a *global reaction*, and an *elementary step*, as one of the components of the mechanism that makes up the overall reaction. In the latter, the chemical equation represents what is believed to happen on a molecular level. For example, the process

$$H+O_2 \rightarrow OH+O \tag{2.21}$$

is an elementary step (which is of considerable importance in combustion) in which a hydrogen atom collides with an oxygen molecule and the three atoms rearrange to give a hydroxyl radical and an oxygen atom. In this case, the rate law may be deduced directly from this concept of the molecular interaction. Doubling the concentration of oxygen molecules in the system would double the frequency of collisions of H atoms with them, and hence the rate of conversion to product species. Thus, the *bimolecular interaction* must have a linear dependence on $[O_2]$, i.e. be first order with respect to it. The *molecularity* of a reaction describes the number of molecules involved in the reactive event. On the basis of a similar argument with respect to the H atoms, the rate law for eqn (2.21) is expressed as

$$\frac{-d[H]}{dt}=\frac{-d[O_2]}{dt}=\frac{d[H]}{dt}=\frac{d[OH]}{dt}=k[H][O_2] \tag{2.22}$$

That is, the rate law is *first order* with respect to each of the reactants and *second order overall*. The reaction rate of an elementary reaction is never dependent on a product concentration. The forward and reverse rate constants of an elementary reaction, k_f and k_r, can be related through the expression $k_f/k_r=K_c$, where K_c is the equilibrium constant defined in terms of reactant and product concentrations.

Most combustion processes take place by a series of elementary steps, as part of a chain reaction. Thus, an overall equation which represents the stoichiometry of the reaction does not necessarily reflect the detailed events as the species react, nor does it give information about the concentration dependences in the appropriate rate law. For example, the overall stoichiometry

$$2H_2+O_2=2H_2O \tag{2.23}$$

may be represented by a global rate law such as

$$\frac{-1}{2}\frac{d[H_2]}{dt}=\frac{-d[O_2]}{dt}=\frac{1}{2}\frac{d[H_2O]}{dt}=k[H_2]^{1.5}[O_2]^{0.7} \tag{2.24}$$

but this cannot be regarded to represent two hydrogen molecules colliding with a single oxygen molecule to form two molecules of water [12].

The global process is an abbreviation for a complex sequence of elementary steps, and a global rate law to represent it is an empirical representation of the dependence of the reaction rate on concentrations which applies over a very limited (specified) range of conditions. It sometimes happens that the detailed reaction mechanism is not completely understood, or that its complexity makes it unsuitable for incorporation into a computer model, and then a global expression has to be used.

For a global reaction, the order must be established by experiment. The fractional reaction orders, as measured, should alert us to the fact that the reaction mechanism is a complex composite of elementary reactions, as would an inverse dependence on a reactant concentration or the intervention of a product concentration dependence in the rate equation.

An elementary step is said to be *unimolecular* if it involves only a single molecule, for example,

$$H_2 \rightarrow 2H \tag{2.25}$$

and it might be expected to obey *first order kinetics* as expressed by the relationship

$$\frac{d[H_2]}{dt} = \frac{1}{2}\frac{d[H]}{dt} = k[H_2] \tag{2.26}$$

While unimolecular reactions are normally first order there are departures at low pressures, which may be important to combustion processes. In fact, the unimolecular reaction is second order in the limit of very low pressures, and this change of overall behaviour (called the 'fall-off' region) signifies that there is a more complex mechanistic substructure to the unimolecular reaction. These types of processes are extremely well understood, and there are sophisticated theories which permit the quantitative interpretation of the behaviour [13].

The units of the rate constant are governed by the overall reaction order (be it a global representation or an elementary reaction), since the units on the right-hand side must equate to those on the left-hand side. Thus, a first-order rate constant has the dimensions of $[\text{time}]^{-1}$, whereas a second-order rate constant has units $[\text{concentration}]^{-1}$ $[\text{time}]^{-1}$. In combustion reactions, the units of time are almost always seconds, whereas convenient concentration units may be (molecules) cm^{-3}, $mol\,cm^{-3}$, $mol\,dm^{-3}$ or $mol\,m^{-3}$. The concentration terms must be unified when data are being taken from different sources, as shown in Table 2.5.

A useful parameter that gives some concept of the timescale on which the chemistry takes place, especially with respect to first order reactions, is the *half-life* of the reaction. This represents the time for 50% of the reactant to be consumed, and is given by $\ln 2/k$. For second or higher order reactions, the corresponding parameter is derived under *pseudo*

Table 2.5 Some conversion factors for concentration and pressure terms (conversion A → B)

	A				
B	mol cm ^{3}	mol dm ^{3}	mol m^{-3}	particle cm^{-3}	Pa[a]
mol cm^{-3}	1	10^{-3}	10^{-6}	1.66×10^{-24}	$1.20 \times 10^{5}/(T/K)$
mol dm^{-3}	10^{3}	1	10^{-3}	1.66×10^{-21}	$120/(T/K)$
mol m^{-3}	10^{6}	10^{3}	1	1.66×10^{18}	$0.12/(T/K)$
particle cm ^{3}	6.023×10^{23}	6.023×10^{20}	6.023×10^{18}	1	$7.23 \times 10^{28}/(T/K)$
Pa[a]	$0.83310^{-6}(T/K)$	$8.33 \times 10^{-3}(T/K)$	$8.33(T/K)$	$1.38 \times 10^{-29}(T/K)$	1

[a] 101 325 Pa = 1 atm = 760 torr.

first order conditions, that is by subsuming certain (constant) reactant concentrations into the rate constant.

2.3.3 Temperature dependence of the rate constant

The temperature dependence of a reaction is incorporated in the rate constant, and is expressed most simply in the Arrhenius form, as

$$k = A \exp^{(-E\ RT)} \tag{2.27}$$

The parameters A and E are independent of temperature in this well-established representation. Although they are derived experimentally, these terms have clear identities even in quite simple interpretations of reaction rate theory. Thus, A is termed the frequency factor with respect to first order reactions, since it may be identified with the rate at which chemical bonds can rearrange in a molecule and relates to a vibrational frequency ($\sim 10^{13} s^{-1}$). In other circumstances, A is called the pre-exponential term and bears some relationship to collision frequencies between the interacting species in elementary reactions. The units of A are also the units of k.

E is the activation energy of the reaction. It may be regarded as a measure of the energy barrier to reaction and is given as J mol^{-1} or kJ mol^{-1} in SI units. The quotient $(E/R)/K$ signifies a temperature coefficient, which testifies to its experimental origin as a quantitative measure of the way in which the rate constant varies with temperature. The coefficient E/R is often quoted in rate data, rather than the activation energy.

The two-parameter representation of the temperature dependence (eqn (2.27)) is satisfactory for most reactions, especially over limited temperature ranges, and is confirmed by the linearity of the relationship

$$\ln k = \ln A - E/RT \tag{2.28}$$

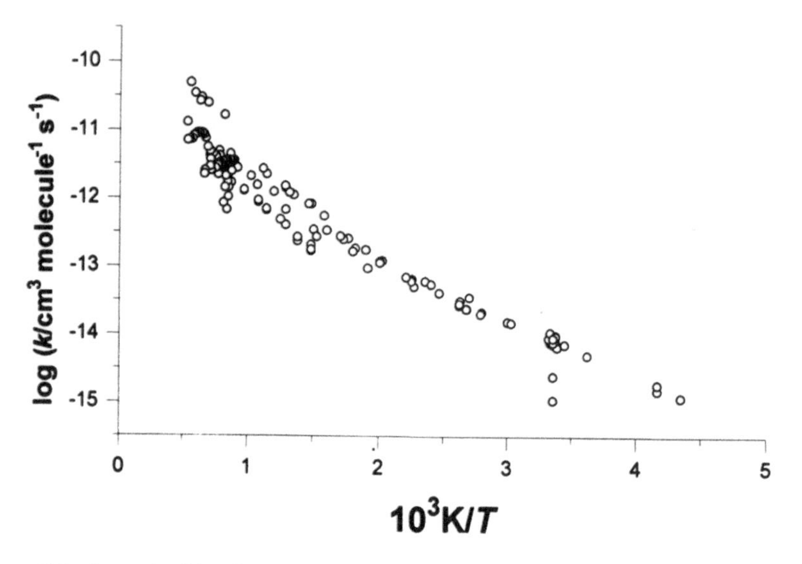

Figure 2.1 A graph of log k versus $1/T$ obtained from the measured rate constant for the reaction $CH_4 + OH$ over the temperature range 300–2000 K. The results from about 50 different experimental studies are shown. Curvature of the Arrhenius function is very clear. The recommended rate constant from these data is $k = 1.57 \times 10^7 T^{2.189} \exp^{(-1400/T)} cm^3 mol^{-1} s^{-1}$ (after Baulch *et al.* [14]).

in the classical Arrhenius plot ln$k$ versus $1/T$. However, in some applications in combustion the rate constant has to be represented over an extremely wide temperature range (500–2500 K, say). Where sufficient experimental measurements are available it may be found that the two-parameter representation is inadequate (Fig. 2.1), which leads to a three-parameter representation of the form

$$k = A' T^n \exp^{(-E/RT)} \tag{2.29}$$

where n is a number of order unity [14]. The product $(A' T^n)$ now has the units of the rate constant.

2.3.4 Termolecular reactions and elements of chain reaction mechanisms

The termolecular reaction

$$H + O_2 + M \rightarrow HO_2 + M \tag{2.30}$$

obeys third-order kinetics in certain conditions

$$v = k[H][O_2][M] \tag{2.31}$$

The symbol M is used to denote any molecule present in the system and its function is to remove some of the energy released by the formation of the new chemical bond, thereby preventing the product from immediately redissociating. M is termed a *third body* or *chaperone* molecule. The rate constant has dimensions of [concentration]$^{-2}$ [time]$^{-1}$ in this case. Third-order reactions often have rates which fall slightly with increasing temperature, which means that the measured 'activation energy' is slightly negative. This has no obvious physical meaning, and signifies that the mechanism of the overall process is not properly represented as a single 'three-body interaction'. Like unimolecular reactions, the rate constants for termolecular reaction also show a pressure dependence: they become second order at a high pressure limit.

The probability of termolecular reactions is so low that these reactions might be expected to be unimportant. However, since they provide virtually the only route for the homogeneous removal of reactive intermediates in gaseous combustion and are responsible for the liberation of considerable amounts of energy they are very important. Such reactions tend to predominate in the post-flame region which follows the main reaction zone of a flame. Termolecular recombination reactions, such as

$$H + OH + M \rightarrow H_2O + M; \qquad \Delta H^{\ominus}_{298} = -468\ \mathrm{kJ\ mol^{-1}} \qquad (2.32)$$

or

$$CO + O + M \rightarrow CO_2 + M; \qquad \Delta H^{\ominus}_{298} = -532\ \mathrm{kJ\ mol^{-1}} \qquad (2.33)$$

fall in this category.

Combustion reactions normally involve a complex mechanism, or sequence of elementary steps, and many examples will be discussed in later chapters. The mechanisms involve chain reactions, in which an active species (usually a free radical or an atom) reacts with a stable molecule to give a product molecule and another active species which can propagate the chain. Thus, in the reaction between hydrogen and chlorine [15], the chain is propagated by the cycle:

$$Cl + H_2 \rightarrow HCl + H \qquad (2.34)$$

$$H + Cl_2 \rightarrow HCl + Cl \qquad (2.35)$$

Such reactions comprise a linear chain because each propagation step leaves the total number of active centres (H and Cl) unchanged.

Although linear-chain propagation is normal, examples are known where chain branching may also occur, that is, in which one active species produces additional species which are capable of continuing the chain. In the oxidation of hydrogen, the reaction

$$H + O_2 \rightarrow OH + O \qquad (2.21)$$

brings about chain branching since both OH and O can react with hydrogen molecules to continue the chain. Branching chain reactions are particularly important in many combustion reactions and their special features will be discussed in Chapter 9.

Typical active species involved in hydrocarbon combustion are H, O, OH, CH_3 and CHO. Reactions of such species have low energy barriers and hence their rates are rapid even though the concentrations of active species are low. With the exception of thermal ignition, in general, reactions involving only molecular reactants are too slow to sustain combustion.

The behaviour of a complex reaction mechanism is described mathematically by a set of simultaneous differential equations, equal in number to that of the chemical species involved. Analytical solution of these equations is usually impossible and, therefore, use is made of the stationary-state approximation in which the rates of change of concentrations of the active centres with respect to time are set equal to zero. In the hydrogen + chlorine example, involving eqns (2.34) and (2.35) amongst other reactions [15], one would write

$$\frac{\mathrm{d}[\mathrm{H}]}{\mathrm{d}t}=\frac{\mathrm{d}[\mathrm{Cl}]}{\mathrm{d}t}=0 \qquad (2.36)$$

This reduces a number of the differential equations to algebraic equations. It does not imply that the concentrations of radicals are invariant in time, but only that they can be related algebraically to the concentrations of stable species, which have finite time derivatives. The algebraic equations can often be solved to give the stationary-state concentration of radical species. In some cases, when the rate of reaction is very high, it is not possible to use the stationary-state assumption, and then recourse must be made to numerical computation.

2.4 Transport properties of gases

In the physics of combustion use is made of the laws of conservation of energy, mass and momentum. Although the applications may be more complicated, the underlying principles are established in the respective transport processes in fluids, namely heat conduction, diffusion and viscosity. The transport of momentum, or viscosity, is involved only indirectly in combustion, for example when it controls the flow velocity profile in a tube.

The conduction of heat is described by Fourier's Law

$$\frac{\dot{q}}{A}=-\kappa\frac{\mathrm{d}T}{\mathrm{d}x} \qquad (2.37)$$

where $\dot{q}$ is the heat flowing in unit time through an area A and κ is the thermal conductivity, which is measured in $\mathrm{W\,m^{-1}K^{-1}}$. The temperature gradient in the medium is dT/dx in the direction of heat flow and the negative sign indicates that the temperature gradient is in the opposite direction to the flow of heat. Thermal diffusivity, a, is defined as $\kappa/\sigma C_p$ and thus has units of $m^2 s^{-1}$, where σ is the molar density and C_p is the molar heat capacity.

For a gas, simple kinetic theory can be used to show that the net heat flux through unit area is $-(1/2)\sigma C_v \bar{c}\lambda(dT/dx)$. In this expression $\bar{c}$ is the mean molecular velocity, and the negative sign signifies that heat flow occurs down the temperature gradient. Comparison with Fourier's Law shows that

$$\kappa = (1/2)\sigma C_v \bar{c}\lambda \tag{2.38}$$

More rigorous treatments result in a similar relationship but with a different numerical factor. Each leads to an expression close to

$$\kappa = (1/3)\sigma C_v \bar{c}\lambda \tag{2.39}$$

The mean free path (λ) is inversely proportional to the gas density and $\bar{c}$ is proportional to $T^{1/2}$, so κ is virtually independent of pressure but increases with temperature very roughly as $T^{1/2}$. The measured value for air at room temperature is about 0.025 $W\,m^{-1}\,K^{-1}$ and hence the thermal diffusivity is about $2\times10^{-5}\,m^2 s^{-1}$.

Diffusion is described by Fick's Law

$$\frac{\dot{n}}{A} = -D\frac{dn}{dx} \tag{2.40}$$

where $\dot{n}$ is the number of molecules per unit time crossing an area, A, (dn/dx) is the concentration gradient, and D is the diffusion coefficient with the dimensions $m^2 s^{-1}$. A simple kinetic theory derivation, analogous to that for thermal conductivity, yields the prediction that the *self-diffusion coefficient* is equal to $(1/2)\,\bar{c}\lambda$. The diffusion coefficient D is inversely proportional to pressure and, in practice, increases with temperature according to about $T^{1.5}$. The measured value of D for a gas molecule such as nitrogen diffusing in air is around $2\times10^{-5} m^2 s^{-1}$.

Derivations of the equations for heat and mass fluxes pertain to the behaviour in quiescent gas or in a laminar flow (i.e. transport through a cross-section which is perpendicular to the imposed gradient as a result of molecular collisions), when there is no forced mixing induced by turbulent motion of the fluid. The expression for the diffusion coefficient comprises the product of a velocity and a scale factor. Similar parameters are used to describe transport owing to turbulent or eddy motion but with values characteristic of the eddies rather than the molecular motions.

The numerical values for the thermal diffusivity and the diffusion coefficient of the same molecules are very similar and the dimensionless ratio, a/D, the *Lewis number* (Le), is close to unity. It represents the ratio of conductive to diffusive fluxes and theoretical solutions to many problems in combustion are obtained by setting this number equal to unity. This implies that both a and D show the same dependence on temperature.

2.5 Grouping of parameters and dimensionless groups

Many of the mathematical equations which represent the physics and chemistry of combustion are written in a non-dimensional form. The equation may not be so cumbersome to write when presented in this way, thereby reducing the chances of error in subsequent algebraic manipulation, but that is only a peripheral advantage. The mode of application may be relatively straightforward, such as the in non-dimensionalisation of a spatial co-ordinate by dividing the distance by a characteristic dimension of the system, e.g. let $\rho = l/r$ where r is the radius of a spherical reaction vessel and l is the distance from the centre to a given position along that co-ordinate. Derivations involving this type of non-dimensionalisation become more general in application. That is, the size of the system need not be specified at the outset, but it may mean also that the equation could be applied to a different shape reactor if a *characteristic dimension* can be specified that replaces the radius r.

In other circumstances, groups of parameters are collected together so that the overall phenomonology, as predicted in a mathematical equation, is described by a more global 'lumped parameter'. It may be necessary to investigate the dependence on a specific physical parameter within such a group at a later stage. Thermal diffusivity, a, which is described in the previous section, is an example of a partial non-dimensionalisation. The Lewis number, which incorporates a, is an example of a non-dimensionalised parameter, or Le could be given its full identity in the form

$$Le = \frac{\kappa}{D\sigma C_p} \tag{2.41}$$

Further reading

Atkins, P.W. (1994). *Physical Chemistry* (5th edn). Oxford University Press, Oxford, UK.

Benson, S.W. (1960). *The Foundations of Chemical Kinetics.* McGraw-Hill, New York, USA.

Benson, S.W. (1976). *Thermochemical Kinetics* (2nd edn). John Wiley, New York, USA.

Bett, K.E., Rowlinson, J.S. and Saville, G. (1975). *Thermodynamics for Chemical Engineers.* Athlone Press, London, UK.

Caldin, E.F. (1958). *An Introduction to Chemical Thermodynamics.* Clarendon Press, Oxford, UK.
Glassman, I. (1986). *Combustion* (2nd edn). Academic Press, New York, USA.
Hirschfelder, J.O., Curtiss, C.F. and Bird, R.B. (1954). *Molecular Theory of Gases and Liquids.* John Wiley, New York, USA.
Laidler, K.J. (1965). *Chemical Kinetics* (2nd edn). McGraw-Hill, New York, USA.
Mulcahy, M.F.R. (1973). *Gas Kinetics.* Nelson, London, UK.
Parsonage, N.G. (1969). *The Gaseous State.* Pergamon Press, Oxford, UK.
Pilling, M.J. and Seakins, P.W. (1995). *Reaction Kinetics* (2nd edn). Oxford University Press, Oxford, UK.
Tabor, D. (1979). *Gases, Liquids and Solids* (2nd edn). Cambridge University Press, Cambridge, UK.
Warn, J.R.W. (1969). *Concise Chemical Thermodynamics.* Van Nostrand Reinhold, London, UK.
Welty, J.R., Wilson, R.E. and Wicks, C.E. (1976). *Fundamentals of Momentum, Heat and Mass Transfer* (2nd edn). John Wiley, New York, USA.

Problems

(1) (a) Use the data in Table 2.1 to compare the maximum thermal energy available from hydrogen, methane, ethane, propane, ethyne (acetylene) ethene and methanol vapour burning in air. The comparison should be made with respect to unit mass of reactants, unit mass of fuel, and unit volume of fuel alone.
(b) Given that the enthalpy of vaporisation of liquid methanol is $39.23\,kJ mol^{-1}$ and its density is $790\,kg\,m^{-3}$, also calculate the thermal energy available from the combustion per unit volume of liquid methanol. (*Note*: Some important conclusions regarding the advantages and disadvantages of different fuels may be drawn from these simple calculations.)

(2) Use the simple formulae which precede eqn (2.15) to estimate mean molar heat capacities at constant pressure and constant volume of carbon dioxide, water vapour and nitrogen. (Remember that the geometrical structures of CO_2 and H_2O are different.) Compare the values obtained with those given by the data in Table 2.2 for temperatures of 298 and 1500 K.

(3) Use the mean molar heat capacities obtained above to estimate adiabatic reaction temperatures at constant pressure and at constant volume for stoichiometric mixtures of propane in air and propane in oxygen. Compare the results with the accurate values in Tables 2.3 and 2.4.

(4) Nitric oxide (NO) is an important combustion-generated pollutant. Use the relation

$$\Delta G_T^{\ominus} = -RT \ln K^{\ominus}$$

to obtain an expression for the equilibrium concentration of NO in air as a function of temperature, assuming that NO, N_2 and O_2 have identical entropies at all temperatures (How good is this assumption?). Use this expression to estimate the mole fraction of NO in air at 300, 1000, 2000 and 5000 K. What implications do these results have for the operation of internal combustion engines?

(5) The following reactions all participate in the combustion of hydrogen:

Reaction	A	E(kJ mol^{-1})
$H_2 + O_2 \rightarrow 2OH$	2.5×10^{9} dm^{3}mol^{-1}s^{-1}	163.0
$H + O_2 \rightarrow OH + O$	2.2×10^{11} dm^{3}mol^{-1}s^{-1}	70.3
$O + H_2 \rightarrow OH + H_2$	1.7×10^{10} dm^{3}mol^{-1}s^{-1}	39.5
$OH + H_2 \rightarrow H_2O + H$	2.4×10^{10} dm^{3}mol^{-1}s^{-1}	21.6
$H + O_2 + M \rightarrow HO_2 + M$	3.8×10^{9} dm^{6}mol^{-2}s^{-1}	0

Evaluate the rate of each reaction for a stoichiometric mixture of hydrogen and oxygen at 3000 K and atmospheric pressure. Assume the active species are each present in concentrations equal to 1% of the total reactants. The first and last of these elementary steps have relatively small rates. Explain why they may still play an important part in the overall reaction. The other three steps form a chain cycle. Which of them would you expect to dominate the combustion behaviour of the hydrogen–oxygen system?

(6) Assume that ethane pyrolyses by the following mechanism:

Reaction	A	E(kJ mol^{-1})
$C_2H_6 \rightarrow CH_3 + CH_3$	$10^{16.9}$ s^{-1}	374.0
$CH_3 + C_2H_6 \rightarrow CH_4 + C_2H_5$	$10^{9.6}$ dm^{3}mol^{-1}s^{-1}	73.2
$C_2H_5 \rightarrow H + C_2H_4$	$10^{14.4}$ s^{-1}	171.0
$H + C_2H_6 \rightarrow H_2 + C_2H_5$	$10^{11.1}$ dm^{3}mol^{-1}s^{-1}	39.2
$C_2H_5 + C_2H_5 \rightarrow C_4H_{10}$	$10^{8.5}$ dm^{3}mol^{-1}s^{-1}	0

By use of the stationary state approximation with respect to the concentrations of H, CH_3 and C_2H_5, show that the reaction will follow a half-order rate law with respect to the C_2H_6 concentration and calculate the global Arrhenius parameters. What will be the major and minor products of this reaction? Calculate the stationary-state concentrations of the radical species, the rate of removal of ethane and the rate of formation of butane at an ethane pressure of 13.33 kPa and a temperature of 800 K.

(7) In order to obtain the equation of 'best fit' to experimental data for the rate constant of the reaction

$$CH_4 + O \rightarrow CH_3 + OH$$

the recommended rate constant over the temperature range 250–2500 K is given by

$$k = 1.57 \times 10^7 T^{1.83} \exp^{(-1400/T)} \mathrm{cm^3 mol^{-1} s^{-1}}$$

Derive the equivalent Arrhenius parameters (A and E/R, in the form $k = A \exp^{(-E/RT)}$) that match the rate constant from the three-parameter expression at 750 K and 1500 K. To what extent does the predicted rate constant from the two-parameter representation differ from the recommended value at the adiabatic flame temperature of methane in air (2250 K)?

(8) Use the expressions

$$\bar{c} = (8RT/\pi M)^{1/2} \text{ and } \lambda = 1/\sqrt{2}\pi n d^2$$

where n is the number of molecules per unit volume, d is the molecular diameter and M is the molar mass, to estimate the dependence of the diffusion coefficient and the thermal conductivity on the pressure, temperature and molar mass of an ideal gas.

(9) Estimate $\bar{c}$, λ, D and κ for air at room temperature and atmospheric pressure, assuming that $d = 0.3$ nm. Hence calculate the Lewis number for air and compare it with the value obtained by using the kinetic theory expressions for a and D.

3 Flames

3.1 Introduction

A flame is caused by a self-propagating exothermic reaction which usually has a luminous reaction zone associated with it. A premixed flame will propagate through a stationary gas at a characteristic velocity termed the *burning velocity*, or it may remain in one place if the reactant gas is forced to move towards the flame front at the same speed. In general, the second type of flame is not particularly stable but may be made so by use of a suitable burner. Apart from initial points of principle, the major part of the present discussion will be concentrated on practical features of laminar, premixed flames. Some contrasting features of diffusion flames are also discussed. In addition, some of the modern techniques that are used for the experimental study of flames are briefly reviewed.

3.2 Mass and energy conservation in premixed flames

The effect of the reaction tube with regard to loss of heat or reactive species, or viscous drag on the flowing gas may be neglected in an ideal propagation of a premixed flame in a tube. The gas flow is considered to be laminar and uniform across the diameter so that the flame front is planar and perpendicular to the flow. The flow rate is also equal in magnitude but opposite in sign to the burning velocity, so that the flame is stationary.

Consider the following conservation equations at two planes perpendicular to the flow, one of which is ahead of and the other behind the flame (Fig. 3.1).

Conservation of mass:

$$\sigma_u S_u = \sigma_b S_b = \dot{m} \tag{3.1}$$

where $\dot{m}$ is the (constant) mass flow rate per unit area through the transition, S is the velocity of the gas stream and σ is its density. The subscripts u and b refer to unburned and burned gas, respectively. This terminology follows convention in which S_u is defined as the burning velocity.

Conservation of momentum (expressed as the flux):

$$p_u + \sigma_u S_u^2 = p_b + \sigma_b S_b^2 \tag{3.2}$$

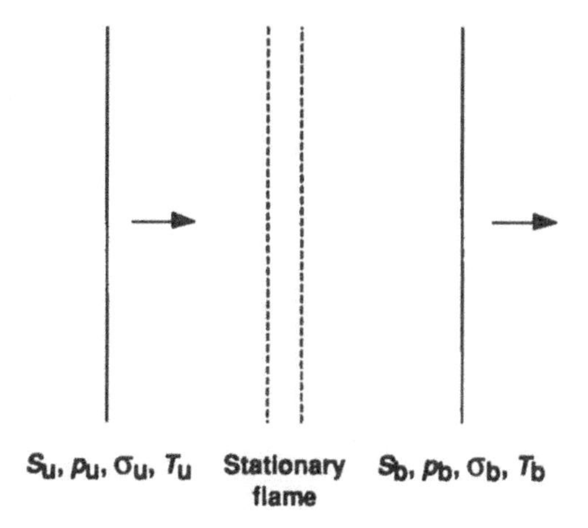

Figure 3.1 Notation used in the description of a stationary flame. The subscripts u and b refer to the unburned and burned gases, respectively.

Rearrangement of these relationships gives

$$\dot{m}^2 = \frac{(p_b - p_u)}{(\sigma_b - \sigma_u)} (\sigma_u \sigma_b) \tag{3.3}$$

$$-\dot{m} = \frac{(p_b - p_u)}{(S_b - S_u)}$$

$$\dot{m} = \frac{(p_u - p_b)}{(S_b - S_u)} \tag{3.4}$$

From eqn (3.3), the pressure and the density must change in the same direction. Thus, there are two types of process possible, namely detonations (Chapter 5), in which both pressure and density increase across the transition, and deflagrations (i.e. combustion waves and flames), in which pressure and density decrease across the transition. Deflagrations are low-velocity expansion waves in which chemical reaction is brought about by heat and mass transport.

According to eqn (3.4) the pressure decrease observed in deflagrations requires there to be an increase in velocity. Thus, the burned gas leaving the flame has a higher velocity, a lower pressure and a lower density than the initial reactant gas (Fig. 3.2 [16]). When the release of chemical energy is taken into account with the consequent rise in temperature, it is found that the pressure change across the flame is so low that it is usually unimportant, apart from being involved in certain distortion effects (see

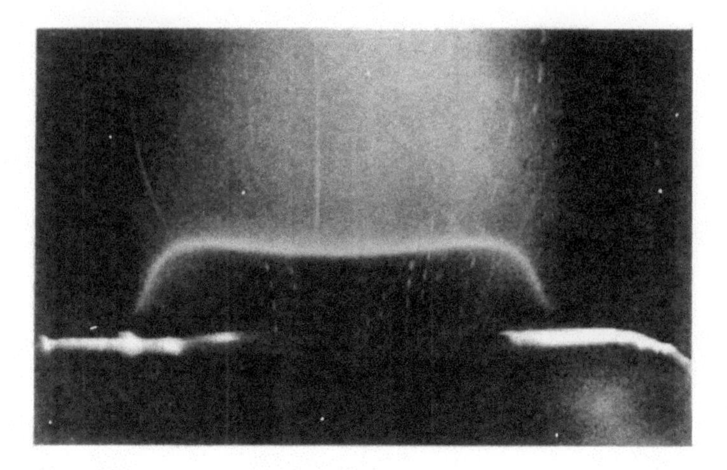

Figure 3.2 Particle tracking (Al_2O_3, ~2 μm dia.) in a button-shaped, premixed laminar flame showing the divergence of flow through the flame front. The particle streaks were obtained by photography with stroboscopic light reflected at right angles to the main beam (from Dixon-Lewis and Islam [16], by courtesy of The Combustion Institute).

section 3.7). In principle, the pressure and density differences across the flame may be used to determine the burning velocity:

$$S_u^2 = \frac{(p_u - p_b)}{\sigma_u((\sigma_u/\sigma_b) - 1)} \tag{3.5}$$

but in practice the method is inaccurate because the differences are so small.

3.3 Structure of the ideal, adiabatic, one-dimensional, laminar, premixed flame

In the detailed structure of the flame, the temperature must increase smoothly from the initial to the final state. The intermediate and product concentrations will increase similarly, whereas the concentrations of fuel and oxidant must show a corresponding decrease (Fig. 3.3). The visible part of the flame is located in the reaction zone and the emission is due largely to electronically excited species, such as CH, CN, C_2, CHO, and also CO_2 emitting light, as they return to their ground state. The origins are discussed in more detail in Chapter 6.

An element of the flowing gas can receive heat in two ways, either from chemical reactions occurring within it or by conduction from the hotter gas ahead of it. Two distinct regions can be recognised, which are attributable to these processes. They are separated by a point of inflexion in the temperature profile. Beginning at low temperatures, for any given cross-section, the heat flow into the cooler region owing to conduction is greater than the corresponding heat loss because the gradient is steeper

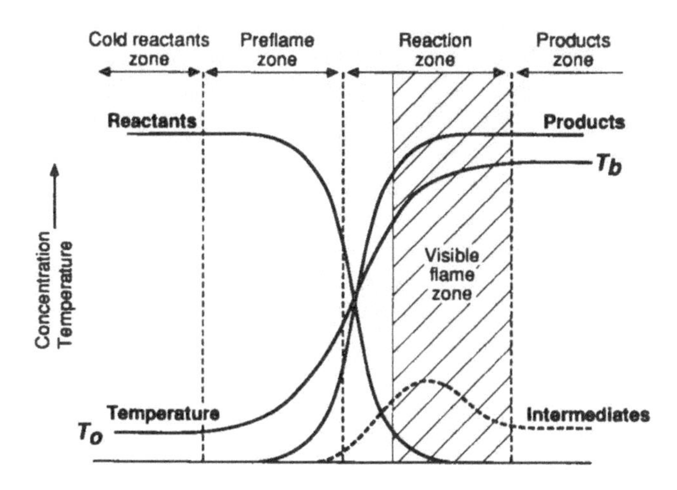

Figure 3.3 Concentration and temperature profiles associated with a one-dimensional, premixed, adiabatic flame.

on the high temperature side. Beyond the point of inflexion the converse is true; that is, the heat loss rate exceeds the heat gain rate. However, at this higher temperature the reaction rate has increased sufficiently for a significant amount of heat to be produced by chemical reaction. The temperature therefore continues to increase through the flame although at a progressively slower rate. It eventually reaches a constant value when all the fuel has been consumed and reaction has ceased. Parallel behaviour also occurs in the reactant concentration profile, the major loss initially being diffusion of fuel into the flame but subsequently being consumption by chemical reaction. The first region is termed the *preheat* or *preflame* zone and the second the *reaction* zone. A third region may also be defined which follows the reaction zone, and is termed the *post-flame* or *product zone*.

The concentration of all species in the post-flame zone must approach that defined by thermodynamic equilibrium at the prevailing temperature (e.g. Tables 2.3 and 2.4). The difficulty in the reaction zone is that species may not reside there for a sufficiently long period for thermodynamic equilibrium to be established. Since the predominant termination reactions are termolecular processes, and therefore they are relatively slow, the species most affected and, by implication, held at concentrations above those at thermodynamic equilibrium, are the propagating free radicals. Since bimolecular interactions tend to be equilibrated, all other intermediate species tend to be enhanced to *super-equilibrium* concentrations (Fig. 3.3). These processes are discussed in more detail in Chapter 6.

3.4 Properties of laminar, premixed flames

Two very important properties of the typical laminar premixed flame are the *burning velocity*, S_u, and the *adiabatic flame temperature*. Examples are listed in Table 3.1, with other important properties. The laminar burning velocity of a given fuel + oxidant mixture is defined in a formal way as the velocity with which a plane flame front moves normal to its surface through the adjacent unburned gas. Despite the difficulty of making very accurate measurement of burning velocity [10, 17], values are known quite precisely for many systems, and are generally between 0.1 and 1.0 m s^{-1}. The term 'flame speed' is used sometimes for S_u. However, it seems preferable to reserve this name for the speed of a non-stationary flame through an initially quiescent mixture. The observed flame speed in such cases includes a component owing to movement of gas ahead of the flame.

The premixed gas composition is usually expressed in terms of an equivalence ratio, φ, which is the actual fuel:oxidizer ratio divided by the fuel:oxidizer ratio corresponding to complete combustion to carbon dioxide and water. The latter is often referred to as the stoichiometric fuel:oxidizer ratio (for which $\varphi = 1$), and the term 'fuel-rich' implies a mixture which contains a higher proportion of fuel. The fuel + oxidant ratio has a marked effect on the burning velocity, which has its maximum value in a marginally fuel-rich mixture, whereas it shows only a small dependence on the pressure and temperature of the reactant gases, usually increasing at reduced pressures or elevated temperatures.

Maximum flame temperatures usually lie in the range 2200–2400 K for hydrocarbon mixtures in air and 3000–3400 K for mixtures in oxygen

Table 3.1 Flammability limits and flame properties for some common gases at atmospheric pressure[a]

	Flammability limits (% fuel by volume)		Stoichiometric composition	Flame temperature	Maximum burning
Reactants	Lower	Upper	(% fuel by volume)	(K)	velocity (m s^{-1})
H_2+O_2	4.0	94	66	3083	11.0
$CO+O_2$	15.5	94	66	2973	1.08
CH_4+O_2	5.1	61	33	3010	4.5
$C_2H_2+O_2$	—	—	—	3431	11.4
H_2+air	4.0	75	28.5	2380	3.1
CO+air	12.5	74	28.5	2400	0.45
CH_4+air	5.3	15	9.0	2222	0.45
C_2H_2+air	2.5	80	7.4	2513	1.58
C_2H_4+air	3.1	32	6.25	2375	0.75
C_2H_6+air	3.1	15	5.4	2244*	0.40
C_3H_8+air	2.2	9.5	3.8	2250*	0.43
n-C_4H_{10}+air	1.9	8.5	3.1	2365*	0.41
C_6H_6+air	1.5	7.5	2.7	2365*	0.41
C_2H_4O+air	3.0	80	7.75	2411*	1.05

[a] The flammability limits (after Coward and Jones [18]) apply to upward propagation in tubes. The flame temperatures refer to stoichiometric mixtures, except those marked* which are maximum values.

(Table 3.1). The cyanogen + oxygen system has an adiabatic flame temperature of 5000 K, which is associated with the stability of the products towards dissociation rather than to an unusually high heat of reaction. Flame temperatures are notoriously difficult to determine and in many cases the quality of the thermodynamic data available is sufficiently high for calculations of the adiabatic flame temperature to be more reliable than experimental measurements [10].

3.5 Flammability limits

Flames can occur only in mixtures within a certain composition range, bounded by the *flammability limits*. These limits refer to the composition range within which ignition and flame propagation can be brought about by the application of an external stimulus, e.g. a spark or pilot flame. The phenomenon is distinct from that of spontaneous ignition, or autoignition, as discussed elsewhere. Flammability limits must be measured under conditions which are not influenced by quenching effects, as discussed in section 3.6, which means that a reaction tube must be of suitably large diameter. The source must also be of sufficient energy to guarantee ignition, otherwise the property under investigation would be that of the limiting ignition energy and not of flammability.

The flammability limits are usually expressed as percentages of fuel by volume (Table 3.1). There are enormous variations between fuels. For

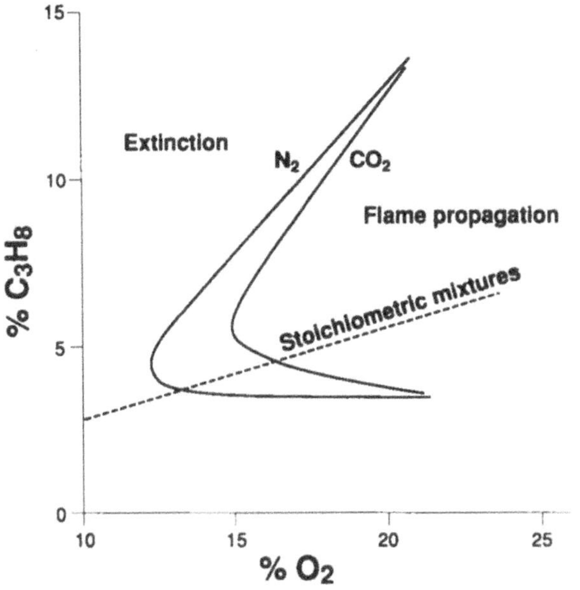

Figure 3.4 Flammability limits for $C_3H_8 + O_2$ mixtures in the presence of added nitrogen or carbon dioxide (after Coward and Jones [18]).

acetylene the range of flammability in air is 2.5–80% while for propane the range is 2.2–9.5%. The lower limit shows less variation for many fuels: it tends to be at about 60% of the stoichiometric composition and takes the same value in both air and oxygen. As progressive amounts of inert gas are added the rich (or upper) and lean (or lower) flammability limits approach each other and eventually coincide (Fig. 3.4 [18]). There is a slightly wider range of flammable compositions for upward rather than downward flame propagation. This is indicative of a role played by heat transport, insofar that a lower density, hot gas plume is able to travel ahead of the flame in upward propagation, which is an aid to development in the preheat zone.

S_u has a maximum fairly close to the stoichiometric composition and falls as the flammability limits are approached. In some systems the limits seem to correlate with a minimum flame temperature (*c.*1400 K for methane), the burning velocity remaining small but finite to the point of extinction. The existence of flammability limits is predicted by theories of flame propagation provided heat loss from the burned gas is included [19]. The heat loss may occur by conduction, convection or radiation and cannot normally be interpreted by a one-dimensional treatment so that the mathematical analysis becomes very complex.

3.6 Flame quenching

A related phenomenon is that of flame quenching close to a solid surface. Although loss of active species may also be involved in quenching, the nature of the surface does not appear to affect its quenching properties and it seems that the flame is extinguished primarily owing to heat loss. Differences in thermal conductivity between various surfaces are unimportant because the heat capacities of solids are high compared with those of gases: they may be regarded as an 'infinite heat sink' as far as a travelling flame is concerned. This is not the case for a stationary flame.

The quenching diameter d_T of a particular gas mixture is the minimum diameter of tube through which a flame in the stationary gas mixture can propagate. The quenching distance d_Q is a related quantity and refers to flame propagation between parallel plates (Table 3.2). The two quantities are related by the expression $d_T = 1.54 d_Q$ [20]. Various theories of flame quenching lead to the conclusion that the dimensionless group (a Peclet number) $d_Q S_u / a$ should be constant. In this expression a is the thermal diffusivity of the unburned gas.

Diffusion and heat transfer are both inversely proportional to pressure, whereas for most flames the burning velocity is largely independent of pressure. Consequently, the quenching distance is also approximately inversely proportional to pressure. Similarly, fast-burning flames have small quenching distances associated with them. This result has been confirmed by experiment, from which values of $d_Q S_u / a$ are generally found to be between 40 and 50 [21].

Table 3.2 Quenching distances for various stoichiometric flames at 101 kPa and 293 K [20]

Reactants	d_Q(mm)	Reactants	d_Q(mm)
$H_2 + O_2$	0.2	H_2 + air	0.6
$CH_4 + O_2$	0.3	CH_4 + air	2.5
$C_2H_2 + O_2$	0.2	C_2H_2 + air	0.5
$C_2H_4 + O_2$	0.1	C_2H_4 + air	1.25
$C_3H_8 + O_2$	0.25	C_3H_8 + air	2.1
		$i\,C_8H_{18}$ + air	2.6
		C_6H_6 + air	1.9

Quenching of flames is important in flame traps. Such devices are used to prevent flame propagation through flammable gases, as in the miner's safety lamp, designed by Sir Humphry Davy, in which the lamp flame was surrounded by a fine copper gauze. The mesh size is crucial in determining whether or not the passage of a particular flame would be stopped. In industrial applications, flame traps frequently comprise an assembly of narrow bore, thin-walled tubes which have a minimal resistance to gas flow while preventing flame propagation.

3.7 Stabilisation of flames on burners

A stationary flame obtained by flowing the premixed gases at the same speed in the reverse direction would have only neutral stability and its position would shift in an uncontrolled manner. In domestic and industrial appliances, flame stability is achieved by attaching the flame to a burner. One of the simplest practical devices is the Bunsen burner, as illustrated in Fig. 3.5. The design is such that its performance is illustrative of both laminar and diffusion flames.

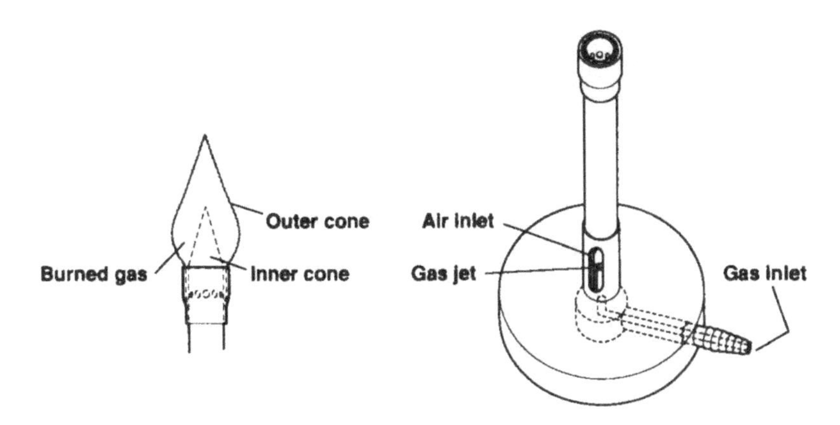

Figure 3.5 The Bunsen burner. The inner cone is a fuel-rich, laminar, premixed flame front. The outer cone is a diffusion flame.

The fuel gas, entering through the nozzle at the base, entrains air, owing to the Venturi effect, and the two gases mix as they travel up the tube. The inner cone is the reaction zone of a premixed flame. At this point the mixture is fuel-rich so the composition of the burned gas does not correspond to complete combustion. The outer cone is then a diffusion flame between the burned gas and the surrounding air. Although the two flames coincide at the burner rim, they can be isolated by use of a Smithells separator [22].

For a typical laminar, premixed flame the burner fulfils three functions:

(i) it permits mixing the fuel and oxidant in the appropriate proportions;
(ii) it provides a suitable section to establish laminar flow; and
(iii) it acts as a heat sink which restricts movement of the flame.

The actual stabilising effect of the burner is provided by the rim alone, and a simple metal ring will perform the same function. The effect of the rim is to remove heat (and possibly active species) from the flame and hence to reduce the burning velocity in its vicinity.

For a laminar flow burner, the flow velocity is very low close to the walls and above the rim, but increases towards the centre of the burner, giving a parabolic velocity profile. At all points within the rim but outside the quenching distance, the flow velocity exceeds the burning

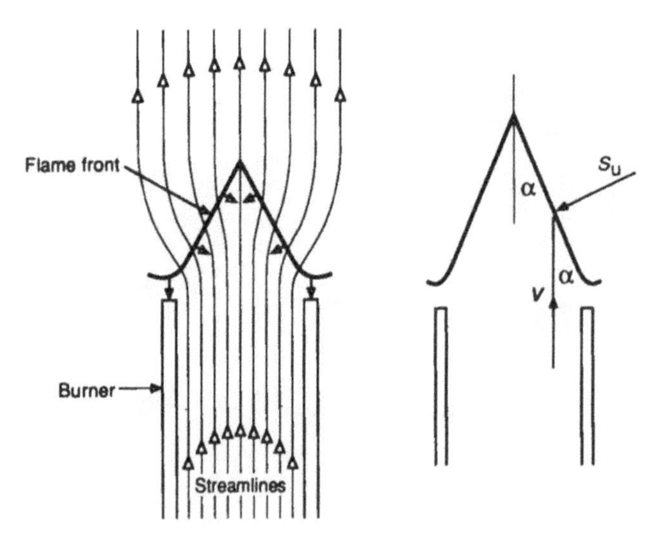

Figure 3.6 Streamlines through a laminar, premixed burner flame. The profile of the open arrows indicates the relative velocities in the burned and unburned region. The burning velocity is shown in relation to the flow velocity and cone angle: $S_u = v \sin\alpha$.

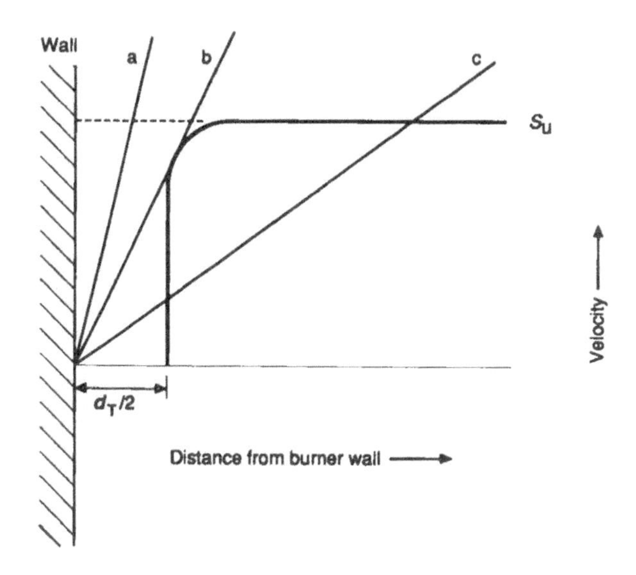

Figure 3.7 Burning velocity and gas velocity inside a burner tube (after Lewis and von Elbe [23]). The lines a, b and c are defined in the text. d_T is the flame quench diameter.

velocity and the flame is inclined upwards from the rim so that the burning velocity equals the component of the flow velocity normal to the flame front. This accounts for the familiar conical appearance of the flame (Fig. 3.6). Measurement of the gas flow velocity and the cone angle using a nozzle designed to give a flat velocity profile forms the basis of one method of measuring the burning velocity.

A flame can be stabilised on the burner only between certain flow velocity limits. If the gas flow is progressively reduced, a point will be reached eventually at which the burning velocity exceeds the gas velocity somewhere across the burner diameter. At the *flash-back limit*, the flame will become unstable and will propagate back down the burner tube, as shown in Fig. 3.7 [23] for various gas flow velocity gradients. If the gas velocity varies according to line *a*, then the flow velocity within the quench zone exceeds the burning velocity and the flame will rise out of the tube until it can stabilise by adopting a conical shape. If the gas velocity varies according to line *c*, S_u exceeds the flow velocity and the flame will flash-back. The critical gradient of flow velocity is given by *b*, which corresponds to the flash-back limit and is approximately equal to $2S_u/d_T$.

A little above the flash-back limit, a tilted flame may occur. The back pressure of the flame allows distortion of the flow, such that the flame may enter the burner in the region where the flow velocity is reduced. The stable, distorted or tilted flame then exists because the constraints of the burner tube lead to less distortion of the flow (Fig. 3.8).

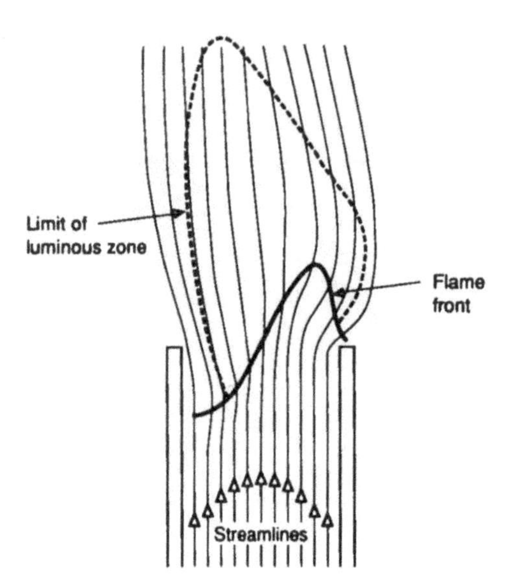

Figure 3.8 Formation of a tilted flame.

A flame can be affected by the entrainment of atmospheric air towards the edge of the burner. This is not particularly important when the flame lies close to the rim but as the flow velocity is increased the flame will rise in order to allow the burning velocity to increase. The dilution by atmospheric air then causes the burning velocity to fall. Therefore, the flame will continue to rise and eventually become unstable at the ***blow-off limit***. This limit is also characterised by a critical velocity gradient.

In general, the values of the gas velocities at the flash-back and blow-off limits will depend on burner dimensions and on the gas composition. For fuel-rich mixtures and high gas velocities, a second phenomenon may also occur. Owing to the entrainment of atmospheric air, the mixture will become progressively more lean above the burner eventually approaching the stoichiometric composition. Because the burning velocity is also increased, a lifted flame is able to form some distance above the burner.

The lifted flame displays two stability limits similar to those of the seated flame. When the gas velocity is reduced, drop-back occurs, the flame taking up its normal position on the burner. At high velocities the flame will blow-out. The velocity–composition regions within which each flame is stable are illustrated schematically in Fig. 3.9. Note that there are conditions in which either flame may occur.

Owing to the phenomenon of quenching described in the previous section, a flame cannot be supported on a burner whose dimensions fall below the quenching diameter, which depends on the gas composition and pressure. Because this critical diameter is approximately proportional to

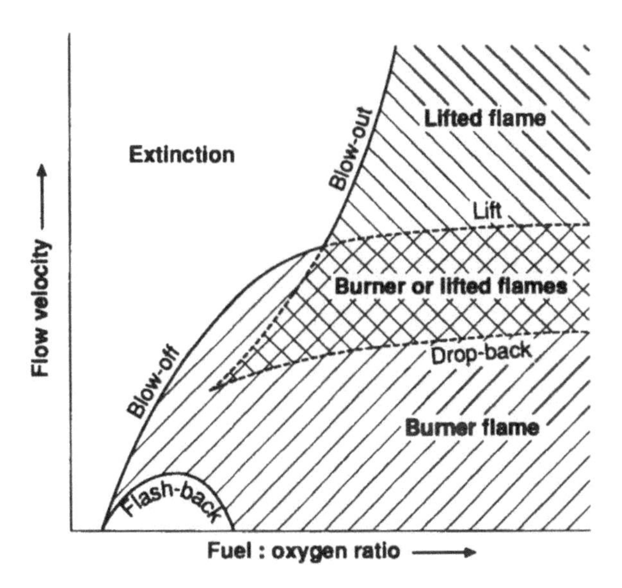

Figure 3.9 Schematic representation of limiting flow velocities for various premixed flame structures as a function of fuel to oxygen ratio.

the reciprocal of the pressure, its value is most easily determined by measuring the pressure at which the flame is extinguished for given sizes of burner. The flow velocity limits for flash-back g_F and blow-off g_B also coincide when quenching occurs.

Clearly, for both industrial and domestic applications, the tube diameter must exceed the quenching distance. It is also found that the gas flow velocity must lie between $2S_u$ and $5S_u$ for a stable cone to form. The flash-back and blow-off behaviour then impose upper and lower limits on the velocity to diameter ratio. Finally, flow up the burner tube must be laminar and therefore the Reynolds number ($Re \equiv du\sigma/\eta$) must not exceed 2300. These constraints are summarised graphically in Fig. 3.10 [24], in which the hatched area indicates the relatively restricted velocity–diameter regime in which a burner can function.

Typical laminar burning velocities for hydrocarbon-air mixtures are about 0.4 m s^{-1} and under turbulent flow conditions are unlikely to be much above 1.5 m s^{-1}. The Bunsen burner will accept flow velocities up to five times the burning velocity so that the maximum flow velocity which can be considered with this design of burner must be about 2 m s^{-1}.

The equivalence ratio has to be optimised for efficient combustion and, in a practical sense is controlled by the ratio of the burner tube area to fuel orifice area. The fuel orifice area must be below the quenching diameter. The momentum of the fuel entering through the orifice area

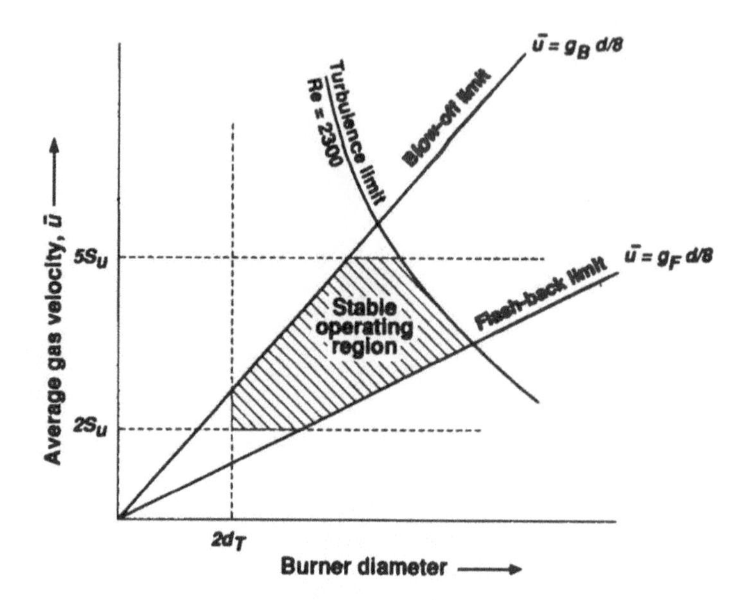

Figure 3.10 The relationship of mean gas velocity to burner diameter for stable operation of a Bunsen burner (after Glassman [24]).

A_F at a velocity u_F is $A_F \rho_F u_F^2$. Entrainment of air then occurs so that momentum is conserved, and

$$A_B \rho_M u_M^2 = A_F \rho_F u_F^2 \tag{3.6}$$

where A_B is the area of the burner tube and the subscript M refers to the mixture. The volumetric fuel to mixture ratio V_F/V_M determines the equivalence ratio, and is given by $A_F u_F / A_B u_M$, or

$$\frac{V_F}{V_M} = \frac{(\sigma_M)^{1/2}}{\alpha \rho_F} \tag{3.7}$$

where $\alpha = A_B/A_F$.

Equation (3.7) may be used if the burner dimensions have to be modified for a different fuel, as happened during the 1960s in the UK when burners operating on coal gas (or town gas), which was primarily a mixture of hydrogen and carbon monoxide, were converted to use natural gas drawn from the North Sea. The dimensions are derived for the correct equivalence ratio, the appropriate fuel orifice having been selected.

3.8 Flame stabilisation at high velocity

The flow velocities in ramjet and turbojet engines (Chapter 13) are about two orders of magnitude higher than in normal burners, so flow stabilisation has to be achieved in a quite different way. Combustion products are recirculated so that they continually reignite the oncoming gas and thus prevent the flame from being 'blown away'. Recirculation may be achieved either by redirecting some of the gas flow so that it travels normal to or even against the main flow (aerodynamic stabilisation) or by inserting a suitable surface which produces eddies (bluff-body stabilisation). Aerodynamic stabilisation is normally associated with the small combustion chambers used in turbojet engines while bluff-body stabilisation is more common in large combustors and is used in the afterburners of gas turbines and in ramjet engines.

3.9 Diffusion flames: properties and chemical background

Combustion occurs at the interface between the fuel gas and the oxidant gas and the burning process, such that it depends more on the rate of mixing in diffusion flames, or jet flames, than on the rates of the chemical processes involved. It is more difficult to give a general treatment of diffusion flames largely because no measurable parameter, analogous to the burning velocity, can be used to characterise the burning process. Although less well characterised because of the difficulties of study and of interpretation, diffusion flames are very important industrially.

Since the dominant physical process is that of mixing, it is possible to make a clear distinction between flames which involve two different flow regions. In slow-burning diffusion flames, typified by candle flames, the fuel rises slowly and laminar flow ensues. The mixing process occurs solely by molecular diffusion, and thus the properties of the flame are determined only by molecular quantities. In industrial burners, and in gas turbines, where the fuel is usually introduced in the form of discrete droplets, burning is rapid, flow speeds are high and the mixing process is associated with the turbulence of the flow. The aerodynamics of the system will then dominate the molecular properties to a great extent.

The burner plays very little part in the stability compared with premixed flames because the burning region is confined to the interface between the fuel and oxidant and the burner serves simply as a nozzle to provide a directed stream of fuel. Flash-back is impossible down to very low nozzle velocities and a very small flame can be stabilised on the nozzle. As the flow velocity is increased, the flame maintains an essentially cylindrical shape but becomes taller. In a cross-section of the flame it is found that the fuel concentration has a maximum on the axis and falls rapidly at the flame boundary (Fig. 3.11). The oxygen concentration also decreases close to the flame and falls approximately to zero at the boundary. By contrast, the concentration of products is a maximum at the boundary, where the major extent of reaction occurs, and falls away both towards the axis and into the surrounding atmosphere. The flame boundary defines the surface at which combustion is complete, but since reaction is normally very rapid it represents the position at which the fuel to oxygen ratio becomes stoichiometric. Since the fuel concentration decreases with height, the position of the boundary moves towards the axis and converges at the tip where all the fuel is consumed. In principle, therefore, the flame should be conical in shape, reaction being confined largely to a thin zone forming the flame boundary which corresponds closely to the reaction zone in a premixed flame.

This model applies only to the case where the fuel and oxygen are moving upwards at the same velocity. Although concentric burners which simulate this model are used, the commonest diffusion flame is that produced when a jet of fuel enters a stagnant atmosphere. The transfer of momentum close to the boundary leads to the formation of eddies. An eddy of combustion products causes the fuel and oxygen streams to become further separated so that diffusion becomes slower. The height of the flame then increases and the shape becomes more cylindrical. If the eddies break away, the fuel and oxygen are brought into more intimate contact so that the burning rate increases and the flame height is reduced. These eddy effects are responsible for the 'flicker' usually associated with diffusion flames. The flow is only partially laminar although that near the axis may be little affected.

The character of the flame depends on the nature of the gas flow which is usually indicated by the Reynolds number. In turbulent flow, the Reynolds number is inversely proportional to the kinematic viscosity, and the value of this parameter increases in the combustion region. Nevertheless, turbulent flow, characterised by a high Reynolds number,

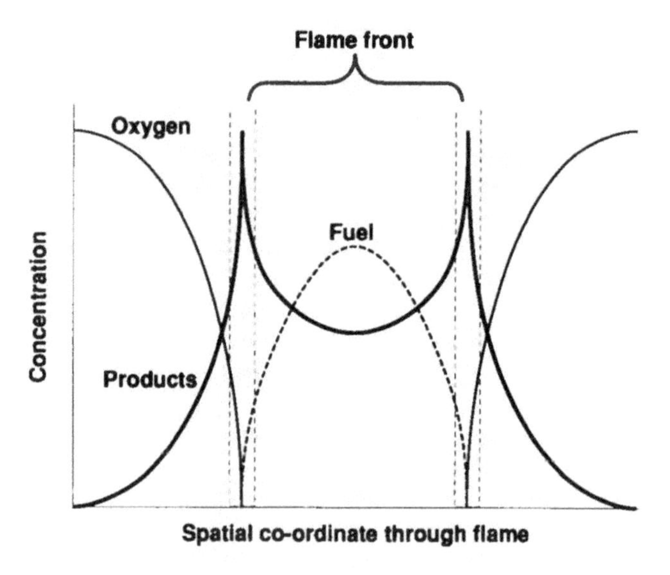

Figure 3.11 Spatial concentration profiles for fuel, oxygen and reaction products through a cross-section of a laminar diffusion flame.

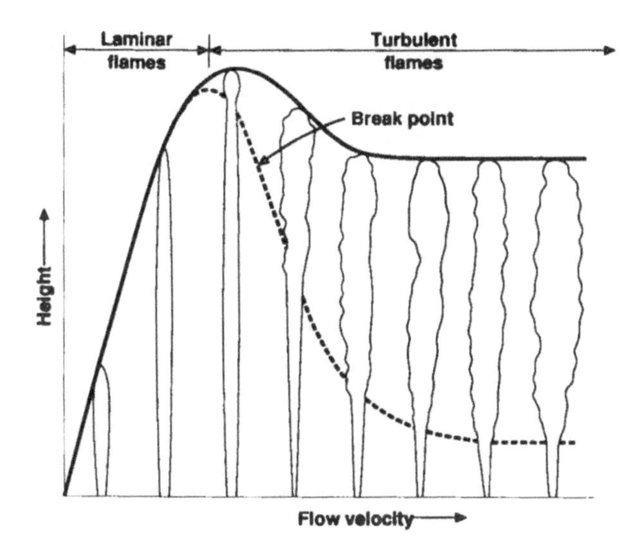

Figure 3.12 Schematic representation of diffusion flame structures as a function of flow velocity showing the transition from a laminar to a turbulent flame (after Hottel and Hawthorne [25]).

may give rise to laminar flow at the flame boundary (and hence gives the typical laminar flow behaviour of tall, cylindrical flames). In these circumstances, mixing is by molecular diffusion rather than by turbulent entrainment. Therefore, the flow criterion must refer to the type of flow in the region in which the flame boundary occurs.

If the nozzle velocity is increased, the laminar flame increases its height almost linearly until turbulent mixing occurs. This appears first at the tip and moves down progressively with increase in velocity. The flame length also falls in this region, the visible boundary spreading outwards. Eventually a stage is reached at which both the height of the flame and the position of the transition, known as the break-point, remain constant irrespective of any additional increase in nozzle velocity (Fig. 3.12 [25]).

In the turbulent flame region, the phenomenon of lift, similar to that observed with premixed flames, may also occur. The reasons are not fully understood but the change in appearance suggests that entrainment of air at the burner rim produces some of the characteristics of the premixed flame and combustion is able to occur when the fuel:oxygen ratio has fallen sufficiently to be within the flammable range. A velocity is eventually reached at which only lifted flames occur and beyond this region there will be a blow-out limit for the lifted flames (Fig. 3.13 [26]).

Diffusion flames can be stabilised in various ways. The usual mode is when a jet of fuel vapour is injected through a burner port or nozzle into

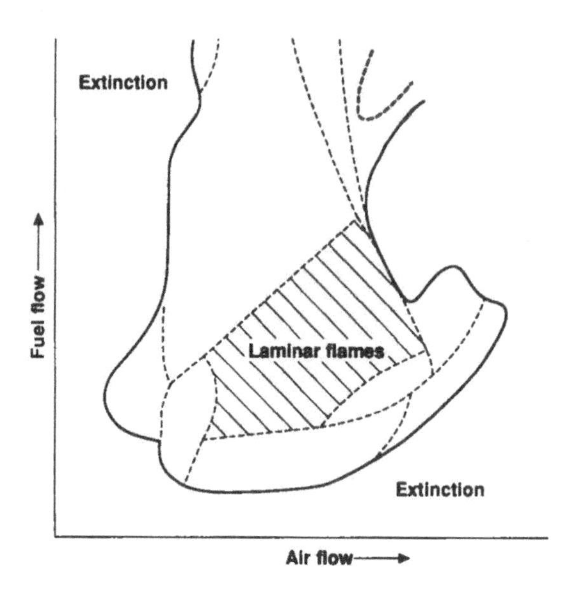

Figure 3.13 Schematic representation of limiting conditions for various diffusion flames. The areas that are not labelled relate to various flame structures that are not discussed here (after Barr [26]).

an oxidising atmosphere, but burners have also been designed in which two parallel streams, one of fuel and one of oxidant, flowing in adjacent ducts, are allowed to come into contact. Such a burner gives a flat flame very suitable for spectroscopic examination. Another special burner giving a flat flame brings opposing flows of fuel and oxidant together giving a counterflow diffusion flame. Flames on wicks also provide examples of diffusion flames; in appearance these flames usually possess an inner blue zone and an outer yellow luminous zone. Samples taken from these zones have demonstrated that the primary chemical processes involve pyrolysis of fuel and that oxidation of these pyrolysis products then takes place at the boundary of the inner flame zone [27, 28].

The concept of a very narrow reaction zone in any diffusion flame is idealised. Considerable reaction occurs on each side of the 'true' flame boundary because heat and mass transfer is possible. As is shown in Fig. 3.11 there is a boundary at which the concentrations of each fall to zero. Reaction takes place largely by pyrolysis of the fuel owing to heat transport and diffusion of active species into the fuel-rich zone, notably OH radicals. The reactions that take place yield intermediates derived from the fuel which then diffuse into the oxidant zone, and they can lead to the appearance of different reaction zones with different physical properties (Fig. 3.14 [10]).

In hydrocarbon flames, the production of carbon particles is characterized by intense luminosity and the shape and behaviour of these

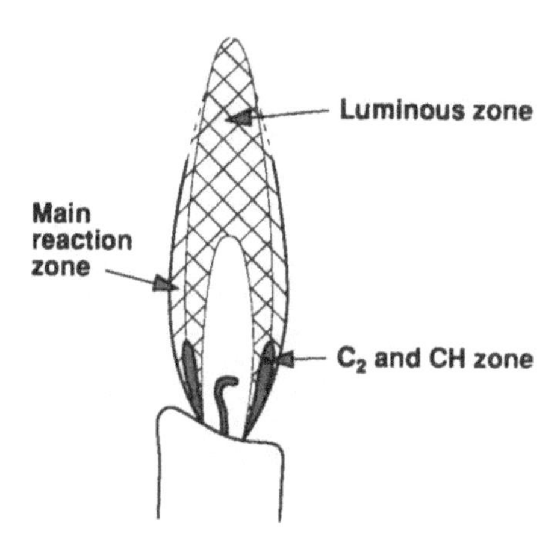

Figure 3.14 Reaction zones in a candle flame (after Gaydon and Wolfhard [10]).

luminous regions can be quite distinct from those of the 'normal' reaction zone. The radiation from diffusion flames corresponds closely to that expected from local thermal equilibrium, indicating that the diffusion of active species into the reactants tends to reduce the probability of obtaining abnormally high local radical concentrations.

3.10 Recent developments in experimental methods for the study of flames

The chemistry, spectroscopy and structure of premixed flames have been studied extensively throughout most of the present century. Flame temperature measurements are foremost, and the techniques used include spectroscopic or other optical methods such as sodium D line reversal, Doppler broadening of translational energies, rotational and vibrational energy distributions in molecules, or emissivity and brightness measurements. Direct measurements with thermocouples have also been made.

Spectroscopic studies in absorption or emission have also contributed substantially to current knowledge about the chemistry of flames. Other methods used to investigate chemical compositions and reaction mechanisms include sampling to mass spectrometers or other analytical equipment, such as gas chromatographs. Mass spectrometry is also used for the detection and measurement of free radicals, by molecular beam sampling from low-pressure premixed, laminar flames [29, 30].

Flame structure is often studied from photography, including Schleiren imaging and interferometry to obtain profiles of density gradients throughout the flame. Ion probes yield data on ion activity, which leads to structural and other information. Gas motion, especially in turbulent conditions, can be characterised by hot wire anemometry or laser Doppler velocimetry (LDV). Burning velocities are notoriously difficult to measure accurately, and a variety of imaginative techniques have been developed over the years to obtain these data. Many experimental techniques used to obtain information on structure, temperature and chemistry are discussed by Fristrom and Westenberg (1965), Gaydon (1974), Gaydon and Wolfhard (1979), Lewis and von Elbe (1956, 1987), and others.

Laser diagnostic methods have been developed over the last two decades, or are still under development, as tools for the investigation of flames in a variety of different environments. Despite the complexity and expense, laser diagnostics have become 'routine' tools in many laboratories, the impetus resulting from an ability to circumvent some shortcomings of the older experimental methods. Major benefits are the non-invasive nature of laser methods and their ability to probe the traditionally difficult environments, such as high pressure, sooty or turbulent combustion systems. Nevertheless, most of the pioneering work has been performed on low pressure, premixed, laminar flames, which is where the more traditional methods have been, and continue to be most used.

In summary, new techniques are required which (i) give *in situ*, spatially and temporally resolved measurements, (ii) probe a small reaction volume deep within the combustion zone or which can be used to obtain instantaneously the variation throughout a two-dimensional plane, and (iii) do not disturb the system by the introduction of a probe. Laser diagnostic methods do not yet satisfy all of these criteria perfectly, but there has been considerable progress towards these goals. The techniques introduced here are laser-induced fluorescence (LIF) and coherent anti-Stokes Raman spectroscopy (CARS). The energetic principles and relevant features are shown in Fig. 3.15. Planar imaging techniques are also briefly surveyed.

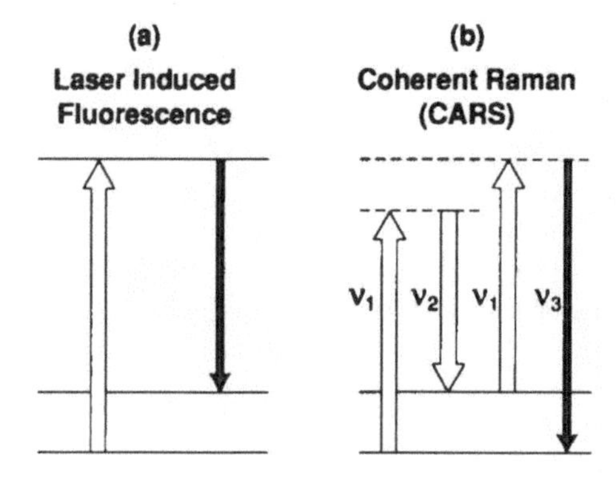

Figure 3.15 Schematic representation of the energy transitions associated with (a) laser-induced fluorescence (LIF), and (b) coherent anti-Stokes Raman spectroscopy (CARS). The broad arrows represent the incident radiation, as described in the text. The solid arrows represent the stimulated signal.

3.10.1 Background principles of laser diagnostics

The radiation sources are pulsed lasers which emit high powers of spectrally pure, coherent radiation. The power that is available, usually delivered in short bursts, i.e. *pulsed radiation*, permits the exploitation of very weak optical processes formerly not accessible with conventional light sources. The coherence ensures efficient delivery of the all of the source power to the probing location. The spectral purity allows high resolution of particular molecular or atomic structural features of significance for chemical identification and measurement of species in the combustion process.

A commonly used source laser, or *pumping laser*, is the solid state, Nd:YAG laser (neodymium ions, Nd^{3+}, doped in yttrium aluminium garnet, $Y_3Al_5O_{12}$) which emits in the infrared range of wavelengths, at 1064 nm. Since most chemical diagnosis or supplementary laser pumping action requires rather higher energies, the output is *frequency doubled* to 534 nm (green emission) by passing it through a suitable crystal. There is sufficient power available at the source to be able to sustain losses associated with the doubling process. Another laser source is the *excimer gas laser*, such as the xenon–chlorine laser, which emits at 308 nm in the ultraviolet range. It relies on the generation of an excited state compound $XeCl^{\bullet}$ for the upper lasing state. The chlorine atoms are generated by passing a discharge through a flow of Cl_2.

The probing of particular species requires operation at a specific wavelength, which is extremely unlikely to correspond to the output of the laser source. The tuning to a required frequency is achieved by pumping a *dye laser*, which gives fluorescent radiation over a broad

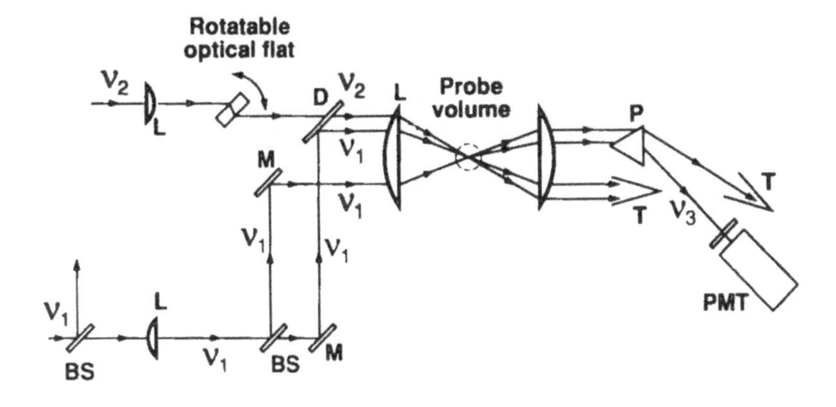

Figure 3.16 A typical optical arrangement required for CARS measurements. BS, beam splitter; M, mirror; D, dichroic filter; L, lens; P, prism; T, light trap; PMT, photomultiplier tube. The phase matched beams, frequencies v_1 and v_2, intersect in the probe volume. The stimulated signal, v_3, is detected by the photomultiplier.

spectral range. A commonly used dye is rhodamine, which flows in a methanol solution through a cell within the laser cavity. Rhodamine B has a fluorescent range of 590–630 nm, for example. This output is also frequency doubled, if necessary, and the wavelength range is scanned by rotation of a diffraction grating to achieve the laser action at a required wavelength. A typical laser system and ancillary equipment used in laser diagnostics, to be described below, is shown diagrammatically in Fig. 3.16.

The laser beam is often split into components, one of which may be used to measure the intensity of the beam. Normally, photomultipliers are used for detection, with a spectrometer, grating or filter to restrict the wavelength received, if necessary. The signals are collected and may be averaged across a number of pulses by use of a boxcar averager with typical gate width down to a nanosecond range, giving essentially 'instantaneous' information even about highly reactive species. The repetition rate of laser pulses may be up to 100 Hz. Laser beam focusing within the region to be probed is essential in order to obtain satisfactory spatial resolution. In techniques involving more than one laser energy, as in CARS, the crossing of the beams is another way of restricting the probe volume, albeit with the loss of intensity of the diagnostic signal. Spatial resolution down to a probe volume of 10^{-3} cm^3 is possible.

3.10.2 Laser-induced fluorescence (LIF)

The quantitative measurement of intermediates which may be present in exceedingly low concentrations in flames may be made by use of LIF spectroscopy. Amongst the species that can be detected are OH, CH, CN, C_2, CH_2O, CH_3O, NH, NO, SO, SH, SO_2 and other nitrogen or sulphur-containing species. Thus access to pollutant derivatives as well as important primary components in the combustion reaction is possible. The laser must be tuned so that the species under investigation can be excited to an upper electronic state, from which there is then a spontaneous emission of radiation, i.e. fluorescence, as the species deactivates to the ground state (Fig. 3.15(a)). *Resonance fluorescence* signifies that the incident radiation is exactly the frequency of the emission, which means that the detection device must be perpendicular to the incident beam, and that care must be taken that reflected signals from the source are not picked up by the detector. The intersection of the illumination and collection beam paths defines the spatial resolution. The fluorescence efficiency of the process must be evaluated in order to take into account losses from the excited state owing to *quenching* by collisions with other species.

Fluorescence spectra usually contain 'fine structure' which arises from a multiplicity of vibrational and rotational states associated with the excited and ground electronic states of the diatomic or polyatomic species, between which transitions may be made. The spectral features may span only a very narrow energy range but they have to be resolved in order to interpret the concentration of species, not least because the probabilities of transitions between different vibrational–rotational states may be temperature dependent.

3.10.3 Coherent anti-Stokes Raman spectroscopy (CARS)

As a chemical diagnostic tool CARS is restricted to a limited range of major components in the combustion system, such as N_2, NO, H_2, CO, O_2 or H_2O, but it can be used for temperature measurements in flames. The measurements can be localised to a small cross-section on a very short timescale, and also in highly luminous or sooting combustion environments. The fine structure of the CARS spectrum for any of these species is quite strongly temperature dependent, and this feature is utilized in the temperature measurements. The procedure is to match the experimental CARS signal, from N_2 say, with computer generated spectra determined at different temperatures. Some instrumental parameters have to be included in the numerical simulation. Normally, the N_2 CARS spectrum is used as the basis for temperature measurements. At the present time, the precision of the measurements is about $\pm$ 40 K at normal flame temperatures. The accuracy is reduced at temperatures below about 1200 K because the fine structure of the spectrum is relatively indistinct.

CARS (Fig. 3.16) requires two, phase-matched laser beams of frequency ν_1 and ν_2 (Fig. 3.15(b)). Commonly the Stokes beam ν_2 is obtained by splitting off part of the pumping laser beam ν_1 to drive a tunable dye laser. The coherent beam ν_3, of frequency $(2\nu_1 - \nu_2)$ is obtained by mixing the pump and Stokes beams appropriately (Fig. 3.16). When the frequency difference $(\nu_1 - \nu_2)$ coincides with a Raman active resonance of the probe molecule (ν_v), the intensity of the radiation at ν_3, then given by $(\nu_1 + \nu_v)$ and termed the anti-Stokes frequency, can become high. This is the signal that is collected and analysed.

3.10.4 Planar imaging techniques

Two-dimensional imaging is evolving as an important tool for the identification of flowfield properties in combustion systems. The developments of laser technology have made possible the sheet illumination of a combustion regime from a tunable dye laser, rather than a single-point diagnosis. The detection of scattered light or a fluorescence signal is made at right angles by use of a solid-state array detector, 100×100

pixels say, coupled to a computer for fast storage, processing and display of signals. The typical sources of information are Mie or Rayleigh scattering in the combustion environment or LIF. The spatial resolution of the signal is determined by the size of the detector pixels, the magnification of the collection optics and the size of the measurement volume at each pixel point. Whereas viewing an open flame may present no particular problem, one of the technical difficulties within closed environments is gaining the optical access to give a sufficiently large field of view. For example, in order to investigate flame propagation or other aspects of combustion in a spark ignition engine, it is necessary to have both a side window above the piston crown at top dead centre for admission of the illuminating beam and a flat, transparent cylinder head for viewing the detection signal.

Further reading

Atkins P.W. (1994). Ch 17, Electronic transitions. In *Physical Chemistry* (5th edn). Oxford University Press, Oxford, UK.

Crosley, D.R. (ed.) (1980). *Laser Probes in Combustion* (ACS Symposium Series 134). American Chemical Society, Washington University Press, USA.

Eckbreth, A.E. (1981). Recent advances in laser diagnostics for temperature and species concentration in combustion. In *Eighteenth Symposium (International) on Combustion*. The Combustion Institute, Pittsburgh, PA, USA, p. 1471.

Eckbreth, A.E. (1988). Laser diagnostics for combustion temperature and species. In *Energy and Engineering Sciences Series* (Vol. 7) (eds A.K. Gupta and D.G. Lilley). Abacus Press, Tunbridge Wells, UK.

Fenimore, C.P. (1964). *Chemistry in Premixed Flames*. Pergamon, Oxford, UK.

Fristrom, R.M. and Westenberg, A.A. (1965). *Flame Structure*. McGraw-Hill, New York, USA.

Gaydon A.G. (1974). *The Spectroscopy of Flames* (2nd edn). Chapman and Hall, London, UK.

Gaydon, A.G. and Wolfhard, H.G. (1979). *Flames: Their Structure, Radiation, and Temperature* (4th edn). Chapman and Hall, London, UK.

Hanson, R.K. (1986). Combustion diagnostics: planar imaging techniques. In *Twenty-First Symposium (International) on Combustion*. The Combustion Institute, Pittsburgh, PA, USA, p. 1677.

Jones J.C. (1993). *Combustion Science, Principles and Practice*. Millennium Books, NSW, Australia.

Kohse-Höinghaus, K. (1994). Laser techniques for the quantitative detection of reactive intermediates in combustion systems. *Prog. Energy Combust. Sci.*, **20**, 203.

Lewis, B. and Von Elbe G. (1987). *Combustion, Flames and Explosions of Gases* (3rd edn). Academic Press, New York, USA.

Lewis, B. and Von Elbe, G. (1956). Combustion waves in non-turbulent explosive cases. In *Combustion Processes* (eds. B. Lewis, R.N. Pease and H.S. Taylor). Oxford University Press, Oxford, UK, p. 216.
Lovachev, L.A. *et al.* (1973). *Combust, Flame*, **20**, 259.
Pilling, M.J. and Seakins, P.W. (1995). *Reaction Kinetics* (2nd edn). Oxford University Press, Oxford, UK.
Wohl, K. and Shipman, C.W. (1956). Diffusion flames. In *Combustion Processes* (eds B. Lewis, R.N. Pease and H.S. Taylor). Oxford University Press, Oxford, UK, p. 365.

Problems

(1) The value of the Peclet number for quenching in most hydrocarbon + air mixtures is about 50. Estimate the quenching distance in stoichiometric mixtures of air + (a) hydrogen, (b) methane, (c) propane, and (d) acetylene at 298 K and a pressure of 1 atm (100 kPa), given the following data:

	H_2	CH_4	C_3H_8	C_2H_2	Air
S_u(m s^{-1})	2.0	0.45	0.43	1.58	—
κ(W m^{-1} K^{-1})	0.151	0.0342	0.0266	0.0213	0.0261
C_p(J mol^{-1} K^{-1})	28.6	35.6	73.0	44.1	29.2

What will be the corresponding values of the quenching diameter? (For the H_2 + air mixture, the experimental value of κ is *c.*0.04 W m^{-1} K^{-1}. Use mean values based on the molar ratios for other properties of the fuel + air mixtures.)

(2) Using the information given in Question (1), calculate the quenching distance in a stoichiometric propane + air mixture at 298 K and 0.1 atm (10 kPa).

4 Flame theory and turbulent combustion

4.1 Introduction

In its experimental realisation, the laminar flame is the laboratory tool which has made possible the study of flame structure and chemistry. Moreover, the theoretical understanding of laminar flames, which is extremely well developed [31, 32], has been the test-bed for the development of many numerical and mathematical tools, and it is the foundation from which many concepts of turbulent combustion emerge. A comprehensive, analytical theory for premixed, laminar, flame propagation is outside the scope of an introductory book, and various levels of its development are to be found in virtually all other general combustion texts (see the further reading section at the end of this chapter). Perhaps Lewis and Von Elbe's book – *Flames, Explosions and Detonations in Gases* – is set apart from most by the extent of the practical experience that accompanies the theory.

The propagation of combustion waves is so central to combustion that it would be remiss to avoid addressing some aspects, but the practical aspects of flames that are discussed in Chapter 3 show that there are significant distortions as real flames depart from the planar ideal. Virtually all practical applications of combustion based on 'flame propagation' involve turbulent processes, premixed or otherwise [33]. Thus current views on the interpretation of structure in expanding flames and turbulent combustion constitute a major component of this chapter.

4.2 Foundations of wave propagation theory

Laminar flame theory is set out to establish the uniqueness of the combustion wave, that is, a single burning velocity for a given set of conditions, and to make the numerical prediction, but it is very complicated in rigorous form. The complexity arises from the need for a simultaneous treatment of both species and thermal diffusion from the reaction zone into the preheat zone, as expressed in the appropriate partial differential equations. Severe approximations and constraints have to be imposed if the treatment takes an analytical form. Only numerical solutions are possible if detailed chemical kinetics are included and each species is treated individually. Detailed and elegant interpretations of flame structure are possible in such treatments [34].

4.2.1 *Diffusional propagation of travelling wavefronts*

Although species diffusion ahead of a flame contributes to its propagation and should not be ignored, the primary driving force in normal flames is the thermal diffusion. Thus, theories of flame propagation that are based purely on species diffusion (e.g. Tanford and Pease [35]) cannot be supported, since isothermal propagation is assumed. However, rigorous insights into *chemical wave* propagation emerge from a solution of the isothermal reaction–diffusion equation, which exemplify the properties of waves in a highly reactive medium [36, 37].

To establish an analogy to flame propagation it is necessary to invoke a rate law that exhibits a highly non-linear feedback. A cubic autocatalysis is used [36, 37], represented in its simplest kinetic form as

$$A+2X \rightarrow 3X;\ v=k[A][X]^2 \tag{4.1}$$

The cubic dependence emerges in the rate expression, and is similar to quadratic autocatalysis (section 9.3). By taking into account mass conservation

$$[A]+[X]=[A]_0 \tag{4.2}$$

and assuming that the diffusion coefficients for A and X are the same (D), the planar propagation of a wave in y may be represented by a single reaction–diffusion equation for one independent concentration.

$$\frac{\partial[X]}{\partial t}=\frac{D\partial^2[X]}{\partial y^2}+k([A]_0-[X])[X]^2 \tag{4.3}$$

The analytical solution of this equation, via its conversion to a moving coordinates with the reference at the mid-point of the wave, and subject to the appropriate boundary conditions, shows that the wave propagates at a rate (r) given by

$$r=(1/\sqrt{2})(Dk[A]_0^2)^{1/2} \tag{4.4}$$

The points of principle, that transpose to *flame propagation*, are that the wave is driven by the chemistry, its rate of propagation being dependent on the square root of the rate constant for the autocatalytic reaction and the concentration of the initial reactant, as well as being dependent on the square root of the diffusion coefficient. Fast chemistry and a high diffusion coefficient are both capable of increasing the speed of the wave. The profile of the wave is also constant throughout the course of its propagation. It does not get 'smeared out' as would occur as a result of diffusion alone.

Similarly, the propagation of a non-isothermal travelling wave is governed by the rate of heat release from the exothermic reaction and its conduction into the reactants ahead of the wave. In most (but not all) cases 'flame' propagation is consolidated by autocatalytic reactions taking place in the reaction zone.

4.3 Non-isothermal wave, or flame propagation

Both species and thermal diffusion have to be taken into account when considering a nonisothermal wave, but at least the chemistry can be simplified, the simplest case being a first order exothermic reaction. In the non-isothermal nature of the process the total mass flow is conserved, so the amount of reactant is expressed as w_A(mol kg^{-1}) of mixture. The stationary state, reaction–conduction and diffusion equations in terms of wave-fixed coordinates take the following form:

$$D\sigma\left(\frac{d^2 w_A}{dz^2}\right) - r\sigma\left(\frac{dw_A}{dz}\right) - \sigma k(T) w_A = 0 \tag{4.5}$$

$$\kappa\left(\frac{d^2 T}{dz^2}\right) - rC\sigma\left(\frac{dT}{dz}\right) + q\sigma k(T) w_A = 0 \tag{4.6}$$

In these expressions σ(kg m^{-3}) is the density, C(J kg^{-1} K^{-1}) is a specific heat capacity, q(J mol^{-1}) is the exothermicity ($=-\Delta H$) and r(m s^{-1}) is the combustion wave propagation rate. The temperature dependence of the rate constant is not specified, but it is likely take the Arrhenius form. The form of the spatial concentration and temperature profiles that these equations represent are shown in Fig. 4.1, subject to the boundary conditions

$$T = T_0,\ w_A = (w_A)_o \text{ as } z \to -\infty,\ T = T_b,\ w_A = 0 \text{ as } z \to +\infty \tag{4.7}$$

and

$$dT/dz = dw_A/dz = 0 \text{ at } z \to \pm\infty$$

In a formal sense the reaction rate is not zero at $-\infty$ because $T_0 \neq 0$. This is often referred to as the 'cold boundary problem' because a mathematical condition is violated. Nevertheless, although not exactly zero in principle, the negligibly small heat release rate that occurs at T_0 preserves the uniqueness of the solution to the energy conservation in practice. Zel'dovich *et al.* (1985) argued that, in fact, very little reaction occurred throughout an appreciable range of temperatures above T_0 such that virtually all of the reaction took place within a narrow range of temperatures above an (arbitrary) temperature, T_r, in the vicinity of T_b. This is a reasonable supposition since the residence time of the fuel in the reaction zone is so short (typically 10^{-5} s) that reaction at T_r must be extremely fast, and so T_r must be quite close to the final flame temperature. This is satisfactory only if the activation energy is appreciable so that the temperature sensitivity of the reaction rate is very high and the flame propagation is predominantly thermal in nature. There is no critical 'onset temperature' for reaction, although the idea is exploited in the analytical treatment.

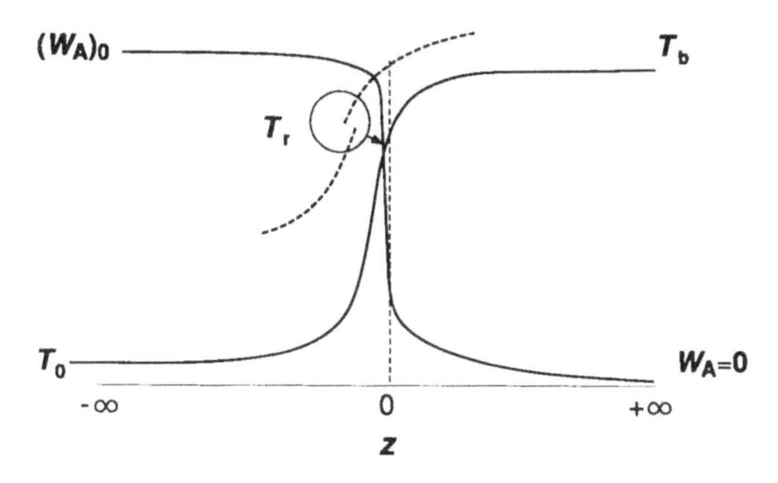

Figure 4.1 Spatial concentration and temperature profiles for a non-isothermal, one dimensional travelling wave with reference to a moving coordinate at the hypothetical reaction onset temperature, T_r, of the wave progressing from right to left. The broken lines indicate the matching of the gradients for the preflame and reaction zones as invoked in the accompanying analysis and also in asymptotic expansion procedures (see text).

By substitution of the temperature excess and extent of reactant consumption, eqns (4.5) and (4.6) may be written as

$$\frac{D\sigma \mathrm{d}^2[(w_A)_o - w_A]}{\mathrm{d}z^2} - \frac{r\sigma \mathrm{d}[(w_A)_o - w_A]}{\mathrm{d}z} - \sigma k(T) w_A = 0 \qquad (4.8)$$

$$\frac{\kappa \mathrm{d}^2[C(T-T_0)]}{C\,\mathrm{d}z^2} - \frac{r\sigma \mathrm{d}[C(T-T_0)]}{\mathrm{d}z} + q\sigma k(T) w_A = 0 \qquad (4.9)$$

In an adiabatic system at $Le = 1$

$$q[(w_A)_o - w_A] = C(T-T_0) \text{ and } D\sigma = \kappa/C \qquad (4.10)$$

so that the species and energy conservation eqns (4.8) and (4.9) may be regarded to be identical in form, and the system can be represented by a single conservation equation.

The definition of T_r implies that there are two regimes associated with the propagating combustion wave. There is a *preheat* or *preflame zone*, in which negligible reaction takes place, so $w_A = (w_A)_o$ throughout, and the temperature profile in this zone is obtained from solutions to a simplification of eqn (4.6) where the heat release rate is neglected:

$$T < T_r, \quad -\infty < z < 0$$

$$\left(\frac{\mathrm{d}^2 T}{\mathrm{d}z^2}\right) - \frac{rC\sigma}{\kappa}\left(\frac{\mathrm{d}T}{\mathrm{d}z}\right) = 0 \qquad (4.11)$$

with boundary conditions

$$T=0, \text{ and } dT/dz=0 \text{ at } z\to-\infty$$

$$T=T_r \text{ at } z=0, \text{ when the origin is located at } T_r \tag{4.12}$$

By contrast, within the *reaction zone* very little of the heat of reaction is used to raise the temperature of the products, so that virtually all of it is lost by conduction and, therefore, at

$$T_r < T < T_b,\ 0 > z > +\infty$$

$$\frac{d^2T}{dz^2}+\frac{q\sigma k(T)w_A}{\kappa}=0 \tag{4.13}$$

The gradients (dT/dz), given by solutions to eqns (4.8) and (4.10) must be continuous at T_r (Fig. 4.1).

The preheat zone

The temperature gradient at T_r, $z=0$, is given by integration of eqn (4.11) and is

$$\left(\frac{dT}{dz}\right)_{z=0}=\frac{rC\sigma(T_r-T_0)}{\kappa} \tag{4.14}$$

Integration of eqn (4.14), subject to the boundary conditions, gives the spatial temperature profile in the form

$$T=T_r\exp(rC\sigma z/\kappa) \tag{4.15}$$

To the extent that C and κ can be adequately represented by mean values, the temperature increases exponentially with distance in the preheat zone. Although, in principle, the temperature starts to increase from $z=-\infty$, a thickness may be arbitrarily defined by considering the zone to begin at $(T-T_0)/(T_r-T_0)=1/100$ (i.e. $\sim T_0+20$ K for a typical $T_r\sim 2000$ K). Replacing r by S_l, the *laminar burning velocity*, gives the thickness of the preheat zone

$$\delta_{pz}=\frac{4.6\kappa}{C\sigma S_l} \tag{4.16}$$

Typical values of δ_{pz} range from 10^{-5} m for a burning velocity of 30 m s^{-1} to 10^{-2} m for a burning velocity of 0.03 s^{-1}.

The reaction zone

The equation to be solved in the reaction zone is eqn (4.13) which, by use of the identity

$$\frac{d}{dz}\left(\frac{dT}{dz}\right)^2=2\left(\frac{dT}{dz}\right)\left(\frac{d^2T}{dz^2}\right)^2 \tag{4.17}$$

may be written as

$$\frac{\mathrm{d}}{\mathrm{d}z}\left(\frac{\mathrm{d}T}{\mathrm{d}z}\right)^2 = \frac{-2q\sigma k(T)}{\kappa} w_A\left(\frac{\mathrm{d}T}{\mathrm{d}z}\right) \quad (4.18)$$

which integrates to

$$\left[\left(\frac{\mathrm{d}T}{\mathrm{d}z}\right)^2\right]_0^{\infty} = \frac{-2q\sigma}{\kappa}\int_{T_r}^{T_b} k(T) w_A \,\mathrm{d}T \quad (4.19)$$

There are three difficulties with eqn (4.19):

(i) the temperature T_r is not known;
(ii) if $k(T)$ is identified as the Arrhenius dependence the equation cannot be solved analytically; and
(iii) w_A is also dependent on temperature.

Following Zel'dovich and Frank-Kamenetskii [38, 39], the ways to cope with these points are as follows.

(i) Since negligible reaction is presumed to occur in the preheat zone, the area under the z–T curve in the range $-\infty < z < +\infty$ is hardly different from the area in the range $0 < z < +\infty$. Therefore, to a good approximation the lower limit on the right- hand side, T_r, may be transposed to T_0.
(ii) The first-order rate constant can be expressed in the approximate form

$$A\exp(-E/RT) = A\exp(-E/RT_b) \times \exp\{(T - T_b)E/RT_b^2\} \quad (4.20)$$

This is a form of the exponential approximation, to be discussed in Chapter 8, in which the flame temperature, T_b, is the reference temperature (cf. eqn (8.18)).
(iii) From eqn (4.10).

$$1 - \frac{w_A}{(w_A)_o} = \frac{(T - T_0)}{(T_b - T_0)}$$

or

$$w_A = (w_A)_o\left\{1 - \frac{(T - T_0)}{(T_b - T_0)}\right\} \quad (4.21)$$

Incorporating the terms discussed in (i)–(iii) into eqn (4.19), with some rearrangement the equation becomes

$$\left[\left(\frac{\mathrm{d}T}{\mathrm{d}z}\right)^2\right]_0^{\infty} = \frac{-2q\sigma A(w_A)_o}{\kappa(T_b - T_0)}\exp(-E/RT_b)$$

$$\int_{T_0}^{T_b}(T_b - T)\exp\{(T - T_b)E/RT_b^2\}\,\mathrm{d}T \quad (4.22)$$

It is easiest to change the variable in order to integrate eqn (4.22) and a convenient form is

$$\Theta = (T - T_b)\frac{E}{RT_b^2} \tag{4.23}$$

so that

$$\left[\left(\frac{dT}{dz}\right)^2\right]_0^{\infty} = \frac{2q\sigma k(T_b)}{\kappa(T_b - T_0)}\left(\frac{RT_b^2}{E}\right)^2 (w_A)_o \int_{-B}^{0} \Theta \exp\Theta d\Theta \tag{4.24}$$

The lower limit is identified as $-B$ because $(T_0 - T_b)E/RT_b^2$ corresponds to the dimensionless adiabatic temperature rise defined with respect to T_b. The solution to this integral is

$$0 + B\exp^{-B} - 1 + \exp^{-B} \sim -1 \text{ for large } B.$$

At $z = +\infty$, $(dT/dz) = 0$, so that eqn (4.24) yields

$$-\left(\frac{dT}{dz}\right)^2 = \frac{-2q\sigma k(T_b)}{\kappa(T_b - T_0)}\left(\frac{RT_b^2}{E}\right)^2 (w_A)_o \tag{4.25}$$

In accord with Fig. 4.1, continuity of the temperature profile is obtained by equating eqn (4.14) and the square root of eqn (4.25), so that

$$\frac{rC\sigma(T_b - T_0)}{\kappa} = \left\{\frac{2q\sigma k(T_b)}{\kappa(T_b - T_0)}\left(\frac{RT_b^2}{E}\right)^2 (w_A)_o\right\}^{1/2} \tag{4.26}$$

With two further substitutions, namely for the term B^2 and also the relationship $(w_A)_o q = C(T_b - T_0)$, and with further rearrangement, the laminar burning velocity S_l is given by

$$S_l = r = \frac{2^{1/2}}{B}[(\kappa/C\sigma)k(T_b)]^{1/2} \tag{4.27}$$

Thus, the laminar burning velocity is proportional to the square root of the thermal diffusivity $(\kappa/C\sigma)$ multiplied by the reaction rate constant (evaluated at the burned gas temperature). This means also that the velocity is independent of the initial temperature. There is no dependence on pressure, but this is the case only for a first order reaction. The thickness of the preheat zone is inversely related to the burning velocity by the thermal diffusivity of the gas (eqn (4.16)).

These important analytical solutions from early flame theory, even for the simplest thermal model, were achieved by making approximations and assumptions. Quantitatively more accurate analytical solutions to the flame equations were obtained in the 1950s by use of other approximate methods [40]. More recently there has been very extensive development and use, for systems in which the chemical reaction is confined to a narrow reaction zone, of a mathematical technique called *high activation energy asymptotics* [32, 41, 42]. The analytical interpretation of many different types of flame structure, such as Bunsen flames, spherical expanding flames, flat flames in stagnation flow or wrinkled flames in non-uniform flow, is made possible by application of high activation energy asymptotics.

According to the Zel'dovich and Frank-Kamenetskii model, the region of rapid reaction is confined to a zone in which $T \approx T_b$ and the preheat zone is essentially an unreactive flow. Furthermore, as the activation energy for the chemical reaction is increased, the length of the reaction zone becomes smaller relative to that of the preheat zone and the temperature change in the reaction zone is also reduced. The reaction zone is infinitely thin and $T_r - T_b = 0$ in the limit $E \rightarrow \infty$. Simplified differential equations may be derived and solved if these limiting conditions are applied. Their properties and gradients at the boundary conditions can then be matched by expanding the solutions in a geometric series with respect to an appropriate small parameter, such as $\varepsilon (= RT_b/E)$ in the present case. The final solution may then be obtained with increasing accuracy by retaining terms of increasingly high order in ε. In fact, the Zel'dovich and Frank-Kamenetskii solution for a first order exothermic reaction represents the leading order term in an asymptotic expansion for the burning velocity [43]. Since, typically, $\varepsilon \approx 0.1$ an expansion in two terms leads to a considerably improved quantitative precision over the leading order term.

4.4 A unifying theme for chemical kinetic / fluid dynamic interactions

The planar wave discussed in the preceding sections is able to retain its integrity because there are no viscous forces in the fluid, and there is no fluid motion either imposed on or induced by the wave propagation. This is an idealisation. There are marked effects on the flame structure and propagation rate if turbulence is created in the gas by supplementary means. The classifications of flame structure under varying conditions of fluid motion may be rationalised in a unifying diagram (Fig. 4.2), from which some of the concepts and terminology associated with combustion - turbulence interactions are introduced. The different phenomena shown are discussed in later Sections.

The fundamental relationships shown in Fig. 4.2 [44] for combustion in premixed fuel + oxidant under flowing conditions are between the ratio of the 'turbulence intensity'/laminar burning velocity (u'/S_l) on the ordinate and the ratio of the 'integral length scale'/laminar flame thickness (L/δ_l) on the abscissa. These terms require some explanation. The *laminar burning velocity*, S_l, is defined as the rate of propagation of a planar flame into the unburned gas under laminar flow conditions, and is a fundamental parameter of a particular fuel + oxidant composition at a given temperature and pressure. The distinction from the burning

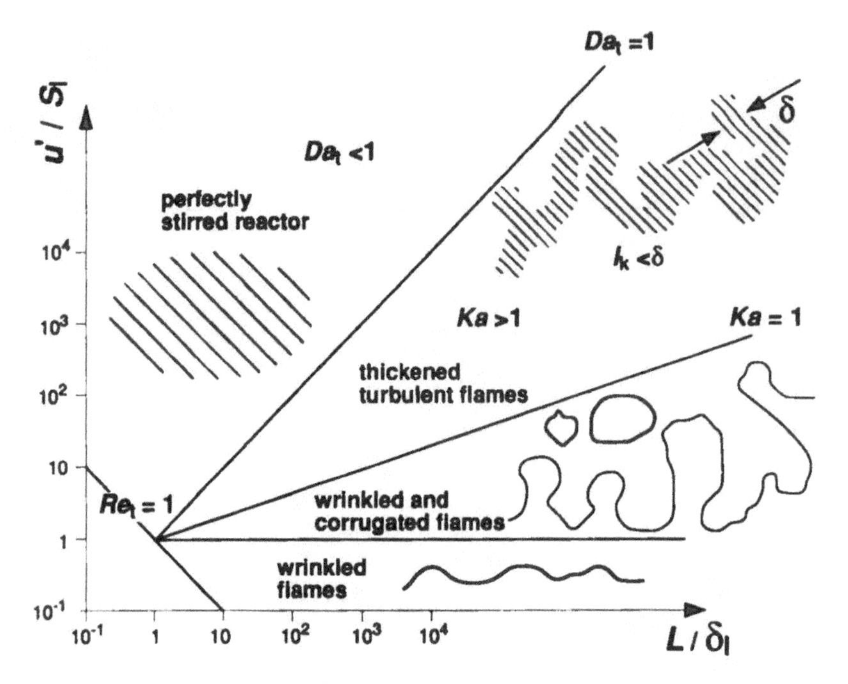

Figure 4.2 A unifying representation of combustion phenomena in well-mixed and turbulent conditions. All terms are defined in the text (after Borghi [44]).

velocity, S_u, will emerge later in connection with stretched flames. The *laminar flame thickness*, δ_l, confers a three-dimensionality to the flame insofar that it represents the depth of the reaction zone in which the bulk of the heat is released. A correction term, called the Markstein number [45], is applied sometimes to δ_l in order to take an effect of flame curvature into account. The laminar flame thickness is often approximated to υ/S_l [46], where υ is the kinematic viscosity. The kinematic viscosity relates to transport of momentum and is given by the dynamic viscosity of the fluid divided by its density (η/σ).

Turbulent flow may be regarded as random local motions superimposed on the uniform motion of the fluid. Terms like eddies, vortices, turbulent spheres and turbulent balls are all employed to describe such motions. In isotropic turbulence the motion is characterised by the *turbulence intensity*, represented as the root mean square (rms) turbulent velocity, u', which signifies an rms of the rate at which a given pocket of gas may be 'tumbling' or 'rotating'. There is no given size of the gas pocket (or the volume within which it is tumbling), but it is possible to define an average size, which is termed the *integral length scale*, L. The largest eddies contain most of the kinetic energy of the turbulence. The Kolmogorov length scale, l_K, is a measure of the smallest structures, through which most of the turbulence energy is dissipated. The Kolmogorov scale is defined as

$$l_K = \frac{\upsilon^3}{\varepsilon} \quad (4.28)$$

in which ε is the dissipation of turbulent energy in the unburned gas. A complexity is that not only do the intensity and length scales have a spectrum of values but also they vary in time. For example, unless there is a supporting generation of energy (as in a combustion process), the viscous forces that prevail in a flowing gas would cause the turbulence to decay.

Turbulent motion appears first as large stable eddies which are anisotropic. Vortex stretching leads to small, faster moving eddies which are more isotropic. Eventually the eddies reduce to the dimensions of molecular motion, the energy being dissipated by viscosity into random kinetic energy of the molecules. Because angular momentum is conserved in this cascade process, the kinetic energy involved increases. This picture of eddy decay refers only to what takes place in a free stream. At the origin of the turbulence, for example, where two gas streams of different velocity meet, the reverse occurs: the shear forces create small-scale vortices which coalesce into larger eddies. A complete description of turbulence requires a small length scale, as well as an integral length scale and in the cascade process it is possible to define a scale parameter λ, *the Taylor microscale*, associated with the small dissipative eddies.

The state of a gas flow is often characterised by the dimensionless Reynolds number, Re, and is defined by

$$Re = \frac{ul}{\upsilon} \quad (4.29)$$

where u is a flow velocity, l is a characteristic length. If, for gas flow in a burner tube, u is chosen as the average flow velocity and l as the burner diameter, then the Reynolds number so obtained can provide a criterion for the transition from laminar to turbulent flow. At $Re < 2300$ the flow is normally laminar, whereas for $Re > 3200$ it becomes turbulent. In the intermediate range, 2300–3200, the behaviour is not so readily predictable; small perturbations are quickly damped while more severe disturbances persist. The Reynolds number is the ratio of inertial to viscous forces in the flow. Sometimes it is convenient to define other Reynolds numbers. For example, $Re_t = u'L/v$ is the *turbulent Reynolds number* based on the length scale L, and is given by the product of terms representing the two axes in Fig. 4.2. The Reynolds number, Re_λ, based on the Taylor microscale, is the ratio of the timescales for large and small eddies.

The four terms, S_l, u', δ_l and L, can be used to define characteristic chemical and turbulence timescales respectively, as follows:

$$t_{ch} = \frac{\delta_l}{S_l} \quad \text{and} \quad t_t = \frac{L}{u'} \quad (4.30)$$

The dimensionless parameter that quantifies the relationship between these timescales is the turbulent Damkohler number, Da_t, and is given by t_t/t_{ch}. The 45° line in Fig. 4.2 corresponds to $Da_t = 1$, which gives a broad division between two important regimes in combustion. The line is not extended to the origin but is cut off at $Re_t = 1$. The regime for the existence of laminar flames lies below the line $Re_t = 1$.

If the turbulence is 'fast' compared with the chemical reaction rate, i.e. $Da_t \ll 1$, then the chemistry is insensitive to effects of heat and mass transport. This introduces the notion of the *perfectly stirred reactor*, to be discussed in section 10.2, within which there are no spatial variations in temperature or concentration because the gas motion is sufficiently vigorous to smooth these out. Often, this idealised condition is assumed in 'zero-dimensional', numerical modelling of combustion processes, even though it may not be very appropriate [4]. The perfectly stirred reaction regime must be restricted to rather low values on the abscissa (Fig. 4.2), because a large L and small δ_l would imply thin highly reactive zones being transported over comparatively large distances. It must also lie relatively high up the ordinate, otherwise there is an implication of a segregation of burned and unburned gas in the wake of a quite well-developed laminar flame in a regime of low turbulence.

The qualitative picture associated with the lower portion of Fig. 4.2, bounded by $Da_t = 1$ and $Re_t = 1$, reveals some of the complexities that develop in flame structure once there are significant departures from laminar flow on the one hand but perfect mixing is not yet approached on the other. It requires also that another criterion, the dimensionless Karlovitz *flame stretch factor*, *Ka*, be defined.

$$Ka = \frac{\delta_l u'}{S_l \lambda} \tag{4.31}$$

This expression incorporates the *turbulent flame stretch rate*, which is defined as u'/λ [46]. The line representing $Ka = 1$ in Fig. 4.2 is derived on the assumption that $\lambda = 0.01L$. *Ka* may also be interpreted as a ratio of the chemical to eddy lifetimes. The terms *flame stretch* and *flame strain* are used in the literature to describe the same flame properties. Although *strain* is a specific property of fluid motion which has its origins in flow non-uniformities and curvature effects, only the term *stretch* is used here.

4.5 Flame stretch and flame stretch rates in laminar flames

The concepts in section 4.3 refer to a planar, laminar flame comprising a very thin reaction zone preceded by a broad, heat and mass transport-dominated zone (Fig. 4.1). It is assumed that $Le = 1$ for the present. If the flame surface is curved or it is non-stationary, as is commonly the case as a result of non-uniform flow and flame expansion, the flame surface suffers the effect of stretch. This concept, introduced by Karlovitz *et al.* [47] and developed by Markstein [45], originated in the need to account quantitatively for the response of flames to velocity gradients in the gas flow and to curvature of the laminar flame surface.

The Karlovitz flame stretch rate (K) that a laminar flame surface undergoes in a non-uniform flow field, or as a result of flame motion or curvature, is a measure of the fractional rate of change of area A of an infinitessimal element of the surface [34].

$$K = \frac{1}{A}\frac{dA}{dt} \tag{4.32}$$

For example, in a spherically expanding laminar flame and in the conical surface of stabilised, axisymmetric Bunsen burner laminar flame the flame stretch rates [48] are given, respectively, by

$$K = \frac{2}{r_f}\frac{dr_f}{dt} \quad \text{and} \quad K = \frac{-u_\infty \sin\theta}{2r_f} \tag{4.33}$$

where r_f is the flame radius at a given axial position in either case, and in addition u_∞ is the uniform upstream velocity and θ is the angle at the apex of a Bunsen burner flame. There is a different stretch rate at different horizontal planes through the Bunsen flame, and also it experiences a negative stretch.

An increasing surface area as a result of flame stretch means that the amount of heat flowing from the reaction zone into the unburned gas is distributed over increasing volumes of gas. This can reduce the burning velocity and flame temperature, and may lead to flame extinction. Flame quenching by the unburned gas is completely different from the 'flame quench' and 'blow-off' phenomena (sections 3.6 and 3.7). A critical value of Ka may be defined for quenching of a stretched flame in the absence of any other external heat sink.

A flame that is expanding spherically in quiescent gas, initiated by a spark at the centre of a reaction vessel say, is a stretched laminar flame and, although uniform over its surface, the stretch rate profile is a function of time. Thus, a measure of its burning velocity is not the (unstretched) laminar burning velocity. To obtain the latter, the flame stretch rate would have to be determined, and the burning velocity extrapolated back from these conditions to that at a zero stretch rate. Dowdy *et al.* [49] have developed a method for the measurement of burning velocities from and characterisation of the stretch in expanding spherical flames. A stationary spherically flame is stretch free.

It is possible also to determine Ka and make the necessary corrections to S_u to obtain S_l, in experiments comprising either a single planar flame in stagnation point flow against a flat surface [50], as shown in Fig. 4.3, or two symmetrical, planar, near-adiabatic flames generated in a counterflow configuration [51]. Numerical computations of flame structure, burning velocity and flame extinction have been made for premixed, counterflow, laminar flames [52]. In the limit of high activation energy the

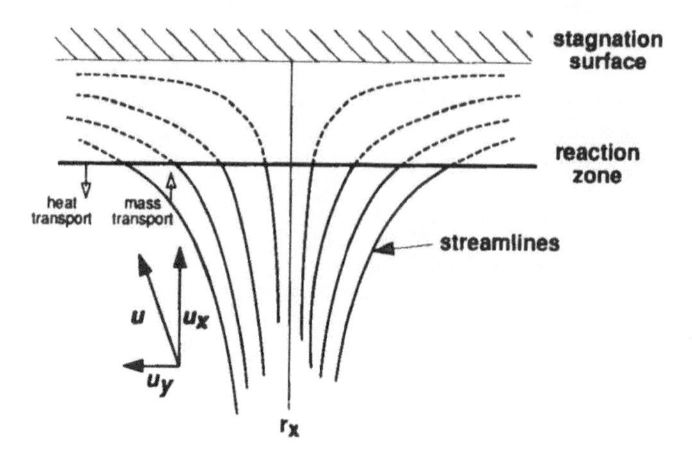

Figure 4.3 A schematic representation of a laminar, premixed flame stabilised against a stagnation surface (after Dixon-Lewis [34]).

reaction may be regarded to be confined to a very thin zone (section 4.3), often termed the flame sheet (Fig. 4.3). Upstream of the reaction zone is a control volume enclosing the transport zone and the divergent streamlines. Convective heat transport occurs along the streamlines, whereas diffusion of heat and mass occurs normal to the flame. The counterflow flame would be represented with the mirror image imposed above Fig. 4.3. A physical surface is not required in order to create a stagnation plane. The stretch rate is uniform over the entire isothermal surface of the flame in these structures, given by $K = -\mathrm{d}u_x/\mathrm{d}x$, (or $-2\mathrm{d}u_x/\mathrm{d}x$ in counterflow).

4.5.1 Dependences of the propagation of stretched flames on Lewis number

Preferential diffusion, that is the favouring of thermal diffusion over mass diffusion ($Le > 1$) or mass diffusion over thermal diffusion ($Le < 1$), can have a very marked effect. These exist in many reactant compositions (e.g. Table 4.1 [53, 54]) although the effect is not great in methane + air. The importance of the magnitude of Le can be illustrated with respect to an adiabatic, laminar flame stabilised in a stagnation flow (Fig. 4.3). The reaction volume loses thermal energy to the external streamlines, while it gains chemical energy from them owing to an increase in the deficient reactant concentration (i.e. fuel from fuel-lean mixtures or oxidant from fuel-rich mixtures). Nevertheless, in a diffusionally neutral mixture ($Le = 1$) the total energy conservation is maintained, and $T_b = T_{ad}$. However,

Table 4.1 Diffusion parameters and Lewis numbers associated with CH_4 + air and C_3H_8 + air mixtures [53, 54*]

Fuel	φ	$a(cm^2 s^{-1})$	$D(cm^2 s^{-1})$	Le
CH_4	0.51	0.213	0.220	1.03
	0.69	0.219	0.219	1.00*
	1.60	0.213	0.207	0.97
C_3H_8	0.51	0.208	0.114	0.55
	0.74	0.212	0.113	0.53*
	2.25	0.197	0.207	1.05

for $Le < 1$, mass gain exceeds heat loss and so combustion is intensified ($T_b > T_{ad}$). Conversely, for $Le > 1$, heat loss exceeds mass gain and so $T_b < T_{ad}$. The flame in stagnation point flow experiences a positive stretch rate, so the burning velocity would be expected to increase as the stretch rate is increased up to the point of flame extinction [48]. However, this effect may be augmented or diminished according to whether Le is greater than or less than unity.

The effects of variations in Le on the behaviour at the apex of a Bunsen flame (the negatively stretched region) are the opposite of those described for the planar flame in stagnation flow. The concave shape with respect to the upstream reactants tends to focus the heat ahead of the flame, whereas it 'defocuses' the reactants approaching the flame.

For spherically expanding flames the situation is more complicated because the (positive) stretch rate diminishes as the flame expands. Thus, diffusionally neutral flames would be expected to show a fall in burning velocity during expansion. Palm-Leis and Strehlow [55] demonstrated that there could be a sufficiently strong compensation by preferential mass diffusion ($Le < 1$) in propane + air flames, which showed very marked increases of the measured flame speed (S_b), in the early stage of development. The effect was greatest in the leanest mixtures ($Le < 1$).

When a smooth flame front is subject to planar disturbances, the resulting wrinkles may grow or decay. Whether the flame is stabilised or destabilised is governed to a great extent by the Lewis number. For $Le > 1$ (favoured thermal diffusion, so that heat loss exceeds mass gain) the combustion is less intense at the 'crest' of the wrinkled flame, whereas combustion is more intense in the 'trough' since it experiences a negative stretch, as is the case at the tip of a Bunsen flame. Therefore, the flame is stabilised insofar that the wrinkles tend to be smoothed out. By contrast, for $Le < 1$, the burning is less intense in the trough than at the crest. This implies that there is an instability of the flame structure, such that the flame front may be fragmented if local extinction occurs in the vicinity of the troughs of flames experiencing stretch [48, 56]. Thus, a conceptual link is established between laminar and turbulent burning.

Turbulent flames may be described by a system of *stretched laminar flamelets*. Since the magnitude of Le must be taken into account in stretched flames, the product $Ka \times Le$ (or the product of Karlovitz stretch factor and Markstein number [56]) becomes important in the interpretation of structure and behaviour in turbulent combustion based on laminar flamelets.

4.6 Laminar flamelet concepts in premixed turbulent combustion

The underlying physical conditions for laminar flamelet concepts is that molecular rates of transport and chemical reaction predominate over the rate of turbulent fluctuations. Apart from the criteria $Da > 1$, $Re_t > 1$ and $Ka < 1$, which represent systems in which chemistry is fast, turbulence is high and flame stretch is relatively weak, the flame thickness, δ_l, must also be small compared with the Kolmogorov scale, l_K. This condition is satisfied below $Ka = 1$, which corresponds to the Klimov–Williams criterion [57, 58] at which $\delta_l = l_K$.

The laminar flamelet regime is associated with these criteria, and the regions of Fig. 4.2 to which they apply may show wrinkled or corrugated flames. The distinction between the two is thought to be the existence of single or multiple flame sheets. The wrinkled flame zone ($u'/S_l < 1$) is a consequence of the self-similar scales of the inertial range, which signifies turbulence at intermediate length scales. There is no turbulent structure within the flame thickness. Although depicted as a sinusoidal shape, the front in the wrinkled flame comprises a series of convex surfaces linked by cusps. The three-dimensional structure in the vicinity of the cusps leads to an enhanced burning velocity.

Expressions have been derived by a variety of routes over the last 40 years, which relate the turbulent burning velocity of stretch-free wrinkled flames (expressed as S_t/S_l) to the turbulence intensity (or 'wrinkling factor' expressed as u'/S_l) [59]. These tend to show only weak departures from linearity but there is variation by a factor of 4 in the gradients of the proposed expressions [59]. At present there is no clear consensus regarding the most satisfactory approach. One way forward, in keeping with modern thinking, is to acknowledge that the wrinkled flame surface is fractal in nature between inner and outer cut-offs governed by l_K and L. The turbulent flame surface area may then be derived via the fractal dimension of the surface and, if stretch-free, the ratio of the surface area to that of the laminar flame (A_t/A) gives S_t/S_l. Since l_K and L are characteristics of the turbulence intensity, this seems to be a convenient basis for relating S_t/S_l to u'/S_u. The fluid mechanical arguments, set out elsewhere [59, 60], yield a relationship of the form

$$\frac{S_t}{S_l} = 1.52 \frac{u'}{S_l} \tag{4.34}$$

The success of this and similar relationships derived in other ways should be judged against experimental measurements, but this is not so readily achieved because turbulent flames are not stretch-free in practice.

4.7 Correlation of measured turbulent burning velocities and turbulence intensity

Experimental relationships between S_t/S_l and u'/S_l for stretched flames were first correlated by Abdel-Gayed and Bradley [61] from an enormous range of flame propagation measurements of hydrocarbon fuels in which the turbulence intensity could be characterised. The techniques included fan-stirring or moving grids in closed bombs, flames stabilised in tubes or ducts, and slot burners [46]. The correlations shown in Fig. 4.4 are based on the product $Ka \times Le$ and, for reasons to be discussed next, related to u'_k/S_l [62]. The relationship in eqn (4.32) for a stretch-free flame is also included, which would suggest that its gradient is about twice the ideal.

The term u'_k represents the effective rms turbulent velocity, and is a variable for expanding flames, such as those initiated by a spark. Eddies that are capable of affecting the flame structure must be smaller than the flame itself. Larger eddies can displace the flame but cannot break it up. Increasingly larger scale eddies contribute to u'_k as the spark initiated flame grows. Therefore u'_k is a function of eddy size and is determined

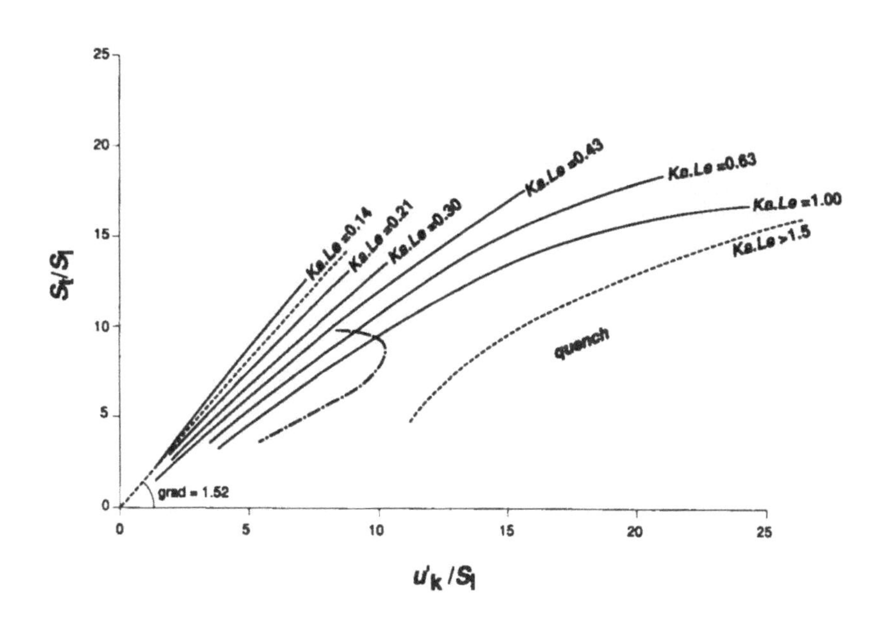

Figure 4.4 Correlations between the turbulent burning velocity (S_t) and effective turbulence intensity (u'_k) expressed in terms of the product *Ka.Le* (after Bradley *et al.* [62]). The broken line passing through the origin represents eqn (4.31). The chain-dot line is a simulation of flame development in a spark ignition engine (Chapter 13).

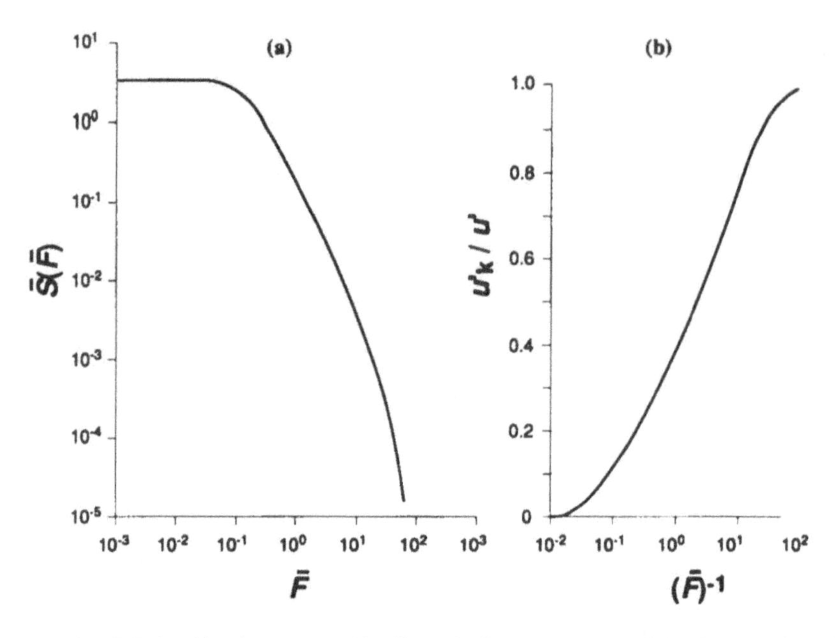

Figure 4.5 Relationships between (a) the dimensionless power spectral density function of the turbulence, $\bar{S}(\bar{F})$, and frequency, $\bar{F}$, and (b) the effective turbulence intensity, u'_k/u', and the integral timescale of the turbulence, $\bar{F}^{-1} \cdot u'_k$ is determined from the area under curve (a) above a limiting frequency set by the size of the developing flame. The threshold frequency diminishes as the flame grows.

from the area under the curve representing the dimensionless power spectral density function ($\bar{S}(\bar{F})$) versus dimensionless frequency ($\bar{F}$). The area is bounded by a low-limit frequency (Fig. 4.5) which decreases as the flame grows [46]. The variation of u'_k with flame growth is determined from the graph u'_k/u' versus $(\bar{F})^{-1}$ (Fig. 4.5). The effective turbulence intensity u'_k approaches u' only at elapsed times that are more than an order of magnitude greater than the integral timescale, the latter being related to the integral length scale through the mean speed of the moving gas.

The non-linearity of the solid lines in Fig. 4.4 may be attributed to the opposed effects of turbulent wrinkling, which increases the flame area and hence S_t, and the effects of flame stretch which reduces S_t. There is a diminishing benefit conferred on S_t as the turbulence intensity is increased. Ultimately, the turbulent flame cannot be sustained, and quenching may occur in the limit of high stretch at the highest turbulence intensities.

Different combustion regimes are also observed with increasing flame stretch in a fan-stirred bomb [63]. For $Ka \times Le < 0.15$, there is a continuous flame sheet. At higher values of $Ka \times Le$ the flame sheet begins to break up, and this process is accelerated by increasing flame quenching at $Ka \times Le > 0.3$. Quenching of the turbulent flame appears to be an inevitability of extinction at high stretch rates, as in the stagnation point, laminar flame. Combustion in spark ignition engines may be quite strongly affected by the dependence of turbulent burning velocity on turbulence intensity and flame stretch (Chapter 13).

4.8 Theory of turbulent diffusion flames

Diffusion flames are more difficult to deal with than premixed flames because there is no simple parameter characteristic of the process, such as the burning velocity. The closest measurable property is the flame height and several treatments have been developed to relate observed flame heights to the properties of the gases

The simplest analysis is due to Jost [64]. The Einstein diffusion equation states that the average square displacement x^2 is given by

$$x^2 = 2Dt \tag{4.35}$$

The height of the flame is taken as the point where the average depth of penetration is equal to the tube radius r. Approximating x^2 by r^2 this gives $t = r^2/2D$, but since the time t is also given by the height y divided by the gas velocity u (taken as constant), then

$$y = \frac{ur^2}{2D} \tag{4.36}$$

The volume flow rate v' is equal to $u\pi r^2$ so that

$$y = \frac{v'}{2\pi D} \tag{4.37}$$

From this simplified approach it is predicted that the flame height at a constant volume flow rate should be independent of burner dimensions. Also, because the diffusion coefficient is inversely proportional to pressure, the height of the flame will be independent of pressure at a constant mass flow rate.

In order to gain insight into the behaviour of turbulent diffusion flames from this approach, the molecular diffusivity D may be replaced by the eddy diffusivity, which is the product of the scale of the turbulence L and a velocity fluctuation term. Since L is proportional to the tube diameter, this leads to a prediction that the height of a turbulent diffusion flame is proportional only to the orifice diameter.

A full mathematical analysis of diffusion flames proves to be very difficult but an approximate procedure may be outlined as follows. Although incorrect since reaction occurs over an extremely wide range of fuel to oxygen ratios, particularly at high temperatures, it is necessary to define the reaction zone as the region in which the fuel and oxygen are in stoichiometric ratio. Another assumption is that the diffusion process is rate-determining so that the reaction rate is related directly to the amounts of fuel and oxygen diffusing into the reaction zone. Although satisfactory in principle, this approximation is difficult to apply because the fundamental parameter required, the diffusion coefficient, varies with both temperature and composition. It is normal also to consider only diffusion in a radial direction [65]. Modifications to the original analysis have resulted in very satisfactory agreement being obtained between predicted and observed flame sizes on several types of burner which are used in domestic and industrial applications [66, 67]. The structure of turbulent non-premixed flames has been reviewed by Bilger [68] in the context of flamelet theories.

Further reading

Bilger, R.W. (1988). The structure of turbulent non-premixed flames. In *Twenty-Second Symposium (International) on Combustion*. The Combustion Institute, Pittsburgh, PA, USA, p. 475.

Bradley, D. (1990). Combustion in gasoline engines. *Internal Combustion Engineering; Science and Technology* (ed J.H. Weaving). Elsevier Applied Science, London, UK, p. 287.

Bradley, D. (1992). How fast can we burn? *Twenty-Fourth Symposium (International) on Combustion*. The Combustion Institute, Pittsburgh, PA, USA, p. 247.

Bradshaw, P. (1971). *An Introduction to Turbulence and its Measurement*. Pergamon, Oxford, UK.

Buckmaster, J.D. and Ludford, G.S.S. (1983). *Theory of Laminar Flames*. Cambridge University Press, Cambridge, UK.

Frank-Kamenetskii, D.A. (1969). *Diffusion and Heat Transfer in Chemical Kinetics* (2nd edn) (Trans. J.P. Appleton). Plenum Press, New York, USA.

Law, C.K. (1988). Dynamics of stretched flames. In *Twenty-Second Symposium (International) on Combustion*. The Combustion Institute, Pittsburgh, PA, USA, p. 1381.

Lewis, B. and Von Elbe, G. (1987). *Flames, Explosions and Detonations in Gases* (3rd edn). Academic Press, Orlando, FL, USA.

Linãn, A. and Williams, F.A. (1993). *Fundamental aspects of Combustion*. Oxford University Press, Oxford, UK.

Peters, N. (1986). Laminar flamelet concepts in turbulent combustion. In *Twenty-First Symposium (International) on Combustion*. The Combustion Institute, Pittsburgh, PA, USA, p. 1231.

Strehlow, R.A. (1984). *Combustion Fundamentals*. McGraw Hill, New York, USA.

Von Karmann, T. and Penner, S.S. (1954). Fundamental approach to laminar flame propagation. In *Selected Combustion Problems* (eds. W.R. Hawthorne and J. Fabri). Butterworths, London, UK, p. 3.

Williams, F.A. (1988). *Combustion Theory* (2nd edn). Addison Wesley, New York, USA.
Zel'dovich, Y.B., Barenblatt, G.I., Librovich, V.B. and Makhaviladze, G.M. (1985). *The Mathematical Theory of Combustion and Explosion*. Consultants' Bureau, New York, USA.

Problems

(1) The burning velocities of stoichiometric hydrocarbon + air mixtures are about $0.4\,m\,s^{-1}$. Using propane as a typical fuel, estimate the average reaction rate and the thickness of the preheat zone. The thermal conductivity of the reaction mixture may be taken as $0.025\,W\,m^{-1}\,K^{-1}$ at 298 K and the mean flame temperature as 1000 K. (Use Table 2.2 to obtain values of C_p and assume that the heat capacity of air is the same as that of nitrogen.)

(2) Estimate the change in burning velocity of a stoichiometric propane + air mixture which would result if the nitrogen in the air is replaced by helium, assuming that the average reaction rate and the mean flame temperature are unchanged. At 298 K, the thermal conductivity of the reactants is $0.036\,W\,m^{-1}\,K^{-1}$.

5 Detonations

5.1 Introduction

The premixed flame or combustion wave (Chapter 3) travels at relatively low velocities, usually between 0.1 and $10\,m\,s^{-1}$. Most explosive mixtures can also display flame front velocities several orders of magnitude greater (2000–$3000\,m\,s^{-1}$). These flames exceed the characteristic sound speed of the gas and are termed *detonations*. Gaseous detonations are characterised by sudden changes in pressure and temperature, by contrast to premixed flames which display a marked temperature change but very little alteration in pressure. The transport of heat or diffusion of reactive species ahead of the flame front into the unburned gas are too slow to explain detonation, and reaction in this case is initiated by a supersonic pressure wave, or *shock* wave. The shock wave heats the reactive gas extremely rapidly and the energy liberated by the subsequent chemical reactions provides the driving force for the shock.

5.2 Shock waves

Shock waves are associated with a large number of physical phenomena both natural and man-made, such as aircraft flying at supersonic speeds, explosions or volcanic eruptions. The shock wave is characterised by a virtually instantaneous pressure rise associated with the rapid release of a substantial amount of energy. If the energy release is not maintained, the shock must be accompanied by a more gradual expansion, or *rarefaction*, which returns the pressure to the ambient value or below it.

A shock wave can occur in any fluid medium and may be defined as a step transition or discontinuity in the physical properties of the medium propagating without change. A sound wave progresses as a sequence of pressure pulses and, provided that the pulses are of small amplitude, the velocity has a constant value characteristic of the medium. If the pulses are of greater amplitude, the velocity increases by an amount which depends on the amplitude, and the wave becomes supersonic. A disturbance which is originally sinusoidal in shape will become progressively more distorted, the region of increasing pressure becoming steeper and that of falling pressure becoming shallower (Fig. 5.1)

An explanation of the way in which a shock wave forms is provided by the following simple model. The steady acceleration of a piston in a cylindrical tube is represented as a series of infinitesimal accelerations

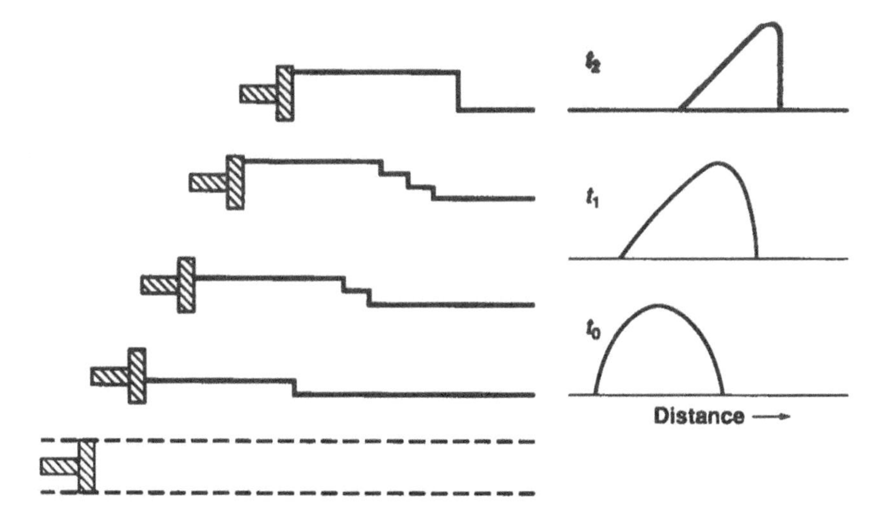

Figure 5.1 Formation of a shock wave from the acceleration of a piston in a tube and the development of a sinusoidal pressure pulse into a shock wave; t_0, t_1 and t_2 refer to successive times.

(Fig. 5.1). Each of these generates in the gas ahead of the piston a pressure pulse which travels at the appropriate sound speed. In addition to imparting movement to the gas, each pulse heats the gas adiabatically thereby raising the sound speed locally in the gas. The sound speed, a, in an ideal gas is given by

$$a = (\gamma p V)^{1/2} = (\gamma RT/M)^{1/2} \qquad (5.1)$$

where γ is the ratio of the principal molar heat capacities (C_p/C_v) and M is the molar mass.

The subsequent impulse travels faster than the one which preceded it and must catch it up. Eventually, these very small pulses coalesce to form a pressure pulse of finite amplitude travelling at a velocity greater than that found in the undisturbed gas. Shock waves may be generated in the laboratory in a manner which resembles that just described (section 5.3).

The gas through which the shock wave has passed is at a higher pressure, density and temperature than the undisturbed gas and moves in the same direction as the shock front itself although at a lower velocity. In a detonation the energy is provided by chemical reaction in the heated gas behind the front, which preserves propagation of the discontinuity.

The properties of the gas on each side of the transition can be considered without reference to the processes occurring in the transition itself since conditions are essentially uniform before and after passage of

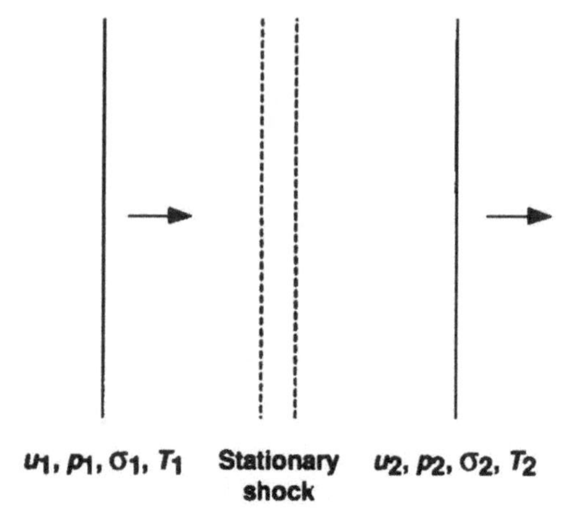

Figure 5.2 Notation used in the description of a stationary shock wave. Subscripts 1 and 2 refer to the pre-and post-shock gases respectively.

the shock and the whole process is considered to be one-dimensional. The system is most conveniently treated by referring it to a co-ordinate system moving with the front (Fig. 5.2). Although the transition is very sharp, since any medium comprises discrete molecules, the shock front thickness must be finite. Measurements show that the transition is very thin, corresponding to a few mean free paths in the gas. This is about 10^{-7} m at atmospheric pressure, which is equivalent to a timescale of nanoseconds (10^{-9} s).

Application of the conservation equations for mass and momentum across the shock front leads to the following expressions:

Conservation of mass (continuity)

$$\sigma_1 u_1 = \sigma_2 u_2 \quad \text{or} \quad u_1/V_1 = u_2/V_2 \tag{5.2}$$

Conservation of momentum

$$p_1 + \sigma_1 u_1^2 = p_2 + \sigma_2 u_2^2 \quad \text{or} \quad p_1 + u_1^2/V_1 = p_2 + u_2^2/V_2 \tag{5.3}$$

where V is the specific volume ($1/\sigma$). These equations take no direct account of the physical state or chemical nature of the medium.

Manipulation of eqns (5.2) and (5.3) leads to the mechanical shock relationships

$$u_1 = V_1 \frac{(p_2 - p_1)^{1/2}}{(V_1 - V_2)} \tag{5.4}$$

$$u_2 = V_2 \frac{(p_2 - p_1)^{1/2}}{V_1 - V_2} \tag{5.5}$$

The *shock velocity*, U_s in stationary co-ordinates, is equal to $-u_1$, so that

$$U_s = V_1\frac{(p_2-p_1)^{1/2}}{(V_1-V_2)} \quad (5.6)$$

and the *particle* or *flow velocity*

$$\dot{W}=(V_1-V_2)^{1/2}(p_2-p_1)^{1/2} \quad (5.7)$$

As expressed, the direction of motion in eqns (5.6) and (5.7) has been reversed to avoid carrying a minus sign. This may be regarded to represent the propagation of the shock and the compressed gas (with positive velocities) into undisturbed gas. A useful dimensionless form for these velocities is obtained by dividing by the sound speed in the undisturbed gas, to give the *Mach number*, M_a. In the limit as p_2 approaches p_1, the shock velocity falls to the sound speed at the conditions in state 1, a_1.

There is a third conservation equation also to be considered.

Conservation of energy

$$e_1+\frac{p_1}{\sigma_1}+\frac{u_1^2}{2}=e_2+\frac{p_2}{\sigma_2}+\frac{u_2^2}{2} \quad (5.8)$$

or

$$e_1+p_1V_1+\frac{u_1^2}{2}=e_2+p_2V_2+\frac{u_2^2}{2}$$

In these expressions e represents the internal energy per unit mass, or specific internal energy. The first two terms represent the enthalpy per unit mass,

$$h_1=e_1+p_1V_1 \quad \text{and} \quad h_2=e_2+p_2V_2 \quad (5.9)$$

and the third term represents the kinetic energy per unit mass. The three conservation equations can be combined to give the *Rankine–Hugoniot relationship*:

$$e_2-e_1=\frac{1}{2}(p_2+p_1)(V_1-V_2) \quad (5.10)$$

or

$$h_2-h_1=\frac{1}{2}(p_2-p_1)(V_1+V_2) \quad (5.11)$$

Energy and enthalpy are properties which depend on the chemical nature of the system so the precise form of these equations will vary from one medium to another. Since e and V (or σ) are themselves functions of p and T, a complete solution for a particular medium requires a knowledge of the appropriate equations of state, i.e.

$$e=f_1(p, T) \quad \text{and} \quad \sigma=f_2(p, T) \quad (5.12)$$

For an ideal gas

$$e = C_v T = pV(\gamma - 1)^{-1} \tag{5.13}$$

Incorporating eqn (5.13) into eqn (5.10) and rearranging yields an expression for the pressure ratio across the shock front, i.e. the ***Hugoniot curve***, in the explicit form

$$\frac{p_2}{p_1} = \frac{(\gamma+1)V_1 - (\gamma-1)V_2}{(\gamma+1)V_2 - (\gamma-1)V_1} \tag{5.14}$$

The specific volume (or density) ratio, V_1/V_2, can be obtained by rearrangement of eqn (5.14) to give

$$\frac{V_1}{V_2} = \frac{(\gamma-1)p_1 + (\gamma+1)p_2}{(\gamma-1)p_2 + (\gamma+1)p_1} \tag{5.15}$$

which, from the continuity eqn (5.2) is also the velocity ratio u_1/u_2. The temperature ratio, T_2/T_1, across the shock front may also be derived by use of the ideal gas equation in the form

$$\frac{T_2}{T_1} = \left[\frac{(\gamma-1)p_2 + (\gamma+1)p_1}{(\gamma-1)p_1 + (\gamma+1)p_2}\right]\frac{p_2}{p_1} \tag{5.16}$$

Equation (5.14) may also be compared with two other important relationships for ideal gases:

for an isothermal change

$$p_2/p_1 = V_1/V_2 \tag{5.17}$$

for an adiabatic change

$$p_2/p_1 = (V_1/V_2)^{\gamma} \tag{5.18}$$

The three relationships in eqns (5.14), (5.17) and (5.18) are shown in a p–V diagram (Fig. 5.3), in which the Hugoniot has a similar significance to the curves which relate states that may be achieved by an isothermal or an adiabatic process. Similar relationships may be derived for real gases given that the equations of state (eqn (5.12)) are available either in numerical or in analytical form, and that the dependence of $\gamma(T)$ is known.

The chord joining the initial (p_1, V_1) and final (p_2, V_2) states is termed the ***Rayleigh line***: if this line makes an angle ψ with the abscissa (Fig. 5.3), eqn (5.6) for the shock velocity can be rewritten:

$$U_s = V_1(\tan\psi)^{1/2} \tag{5.19}$$

In the real shock limit, at which the shock velocity reduces to the sound velocity, the Rayleigh line, the tangent, the adiabatic curve and the Hugoniot curve all coincide.

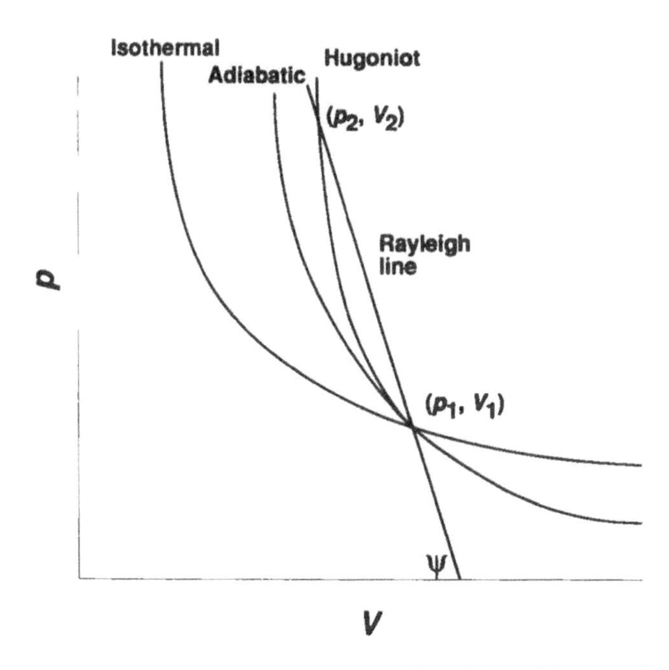

Figure 5.3 Loci of points attainable by shock (Hugoniot), adiabatic and isothermal transitions. The slope of the Rayleigh line connecting the points describing the initial (p_1, V_1) and final (p_2, V_2) states gives a measure of the velocity.

The same analysis applies to condensed phases, for which the compressibility, $(\partial p/\partial \sigma)_s$, is much lower and the sound speed is correspondingly greater. Since the derivative (expressed under isentropic conditions) represents the square of the sound speed in the undisturbed fluid

$$a^2 = \left(\frac{\partial p}{\partial \sigma}\right)_s \tag{5.20}$$

Table 5.1 Typical shock wave properties in air and water [69]

Medium	Sound speed ($\mathrm{m\,s^{-1}}$)	Shock Mach number	Flow Mach number	Temperature increase (K)	Pressure ratio
Air	347	4.12	3.29	900	20
		9.03	7.87	3440	100
		12.6	11.3	5090	200
		19.7	18.1	7940	500
Water	1532	1.15	0.07	5	2000
		1.52	0.28	36	10000
		1.88	0.45	70	20000
		2.62	0.81	152	50000

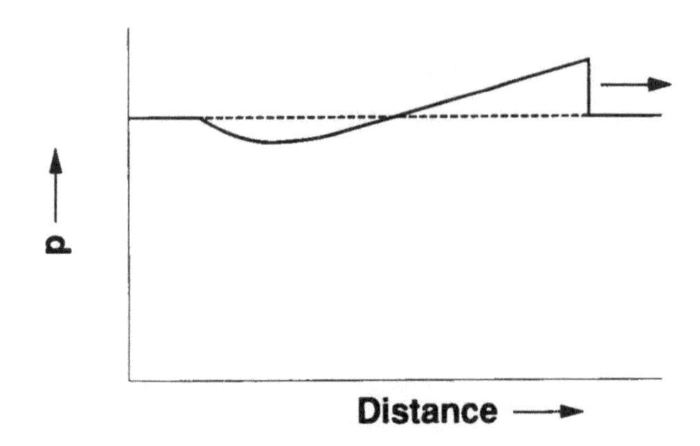

Figure 5.4 Pressure profile of a blast wave.

the relevant equations of state also reflect these differences in compressibility. Some typical shock parameters for air and water are listed in Table 5.1 [69].

Shock waves are generated by explosions. Since the source is a single event, energy is not provided continuously and the shock front in the surrounding atmosphere is immediately followed by an *expansion* or *rarefaction wave* which degrades the front. The pressure profile therefore changes shape as the wave moves further from the source, taking a form which is normally called a *blast wave* (Fig. 5.4).

5.3 Application of shock tubes in kinetic and combustion studies

In order to understand the chemical processes in flames in a quantitative way, it is essential to obtain kinetic data on elementary reactions over very wide ranges of temperatures. Sometimes these data can be derived from flame studies, but another source of data at high temperatures is the study of reactions behind a shock wave under controlled experimental conditions in a *shock tube*.

A high temperature is readily attained in a shock heated gas, governed ideally by the pressure ratio across the shock front and the ratio of heat capacities of the gases (eqn (5.16)). The discontinuity means that a reactant gas is raised instantaneously to T_2. The time interval available before a rapid cooling occurs, 10–1000 μs say, is short, but this is kinetically significant at high temperatures. The system is virtually adiabatic over the reaction interval. These are ideal circumstances in which quantitative kinetic measurements may be made.

A plane shock wave is produced in a long, closed tube by the sudden bursting of a diaphragm (e.g. aluminium foil) which separates one gas at high pressure, the *driver gas*, from another at low pressure, the *test gas*. The reactants in the test gas are usually heavily diluted with inert gas (> 95%). A diagrammatic representation of a shock tube and how some

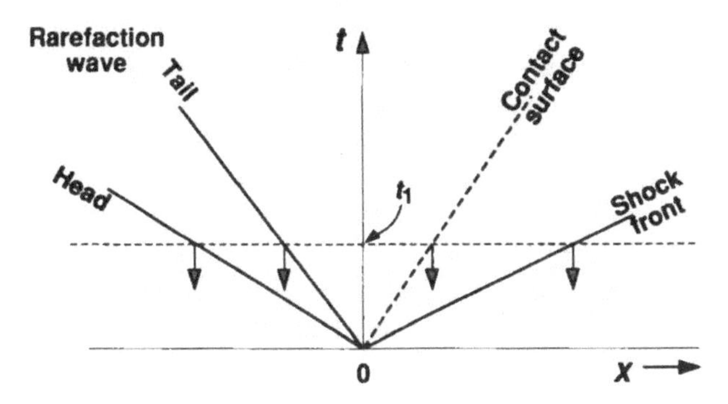

Figure 5.5 An (x, t) diagram showing the development of different regions during the propagation of a shock wave in a tube following the bursting of a diaphragm at $x = 0$, $t = 0$. The arrows signify the displacement of the shock front, the contact surface and the rarefaction fan at time t_1.

of the events infold in time are shown in Fig. 5.5. The contact surface represents the interface between the driver gas and the test gas. It moves rapidly along the tube but only at a subsonic velocity as opposed to the supersonic velocity of the shock front. The driver gas at the *contact surface* is cold, and so when it envelopes the test gas, any high temperature reactions are quenched instantly. The reaction time is represented by the interval between the arrival of the *incident shock* front and the arrival of the contact surface. Its duration is determined by the point in the tube at which the kinetic measurements are being made. The *rarefaction fan*, which is driven back into the driver gas, is an expansion wave and it has a divergent character because the velocity of its 'tail' is governed by gas that has already been cooled by the rarefaction head.

These are the principles, but to obtain meaningful kinetic information many corrections have to be made. These are governed largely by the shock tube design and gas properties, and include viscous and boundary layer effects that affect the shock wave, and movement of the test gas along the tube which affects its reaction time. The correction methods and procedures for deriving the shocked gas temperature and the shock velocity are very well established.

The simplest form of shock tube is closed at both ends. In such a case the shock would be reflected from the test gas end, so that the reactants may experience a double heating as the *reflected shock* passes through them. This means that exceedingly high temperatures are attained before the contact surface arrives. Unless care is taken to *tailor* the gas mixtures, so that the reflected shock passes through the contact surface, it will be reflected back again, which may destroy the control of heating and quenching that is being sought for kinetic measurements. Another effect is that the rarefaction fan would be reflected from the end wall of the driver section, so that it also follows the shock and the contact surface. In fact the expansion wave moves at the local sound speed, which is faster than the contact surface velocity so, with a sufficiently long test section and sufficiently short driver section, the rarefaction may overtake the contact surface, and become the major gas cooling factor.

Some tubes, termed *chemical shock tubes*, have been designed so that conventional chemical analysis can be performed on the reaction products, by use of gas chromatography or related techniques. However, in most cases kinetic data are derived from spectroscopic studies of the reactants or products via windows in the shock tube walls. Spectroscopic methods are also used to identify the occurrence of ignition when ignition delays are being measured. Typically, the shocked gas temperatures for kinetic studies [70] or ignition delay times [71] fall in the range 1500–2500 K, but very much higher temperatures can be generated if necessary. Lower temperatures are not often used for kinetic studies because non-idealities become more of a problem and also because other experimental techniques become accessible, but they are of interest for ignition delay measurements of hydrocarbons [72].

5.4 One-dimensional structure of detonation waves

The shock wave behaves strictly as a one-dimensional phenomenon and, although not fully justified, the same restriction is applied to detonations. In writing the equation of state $e=f(p, T)$ for the gas, it was assumed that no chemical reaction had taken place. In determining the final state (p_2, T_2) for an exothermically reacting system, an allowance must be made for the heat release. The effect is to displace the Hugoniot curve to higher values of p and V so that it no longer passes through (p_1, V_1) (Fig. 5.6). The Hugoniot still represents the solution to eqns (5.10) or (5.11) but defines the states which are mathematically possible from a given initial state by a shock transition accompanied by exothermic reaction. Not all these states are physically real, however. Between the points B and C (Fig. 5.6), $(p_2-p_1)/(V_1-V_2)$ is negative so that the value of U_s given by eqn (5.6) is imaginary: a transition from (p_1, V_1) to any point between B and C therefore cannot correspond to a realisable situation. There are thus two quite separate regions, which correspond to $p_2>p_1$, $V_2<V_1$ (AB) and $p_2<p_1$, $V_2>V_1$ (CD).

In the region CD to the right of (p_1, V_1), the slope of the Rayleigh line is less than the angle made by the tangent at (p_2, V_2), so that the combustion is subsonic. Furthermore, the pressure and density decrease across the transition and the gas leaves it at a greater velocity than that at which it enters. This follows from eqns (5.6) and (5.7) and describes a deflagration (Chapter 3). Points on the curve AB to the left of (p_1, V_1) describe the detonation process.

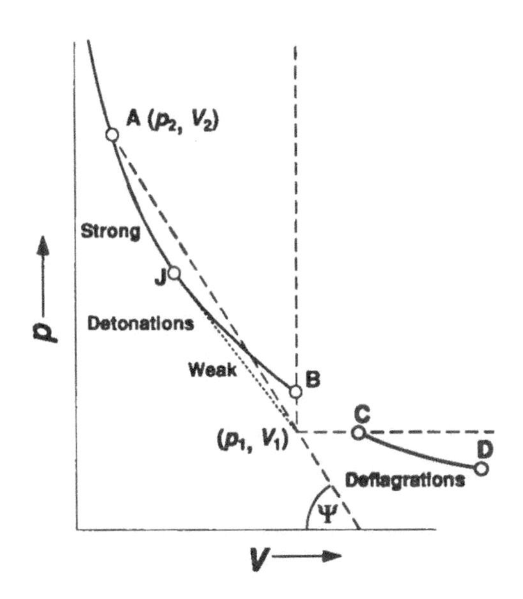

Figure 5.6 Hugoniot curve for the transition involving exothermic chemical reaction.

The Rayleigh lines drawn through (p_1, V_1), which make a tangent to the curve, correspond to the minimum attainable detonation velocity: all other Rayleigh lines through (p_1, V_1) have steeper slope and correspond to higher velocities. Those lines also indicate two possible final states. The detonation with the minimum velocity is termed a Chapman–Jouguet detonation and the final state described by point J to which this detonation refers is the Chapman–Jouguet or C–J state [73–76]. Detonations associated with final states to the left of the C–J state in Fig. 5.6 are referred to as *strong* detonations, while those corresponding to points to the left are termed *weak* detonations.

On the strong detonation section of curve AB, the tangent at (p_2, V_2) has a steeper slope than the corresponding Rayleigh line. The front is then subsonic with respect to the gas behind it and any perturbations generated behind the front will catch up with it and hence cause the detonation to be attenuated. Strong or over-driven detonations may occur, for example when initiated by a very strong shock wave, but unless they are supported by energy additional to that of the reaction they are unstable and decay to C–J detonations. On the weak detonation section of curve AB, the detonation is supersonic with respect to the following gas. This means that the whole of the reaction energy cannot contribute to the motion of the front. Thus, while all final states on AB are allowed by the conservation equations, only the C–J condition corresponds to a stable self-supporting detonation.

Since the Rayleigh line to the C–J point is tangential to the Hugoniot, the sound speed a_2 in the burned gas is given by eqn (5.20) in the form

$$a_2 = V_2 \left(\frac{p_2 - p_1}{V_1 - V_2} \right)^{1/2} \tag{5.21}$$

Combining this with eqns (5.6) and (5.7) leads to the relationship

$$D = u_1 + a_2 \tag{5.22}$$

The detonation velocity D is thus equal to the sum of the particle, or flow, velocity and the sound speed in the burned gas. This may be regarded as a mathematical statement of the *Chapman–Jouget postulate*. In physical terms it means that the detonation front moves at the sound velocity with respect to the gas behind it.

The importance of the C–J postulate is that it enables the detonation velocity, and hence the associated properties, to be predicted from the thermodynamic properties of the burned gas. In principle, a detonation should be possible for all exothermic reactants but, in practice, kinetic factors are involved and composition limits are imposed on detonable mixtures. Some properties of gaseous detonations are given in Table 5.2.

This picture of the detonation wave is oversimplified because it assumes that chemical reaction proceeds to completion instantaneously. A better (ZND) representation, suggested independently by Zeldovich [77], Von Neumann [78] and Doring [79], is to consider a family of Hugoniot curves corresponding to successive fractions of reaction (Fig. 5.7). As the detonation is a steady phenomenon, all states through which the system passes must lie on the single Rayleigh line corresponding to the detonation velocity. The initial shock transition is to a point A on

Table 5.2 Typical detonation wave properties in gaseous mixtures. The detonability limits apply to atmospheric pressure [23][a]

Reactants	Detonability limits (% by volume)		Detonation velocity ($m\,s^{-1}$)	Detonation temperature (K)	Detonation pressure (MPa)
	Lower	Upper			
$H_2 + O_2$	15	90	2825	3700	1.805
$CO + O_2$	38	90	1760	3500	1.86
$CH_4 + O_2$	—	—	2322	3700	—
$C_2H_2 + O_2$	35	92	2350	4200	4.4
$NH_3 + O_2$	254	75	2400	—	—
$C_3H_8 + O_2$	3.1	37	2350	—	—
H_2 + air	18.2	59	1940	2950	1.56
C_2H_2 + air	4.2	50	1900	3100	1.9
CH_4 + air	—	—	1800	2740	1.72
C_3H_8 + air	—	—	1800	2820	1.83

[a] The other figures, for stoichiometric mixtures at atmospheric pressure, are from a variety of sources using different methods and should only be considered as approximate.

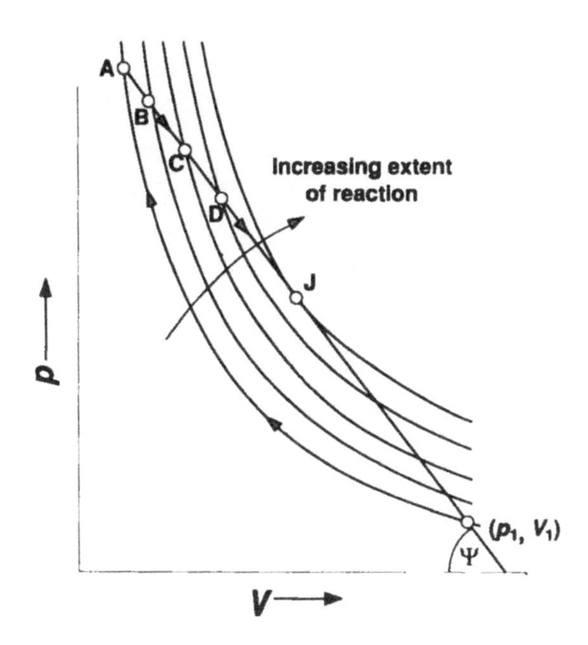

Figure 5.7 A representation of a finite reaction zone by a family of Hugoniot curves corresponding to different extents of reaction.

the non-reactive Hugoniot. The state of the gas then changes along the Rayleigh line in the direction of progressively greater extents of reaction until it reaches the C–J point (J) on the fully reacted Hugoniot.

In this ZND model, the detonation is visualised as a simple shock transition followed by a reaction zone of finite length within which the pressure and density fall and the temperature rises. The reaction zone is expected to have a width of about a millimetre or less, corresponding to a time of about 0.5 µs, at atmospheric pressure. It is very difficult to measure reaction zone thicknesses but indirect methods have suggested reaction times of 0.3–0.5 µs in benzene–oxygen mixtures [80]. A sharp maximum in the pressure and density profiles is predicted and is referred to as the Von Neumann spike (Fig. 5.8), which has been observed experimentally [81, 82]. The whole region between the front and the C–J plane represents a steady flow process and propagates without change.

The burned gas which leaves the C–J plane is moving in the same direction as the detonation but at the particle velocity and, in the absence of an additional source of energy, it must eventually stop. If the detonation is initiated at the closed end of a long tube, the gas close to the end is forced to remain motionless and so an expansion wave must follow the detonation front, cool the burned gas, and accelerate it in the

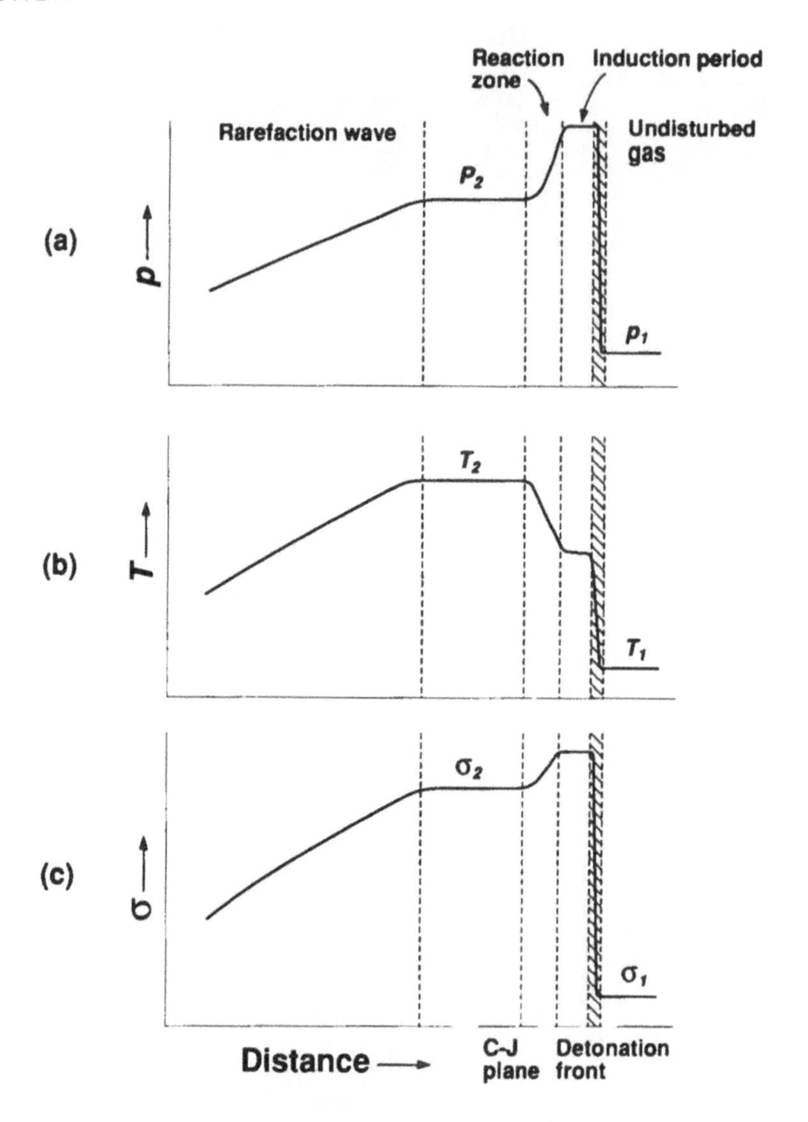

Figure 5.8 Variation of (a) pressure, (b) temperature, and (c) density, through a detonation.

reverse direction. An expansion wave is isentropic and therefore travels at a velocity equal to the sound velocity of the gas superimposed upon the particle velocity. This is precisely the condition at the C–J plane to which the head of the expansion wave is therefore 'tied'. The rarefaction region is non-steady since the profiles there change with time as the detonation propagates through the medium, much like a blast wave. Steady motion of the detonation front is possible only if the flow behaviour of the burned gas matches that of the expansion head.

5.5 Mathematical treatment of detonation

A complete solution of the Rankine–Hugoniot equations to give the C–J detonation velocity is much simpler than the corresponding calculation of the burning velocity of a premixed flame because it depends only on the initial and final states of the gas. To obtain an analytical solution, the fluid is assumed to behave as an ideal gas with a constant specific heat capacity and a fixed molar mass. Chemical reaction is accompanied by a heat release q per unit mass of gas. The following thermodynamic relations then apply

$$\frac{p_1 V_1}{T_1} = \frac{p_2 V_2}{T_2} = R' \tag{5.23}$$

$$h_1 = C'_p T_1 \qquad h_2 = C'_p T_2 \tag{5.24}$$

$$\gamma = C'_p / C'_v \qquad C'_p - C'_v = R' \tag{5.25}$$

The enthalpies, specific heat capacities, specific volumes and the gas constant R', like the heat release, are referred throughout to unit mass of reacting gas. The Rankine–Hugoniot relation (eqn (5.11)) may then be written

$$\frac{\gamma}{(\gamma-1)}(p_2 V_2 - p_1 V_1) - q = \frac{1}{2}(p_2 - p_1)(V_2 + V_1) \tag{5.26}$$

Making the following substitutions for the pressure and specific volume ratios, and for a dimensionless representation of the exothermicity

$$\pi = p_2/p_1; \qquad \Gamma = V_2/V_1; \qquad B = 2(\gamma - l)q/p_1 V_1 \tag{5.27}$$

eqn (5.26) becomes

$$2\gamma(\pi\Gamma - 1) - B = (\gamma - 1)(\pi - 1)(\Gamma + 1) \tag{5.28}$$

The (negative) slope of the tangent to the Hugoniot curve with co-ordinates (π, Γ) is obtained by differentiating eqn (5.28) to give

$$\frac{-\mathrm{d}\pi}{\mathrm{d}\Gamma} = \frac{(\gamma+1)\pi + (\gamma-1)}{(\gamma+1)\Gamma - (\gamma-1)} \tag{5.29}$$

The C–J condition is satisfied by equating this derivative to the (negative) slope of the Rayleigh line, that is, $(\pi - 1)(1 - \Gamma)$ in these generalised co-ordinates. From this equality, one obtains the relation

$$\pi = \frac{\Gamma}{(\gamma+1)\Gamma - \gamma} \tag{5.30}$$

Expressing the detonation velocity expressed as the dimensionless Mach number with respect to the sound speed in the cold gas, a_1 (eqn (5.1)) gives

$$M_a = \frac{D}{a_1} = \frac{D}{(\gamma p_1 / V_1)^{1/2}} \tag{5.31}$$

Replacing the terms in eqn (5.6) with the pressure and specific volume ratios (eqn (5.27)) gives

$$M_a^2 = \frac{1(\pi - 1)}{\gamma(1 - \Gamma)} \tag{5.32}$$

The above relationships may be used to derive the dimensionless Hugoniot equation (eqn (5.28)) in terms of M_a and γ, as follows. Rearrangement of eqns (5.30) and (5.32) give, respectively,

$$\gamma(\Gamma - 1) = (\Gamma/\pi)(1 - \pi) \tag{5.33}$$

and

$$\gamma(\Gamma - 1) = (1 - \pi)/M_a^2 \tag{5.34}$$

from which

$$\Gamma = \pi/M_a^2 \tag{5.35}$$

Substituting eqn (5.35) in to eqn (5.30) gives

$$\pi = \frac{1 + \gamma M_a^2}{(\gamma + 1)} \tag{5.36}$$

and from eqns (5.35) and (5.36)

$$\Gamma = \frac{1 + \gamma M_a^2}{(\gamma + 1)} \tag{5.37}$$

Substituting eqns (5.36) and (5.37) in to the Hugoniot equation (eqn (5.28)) yields the following quadratic:

$$(M_a^2)^2 - CM_a^2 + 1 = 0 \tag{5.38}$$

where

$$C = [B(\gamma + 1)/\gamma] + 2 \tag{5.39}$$

Taking the upper root of eqn (5.38), which applies to the detonation branch in Fig. 5.6,

$$2M_a^2 = C + (C^2 - 4)^{1/2} = C + C(1 - 4/C^2)^{1/2} \tag{5.40}$$

The term in the brackets may be expanded using the series

$$(1 + x)^n = 1 + nx + n(n - 1)x^2/2 + \ldots. \tag{5.41}$$

If $C \gg 1$, as is usually the case, only the first term need be considered so that

$$M_a^2 \approx (\gamma + 1)B/\gamma = 2q(\gamma^2 - 1)\gamma p_1 V_1 \tag{5.42}$$

From eqns (5.31) and (5.32), the detonation velocity may be expressed as a function of exothermicity as

$$D = [2q(\gamma^2 - 1)]^{1/2} \tag{5.43}$$

The particle velocity follows from eqn (5.7) as

$$W = (1 - \Gamma) D \tag{5.44}$$

The temperature ratio across the detonation is given by

$$\frac{T_2}{T_1} = \pi\Gamma \tag{5.45}$$

and by use of eqns (5.36) and (5.37), for conditions in which $\gamma M_a^2 \gg 1$

$$\frac{T_2}{T_1} = \frac{\gamma^2 M_a^2}{(\gamma + 1)^2} \tag{5.46}$$

An alternative expression, obtained by substituting for M_a^2, is

$$T_2 = 2q\gamma / C_v(\gamma + 1) \tag{5.47}$$

At the same level of approximation, the adiabatic, constant volume temperature rise ($T_v = q/C_v$) gives

$$T_2/T_v = 2\gamma/(\gamma + 1) \tag{5.48}$$

In view of the number of approximations made, the calculated detonation properties for fuel–air mixtures turn out to be surprisingly close to those listed in Table 5.2. For fuel–oxygen mixtures, this simplified treatment yields less satisfactory results, but when allowance is made for dissociation of products at the higher temperatures involved, the agreement between the calculated and experimental detonation parameters is greatly improved [83].

5.6 Three-dimensional structure of detonations

Experimental measurements of detonation velocities agree well with the predictions based on the Chapman–Jouguet theory, typically to within ±2% except for small diameter tubes or close to the detonability limits when deviations of 10% or more are observed. In such cases, three-dimensional effects can be seen, the most notable being *spinning detonation* in which the front is tilted and rotates about the axis as it travels down the tube. If longitudinal instability occurs then *galloping detonations* are observed in which the detonation wave repeatedly decays and then jumps back to its original velocity. The occurrence of exothermic reaction in the reaction zone renders the one-dimensional system unstable. All stable detonations appear to show complex, three-dimensional structure.

One way of visualising the origin of instability is as follows. Consider the detonation front to be represented by a square wave model in which the non-reactive shock is followed by an induction period at the end of which the chemical energy is released instantaneously (Fig. 5.9(a)). If the

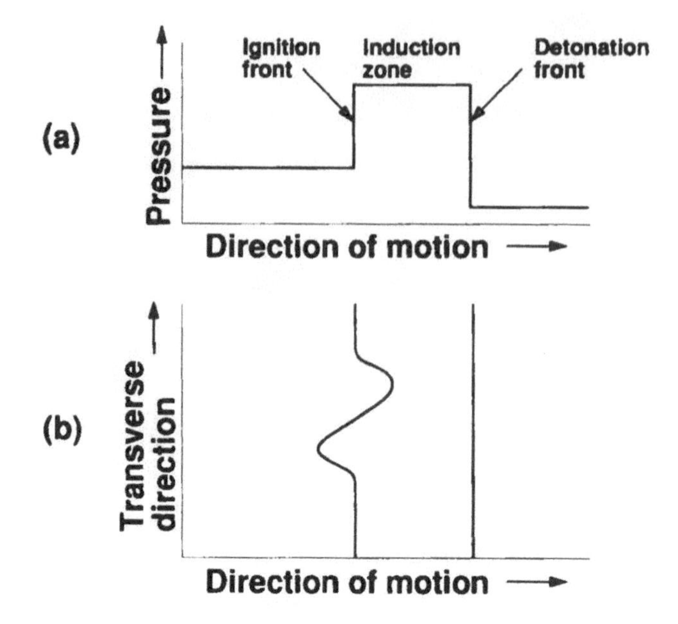

Figure 5.9 (a) A model showing the origin of transverse instabilities in a detonation wave. (b) Properties perpendicular to the front (after Shchelkin and Troshin [84]).

ignition front is given a minor perturbation so that it becomes distorted in a direction perpendicular to the flow (Fig. 5.9(b) [84]), small regions of high and low pressure now co-exist so that the high-pressure region expands sideways to damp out the disturbance. However, in the detonation the expansion of the gas causes it to cool isentropically and the ignition delay (or induction period), being temperature dependent, increases significantly. Thus the perturbation is amplified. This analysis underestimates the instability of detonations, but it serves to show that three-dimensional effects are likely to be greater for reactions with higher activation energies and longer reaction zones, which tend to be associated with marginal conditions.

The three-dimensional structure of a steady detonation arises from contact with a rigid surface. The gas adjacent to the surface must be brought to a standstill with respect to motion normal to the wall, and a shock is reflected into the oncoming gas behind the incident shock (Fig. 5.10(a)). If the surface is inclined at an angle to the flow, the wave pattern in Fig. 5.10(b) is established. The point of intersection of the incident and reflected shocks travels upwards along the wall. Its velocity and angle satisfy the constraint for motion normal to the wall. This is termed *regular reflection*.

As the surface becomes further inclined to the flow, the velocity of the reflected shock increases until a stage is reached at which the point of

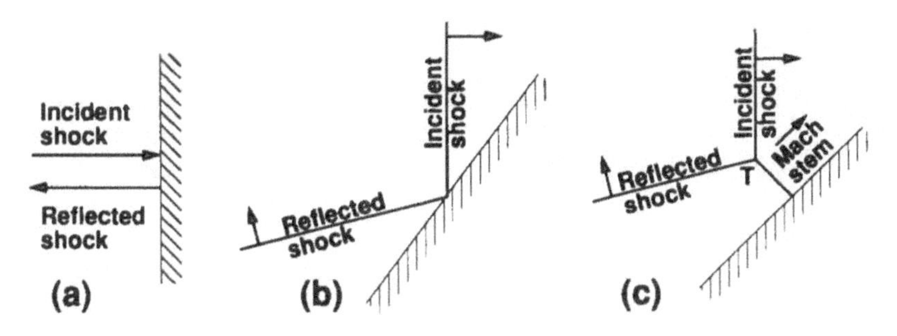

Figure 5.10 Shock reflection (a) normal to a plane, (b) at an angle to a plane, and (c) formation of a Mach stem.

intersection is forced to leave the wall. A third shock, known as the *Mach stem*, perpendicular to the wall, forms at the point of intersection, which is now termed the *triple point*. The gas immediately adjacent to the wall passes through the single shock while that further away is processed by two shocks. This creates a mismatch in the velocities of the shock-heated gas streams and leads to a *slip line*. This behaviour is called *Mach reflection* (Fig. 5.10(c)).

These features can be used to interpret the cellular structure of detonation, given here as a two-dimensional description. The detonation comprises alternating convex shocks and Mach stems connected by triple points, as described above. Leading back from each triple point is a reflected shock, termed a *transverse wave*. As the detonation front propagates the transverse waves move into the regions processed by the curved incident shock. When the transverse waves collide, the pattern is effectively inverted so that the rapid reaction initiated by the collison leads to a new Mach stem and a new pair of transverse waves moving outwards into the now weakened incident shock.

This complex wave behaviour can be studied relatively easily because the velocity shear associated with the slip at the triple point causes it to describe a pattern on a soot-coated surface (Fig. 5.11 [85]). It seems that

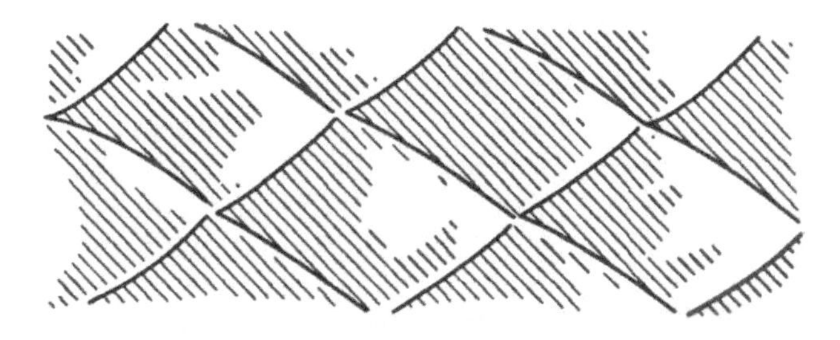

Figure 5.11 Smoke track records showing cellular detonation structure (after Strehlow [85]).

the cell structure is coupled to the reaction zone length and the geometry of the tube, so that the transverse wave spacing and the cell length decrease as the the strength of the detonation is increased. In view of this complex three-dimensional structure one might expect the one-dimensional C–J theory not to be valid [86]. Nevertheless, C–J calculations are universally employed and are often more reliable than experimental measurements.

5.7 Initiation of detonation and the deflagration to detonation transition

For the majority of exothermic reaction mixtures, two quite stable combustion waves are possible, a propagating flame and a detonation. Since detonations can cause severe damage, it is important to understand how detonation may be initiated. Three distinct modes of initiation are by shock wave, electric spark, or flame. In each case, the degree of confinement or the geometry of the container plays a dominant role.

A shock wave passing through a combustible mixture is reflected from the closed end of a tube. Behind the shock wave there is an induction interval within which there is little change in the gross properties of the system, although some chain-branching reaction will have begun. This zone is followed by the liberation of heat to the gas as the chemical reactions approach completion. Thus a reaction wave follows the shock wave at a constant distance from it. However, at the end wall the release of energy causes an adiabatic explosion in the stationary gas which generates a reaction shock wave travelling in the induction zone ahead of the reaction wave. The effect of the reaction shock is to reduce the length of the induction zone so that the two waves travel together until they overtake the original reflected shock, at which the steady detonation forms. A reaction wave of this kind will always follow a shock wave in a combustible mixture. Any perturbation associated with this reaction wave which leads to a weak shock or compression wave can then cause the reaction wave to accelerate and eventually overtake the front.

Initiation by electric spark can lead either to a normal premixed flame or to a detonation depending on conditions. The reason is that the electric spark generates a spherical shock wave followed by a kernel of spark-heated gas. If the temperature and pressure behind the expanding shock wave remain sufficiently high to cause rapid reaction, i.e. to generate a reaction wave, then detonation can follow from this process. If a detonation condition is not met, then it is still possible for the hot kernel to ignite a flame by heat transfer to the surrounding gas.

Initiation by flame is the most complex process of all. In this case also, the behaviour depends critically on the nature of the container. When the flame is ignited at the closed end of a long tube, the transition to detonation often occurs only at the end of a long predetonation run. The burned gas behind a flame has a lower density, and hence occupies a larger volume, than that of the gas ahead. Thus the flame front leaving the closed end of the tube will travel at several times the laminar burning velocity and will generate pressure waves in the unburned gas ahead. This causes preheating of the gas before it enters the flame and hence when the gas actually enters the flame even more rapid reaction ensues. The planar flame front soon becomes unstable and turbulent motion further increases the rate of burning. Pockets of unburned gas become trapped in the burning zone and are compressed and heated there until self-ignition occurs and a detonation eventually forms.

Further reading

Bradley, J.N. (1962). *Shock Waves in Chemistry and Physics.* Methuen, London, UK.
Cook, M.A. (1958). *The Science of High Explosives.* Reinhold, New York, USA.
Fickett, W. and Davies, W.C. (1979). *Detonation.* University of California Press, Berkeley, CA, USA.
Gaydon, A.G. and Hurle, I.R. (1963). *The Shock Tube in High-Temperature Chemical Physics.* Chapman and Hall, London, UK.
Greene, E.F. and Toennies, J.P. (1964). *Chemical Reactions in Shock Waves.* Edward Arnold, London, UK.
Lewis, B. and Von Elbe, G. (1987). *Combustion, Flames and Explosions of Gases* (3rd edn).
Academic Press, New York, USA.
Sokolik, A.S. (1963). *Self-Ignition, Flame and Detonation in Gases.* Israel Program for Scientific Translations, Jerusalem, Israel.
Strehlow, R.A. (1984). *Combustion Fundamentals.* McGraw Hill, New York, USA.
Zel'dovich, Ya.B. and Raizer, Yu.P. (1968). *Element of Gas Dynamics and the Classical Theory of Shock waves* (eds W.D. Hayes and R.F. Probstein). Academic Press, New York, USA.

Problems

(1) Calculate the final pressure and temperature reached in the adiabatic compression of an ideal gas from 100 kPa and 290 K, at a compression ratio of 10:1. Calculate the temperature that would be reached by the same gas if it experienced shock heating from the same initial conditions to the same final pressure.

(2) Use the approximate formulae to estimate the detonation temperature velocity and pressure for stoichiometric hydrogen + air and methane + air mixtures initially at 100 kPa and 290 K.

6 High temperature and flame chemistry ($T > 1000\,K$)

6.1 Background and kinetic overview

This chapter and the following one are concerned with the chemical processes involved in combustion, and especially those connected with hydrocarbon fuels. A division is made at approximately 1000 K between the 'low temperature' and 'high temperature' regimes, because there is a very clear distinction between the types of reactions that dominate the overall combustion processes as temperatures rise from below 850 K to beyond 1200 K. The division itself is artificial; it does not imply the sudden switching off of one set of reactions and the switching on of another, and the choice of 1000 K as the threshold is one of convenience within a broad temperature range.

Apart from the qualitative structure of the mechanisms involved in the high temperature regime, it is important to gain some insight into the relative importance of different processes under different conditions. Combustion reactions must be reasonably rapid since, in general, appreciable extents of reaction must occur in times within the range 10^{-6} to 10^{-2} s (1 μs to 10 ms).

The interpretation of combustion processes by modelling of detailed kinetic mechanisms has become possible with the widespread availability of computers [87], which requires a quantitative knowledge of the parameters of the rate constant for each elementary reaction involved in the combustion process, given as either

$$k = A\exp(-E/RT) \text{ or } k = A\,T^n\exp(-E/RT) \tag{6.1}$$

The parameters for a wide range of elementary reactions have been generated from experimental studies. Moreover, there have been a number of ongoing 'rate data assessment' programmes, from which recommendations of the 'best' data for many elementary reactions have been made [14]. Without these contributions many of the parameters would have to be estimated, or they may have too great an uncertainty for numerical computations to be very meaningful.

Many elementary reactions have not yet been studied experimentally in isolation, and in other cases the precision of the experimental data may not be as satisfactory as is required. Recourse may then be made to theoretical derivations, often based on analogy to other 'known' reactions. Predictive approaches are derived from group additivity rules for thermochemical properties [8, 88].

6.1.1 Relative rates of elementary reactions

The free radical chain initiation processes are not normally predominant in the control of events, and the development of combustion falls to the chain propagation and branching characteristics of the reaction. Nevertheless, the alternative modes for the initiation of alkane combustion illustrate very well some of the underlying issues.

Consider the possibilities:

$$RH + O_2 \rightarrow R + HO_2; \quad k_{6.2} = 1 \times 10^{14} \exp(-25000/T)\,cm^3\,mol^{-1}\,s^{-1} \quad (6.2)$$

$$RH \rightarrow R_1 + R_2; \quad k_{6.3} = 5 \times 10^{16} \exp(-42000/T)\,s^{-1} \quad (6.3)$$

In the generalised representation of eqn (6.2), the oxidation of the fuel (RH) by hydrogen atom abstraction may yield a hydroperoxy radical HO_2 and an alkyl radical (R) in which the carbon backbone of the alkane remains intact. By contrast, the unimolecular decomposition of the fuel (eqn (6.3)) may yield two alkyl radicals (R_1 and R_2), as a result of which the carbon backbone of the alkane molecule is severed. Typical rate constants for these types of process are given in the simple Arrhenius form, and the relative rates of initiation are thus

$$(v_{6.2}/v_{6.3}) = 2 \times 10^{-3} \exp(17\,000/T)[O_2] \quad (6.4)$$

Temperature has the greatest effect on which of these reactions dominates, through the relative magnitudes of the activation energies, but the concentration of oxygen is also important. Consider normal atmospheric conditions, for which the concentration of O_2 is approximately $2.5 \times 10^{-6}\,mol\,cm^{-3}$ at 1000 K, and is inversely proportional to temperature. The relative rates of initiation, given by eqn (6.4) in the temperature range 800–1400 K in 200 K increments, are then 10.8, 0.12, 0.006 and <0.001, respectively. Thus the oxidation route to initiation would predominate at temperatures somewhat below 1000 K, and the carbon backbone structure of the fuel would remain intact. However, there is an increasing importance of the degradation route as the temperature is raised, and it must be regarded to be the virtually exclusive mode of initiation above about 1400 K.

This calculation illustrates how, at high temperatures, reactions with a high activation energy and high pre-exponential factor tend to be more important than those with a lower activation energy but lower pre-exponential term. Many of the simpler bimolecular reactions have similar pre-exponential factors and differ only in their activation energies. Since the exponential term approaches unity at high temperatures, many of these reactions have similar rate constants at flame temperatures and so are of comparable importance. For this reason, the reactions associated with flames should be represented as reversible processes since, at $T > 1500$ K say, there may be a sufficiently high reaction rate for both the forward and reverse processes that an equilibrium is established.

At lower temperatures, in many cases the equilibrium is displaced so far to the right hand side that the reverse process may be disregarded. These circumstances pertain when high activation energies for the reverse reaction are involved, as in eqns (6.6) and (6.8) given below. The rate parameters are given in the two-parameter Arrhenius form.

Amongst bimolecular reactions, the most important ones are metathetical reactions involving an atom or simple radical $A + BC \rightarrow AB + C$. Such reactions do not usually involve complex rearrangements and therefore have high pre-exponential factors. The most common reactions in this category are H atom abstractions, for example,

$$H + CH_4 \Leftrightarrow H_2 + CH_3; \quad k_f = 8.49 \times 10^{14} \exp(-8205/T)\,\mathrm{cm^3\,mol^{-1}\,s^{-1}} \tag{6.5}$$

$$k_r = 5.08 \times 10^{13} \exp(-11270/T)\,\mathrm{cm^3\,mol^{-1}\,s^{-1}}$$

$$OH + CH_4 \Leftrightarrow H_2O + CH_3; \quad k_f = 6.12 \times 10^{13} \exp(-3937/T)\,\mathrm{cm^3\,mol^{-1}\,s^{-1}} \tag{6.6}$$

$$k_r = 6.90 \times 10^{12} \exp(-12400/T)\,\mathrm{cm^3\,mol^{-1}\,s^{-1}}$$

The nature of the species involved depends on the particular fuel-oxidant system but in virtually all systems in which the elements H and O occur, the OH radical is an important and frequently dominant species causing oxidation. Reactions involving HO_2, H and O are also significant. The principal route to water as a final product of hydrocarbon combustion is from abstraction processes such as the reaction shown in eqn (6.6).

There are important shifts of emphasis of the main chain propagating radicals as the temperature is increased. In the 'high temperature' range, this may be attributed mainly to the competition between the chain branching and non-branching reaction modes

$$H + O_2 \Leftrightarrow OH + O; \quad k_f = 2.00 \times 10^{14} \exp(-8455/T)\,\mathrm{cm^3\,mol^{-1}\,s^{-1}} \tag{6.7}$$

$$k_r = 1.46 \times 10^{13} \exp(-252/T)\,\mathrm{cm^3\,mol^{-1}\,s^{-1}}$$

$$H + O_2 + M \Leftrightarrow HO_2 + M; \quad k_f = 2.30 \times 10^{18}\, T^{-0.8}\,\mathrm{cm^6\,mol^{-2}\,s^{-1}} \tag{6.8}$$

$$k_r = 2.80 \times 10^{15} \exp(-23000/T)\,\mathrm{cm^3\,mol^{-1}\,s^{-1}}$$

The relative rates of the forward reactions may be expressed as

$$v_{6.7}/v_{6.8} = 8.7 \times 10^{-5}\, T^{0.8} \exp(-8455/T)\,[M]^{-1} \tag{6.9}$$

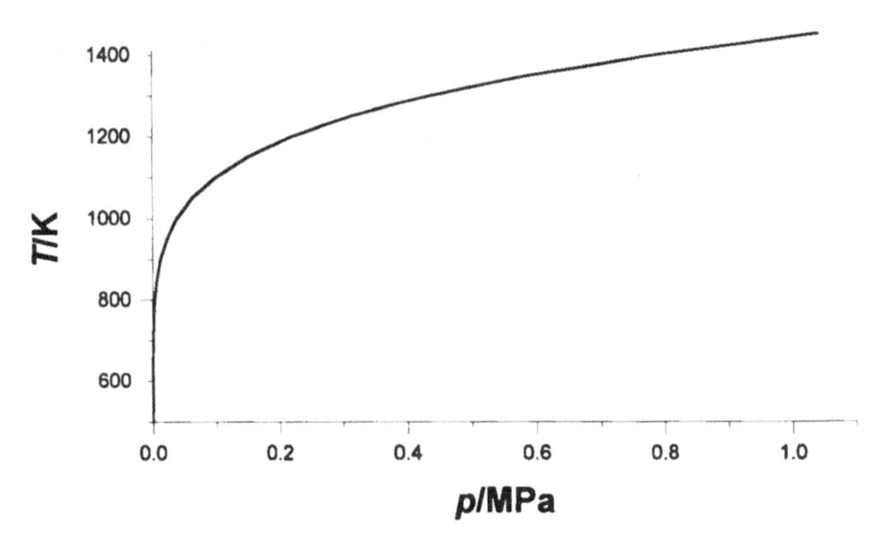

Figure 6.1 Pressure and temperatures at which the rate of the reaction $H + O_2 \rightarrow OH + O$ (eqn (6.7)) is equal to that of $H + O_2 + M \rightarrow HO_2 + M$ (eqn (6.8)), given by eqn (6.9). The third body efficiency of M in eqn (6.8) is taken to be that of air.

and the pressure–temperature relationship at which the rates of these two reactions are equal is shown in Fig. 6.1. The efficiency of the 'chaperone species' M in eqn (6.9) is assumed to be that of air, but [M] represents the total concentration of species in the system, and the pressure for an ideal gas is derived from this. The formation of OH and O predominates at temperatures above the line in Fig. 6.1 and occurs at $T < 1000$ K at pressures below 100 kPa. A temperature in excess of 1400 K is required at pressures exceeding 1 MPa, as is the case in reciprocating engine combustion.

The comparable rate parameters for the branching reaction shown in eqn (6.7) and that of a typical abstraction process shown in eqn (6.5) indicate that the reaction of H atoms with oxygen is as important as the abstraction process, over the temperature range shown in Fig. 6.1, except in very fuel-rich conditions. The importance of the switch in propagating species from HO_2 to OH brought about by the competition between eqns (6.7) and (6.8) is discussed below, but first more detail of the chemistry associated with hydrocarbon combustion is required in order to substantiate the role of reactions involving H atoms.

6.2 Mechanisms of alkane oxidation

The underlying mechanistic structure for the high temperature combustion of methane and ethane is shown in Fig. 6.2 [89]. In this scheme the main propagating free radicals are shown as H, O, OH, HO_2 and CH_3. There are a number of key features.

(i) The mechanisms of the two primary fuels are linked at an interplay between CH_3 and C_2H_6, C_2H_5 or C_2H_4. C_2-containing species are detected in methane flames, especially under fuel-rich conditions.

(ii) Formaldehyde (CH_2O) and formyl radicals (CHO) are the main partially oxygenated products of the C_1 and C_2 hydrocarbon fragments. Decomposition of CHO can be a source of H atoms, via

$$CHO + M \rightarrow H + CO + M;$$

$$k = 4.65 \times 10^{13} \exp(-7167/T)\,\mathrm{cm^3\,mol^{-1}\,s^{-1}} \quad (6.10)$$

but its competitive oxidation is also a major source of HO_2 at all temperatures

$$CHO + O_2 \rightarrow CO + HO_2;\ k = 3.01 \times 10^{12}\,\mathrm{cm^3\,mol^{-1}\,s^{-1}} \quad (6.11)$$

(iii) Carbon monoxide (CO) is the end product of virtually all of the chain sequences. This signifies that carbon monoxide is almost invariably a precursor to carbon dioxide, as the final product of combustion.

The combustion of higher alkanes is closely connected to the sequence of reactions shown in Fig. 6.2 [89], insofar that these processes are the final stages of the reaction chains involved, regardless of the molecular mass of the primary fuel. Degradation of the molecular structure of the fuel is favoured at high temperatures, yielding a predominance of free radicals and molecular intermediates which contain only small numbers of carbon atoms. In particular the formation of alkyl radicals in any free radical abstraction process, for example (X=O, OH, etc.)

$$n\text{-}C_7H_{16} + X \rightarrow n\text{-}C_7H_{15} + HX \quad (6.12)$$

is followed by

$$n\text{-}C_7H_{15} \rightarrow CH_3 + n\text{-}C_6H_{12} \text{ (alkene)} \quad (6.13)$$

$$\rightarrow C_2H_5 + n\text{-}C_5H_{10}$$

$$\rightarrow n\text{-}C_3H_7 + n\text{-}C_4H_8$$

$$\rightarrow n\text{-}C_4H_9 + C_3H_6$$

The alkyl radicals ($C > 2$) are themselves susceptible to a similar sequence of decomposition processes, and the molecular alkenes are susceptible to free radical attack (O, OH, HO_2) at the double bond which leads to further degradation of the carbon structure. A typical activation energy for reactions of the type shown in eqn (6.13) is 120 kJ mol^{-1}, but because these are unimolecular reactions there is a high frequency factor associated with the rate constant. Thus, high temperatures tend to favour the unimolecular decomposition of the alkyl radicals rather than their

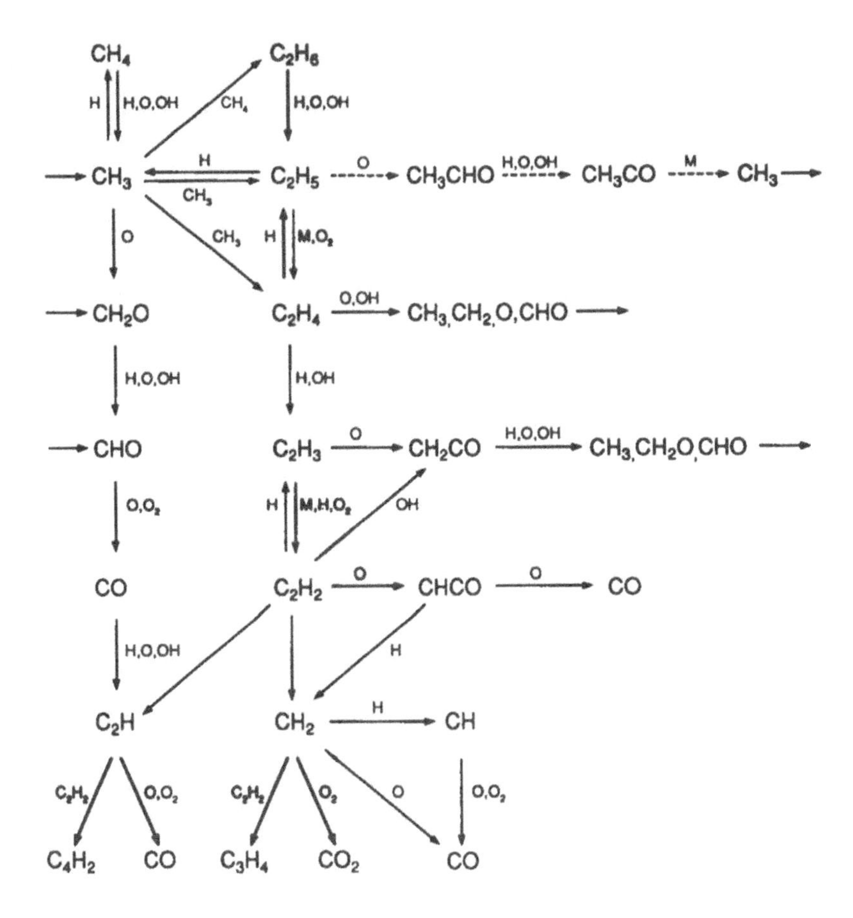

Figure 6.2 Schematic representation of a mechanism for methane oxidation at high temperatures (after Warnatz [89]).

bimolecular interaction with oxygen. Hydrogen atoms may also originate from alkyl radical decomposition via

$$RCH_2CH_2^{\bullet} \rightarrow H + RCH{=}CH_2 \quad \text{(alkene)} \tag{6.14}$$

in which R represents a unit comprising C_nH_{2n+1}. Some typical Arrhenius parameters for the unimolecular decomposition of alkanes and alkyl radicals are given in Table 6.1 [90, 91].

6.2.1 Principal propagating free radicals and reactions

At premixed flame temperatures ($T > 2000$ K) the branching channel (eqn (6.7)) to O and OH is the predominant reaction between H atoms and

Table 6.1 Arrhenius parameters for some unimolecular reactions [90, 91]

Reaction	A(s^{-1})	E(kJ mol^{-1})
Molecular decomposition		
$C_2H_6 \rightarrow 2CH_3$	5.0×10^{16}	370
n-$C_4H_{10} \rightarrow 2C_2H_5$	1.9×10^{17}	342
i-$C_4H_{10} \rightarrow CH_3 + i$-$C_3H_7$	6.3×10^{16}	345
$(CH_3)_4C \rightarrow CH_3 + t$-$C_4H_9$	6.3×10^{16}	343
$C_3H_6 \rightarrow CH_3 + C_2H_3$	1.3×10^{16}	359
Free radical decomposition		
$CH_2CH_2CH_3 \rightarrow C_2H_4 + CH_3$	1.6×10^{14}	136
$CH_2CH_2CH_2CH_3 \rightarrow C_2H_4 + C_2H_5$	2.5×10^{13}	120
$CH_3CHCH_2CH_3 \rightarrow C_3H_6 + CH_3$	7.3×10^{14}	137
$CH_2CH(CH_3)_2 \rightarrow C_3H_6 + CH_3$	2.8×10^{12}	130

O_2, regardless of the pressure at which combustion takes place. This particular reaction, coupled to

$$O + RH \rightarrow OH + R \tag{6.15}$$

and the principal propagation

$$OH + RH \rightarrow R + H_2O \tag{6.16}$$

are crucial to the branching chain-thermal interaction that gives rise to the self-sustaining properties of hydrocarbon flame propagation.

In pre-ignition stages, which may be at the beginning of spontaneous ignition or may be in the preflame zone of a propagating flame, the temperature and pressure govern whether or not the branching channel predominates. At temperatures below the line in Fig 6.1 the HO_2 radical formed in the reaction shown in eqn (6.8) becomes the principal propagating species, which reacts as follows:

$$RH + HO_2 \rightarrow R + H_2O_2 \tag{6.17}$$

$$H_2O_2 + M \rightarrow 2OH + M \tag{6.18}$$

The temperatures at which this particular sequence is important are in the range 800–1200 K (Fig. 6.1). Other reactions become important below 800 K (Chapter 7).

The reactions shown in eqns (6.17) and (6.18) constitute a chain-branching sequence, but with some qualification. The rate of development of reaction is moderated by the appreciably slower propagation via HO_2 radicals since the activation energies associated with reactions of the type (e.g. eqn (6.17)) fall in the range 40–60 kJ mol^{-1}. Secondly, the branching rate is governed by the half-life of hydrogen peroxide. Although this is short ($t_{1/2} < 1$ ms) at temperatures above 1000 K, it does not match the characteristic times ($\tau \approx \mu s$) associated with the radical branching route (eqn (6.7)). This means that there is a much slower acceleration

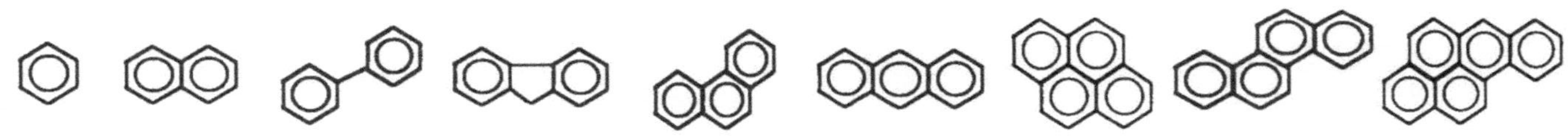

	benzene	naphthalene	diphenyl	fluorene	phenanthrene	anthracene	pyrene	benzophenanthrene	benzopyrene
Formula	C_6H_6	$C_{10}H_8$	$C_{12}H_{10}$	$C_{13}H_{10}$	$C_{14}H_{10}$	$C_{14}H_{10}$	$C_{16}H_{10}$	$C_{18}H_{12}$	$C_{20}H_{12}$
M_r =	78	128	154	166	178	178	202	228	252
M Pt / °C =	5.5	80	71	116	101	216	156	255	176
B Pt/ °C =	80	218	256	293	340	340	393	448	

– – – – – tar – – – – →

Figure 6.3 Structural representations, relative molecular masses (M_r) and physical properties of some simple polynuclear aromatic hydrocarbons.

of rate when HO_2 propagation predominates, compared with that at higher temperatures.

6.3 Mechanisms of aromatic hydrocarbon oxidation

This class of compounds begins with benzene and its derivatives comprising the benzene ring with substituted alkyl side-chains of differing size or number (e.g. toluene, $C_6H_5CH_3$, and the xylene isomers, $C_6H_4(CH_3)_2$). There can also be an increase in the number of aromatic rings; naphthalene, diphenyl and fluorene contain two, phenanthrene and anthracene contain three, and so on into polycyclic aromatic hydrocarbons (PAH) of increasing complexity (Fig. 6.3). A detailed understanding of the oxidation chemistry of aromatic compounds has begun to emerge only relatively recently [92], prompted by the high quantities of toluene and the xylenes in unleaded petrol (gasoline). There are also supplementary concerns about the formation of PAH in combustion systems.

The oxidation of aromatic hydrocarbons is confined essentially to the high temperature regime. Unlike most aliphatic hydrocarbons, the aromatics do not undergo vigorous oxidation or give rise to spontaneous ignition at temperatures below about 800 K. The benzyl and phenyl radicals, that are the primary propagating chain species, are relatively unreactive (Fig. 6.4 [92]). This arises in each case because the electron associated with the free radical centre is able to become delocalised across the electron-deficient, aromatic ring structure. There are consequences of this for the reactivity of fuel mixtures which include benzene, toluene or the xylenes (Chapter 13). The aromatic ring structure appears to be broken down via the formation of the unsaturated, cyclopentadienyl radical before fragmentation into smaller carbon-containing units.

6.4 The relevance of reactions in hydrogen and carbon monoxide oxidation

Hydrogen is the simplest fuel that is able to undergo oxidation, and much more is known about 'the H_2+O_2 reaction' than any other comparable combustion system [2]. H atoms and OH, O and HO_2 radicals are formed in hydrocarbon combustion processes. These species are also central to the development of hydrogen oxidation, and the understanding of their elementary reactions in that system underpins the interpretation of chain propagation and branching in the oxidation of all C–H–O containing fuels. The relevance of carbon monoxide oxidation is that the formation of virtually all CO_2 is via CO as its precursor.

6.4.1 Foundation to the combustion of hydrogen

Premixed flames of hydrogen in oxygen or air are quite typical except that they show little or no visible radiation, that observed normally being due to trace impurities. A considerable intensity of emission from the OH

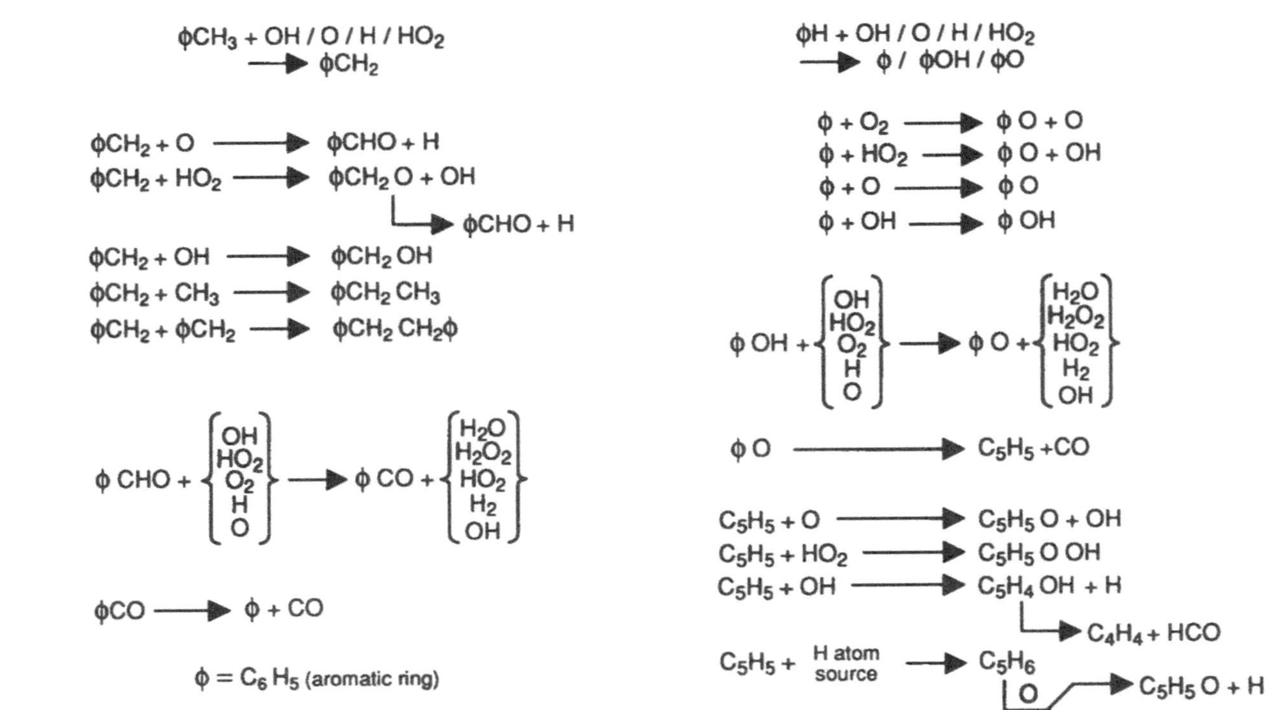

Figure 6.4 A schematic representation of the reactions that are believed to be involved in the early stages of oxidation of toluene and benzene at temperatures of about 1200 K (after Emdee *et al.* [92]).

radical can be detected in the ultraviolet region of the spectrum (section 6.6). The highest adiabatic flame temperature ($\approx 3100\,K$) occurs in a $2H_2 + O_2$ mixture and the burning velocity is very close to its maximum, and the proportion of water in the burned gas is about 57%, with about one-quarter of the composition remaining as free radicals [10].

However, the stoichiometric mixture does not necessarily represent the most reactive composition in all respects. Consider the elementary reaction sequence,

$$H + O_2 \rightarrow OH + O;$$

$$k = 2.00 \times 10^{14} \exp(-8455/T)\,cm^3\,mol^{-1}\,s^{-1} \text{ branching} \qquad (6.7)$$

$$O + H_2 \rightarrow OH + H;$$

$$k = 2.126 \times 10^{14} \exp(-6860/T)\,cm^3\,mol^{-1}\,s^{-1} \text{ branching} \qquad (6.19)$$

$$OH + H_2 \rightarrow H_2O + H;$$

$$k = 5.94 \times 10^{13} \exp(-3880/T)\,cm^3\,mol^{-1}\,s^{-1} \text{ propagation} \qquad (6.20)$$

A complete cycle of these three elementary reactions begins with one hydrogen atom and leads to the production of three hydrogen atoms according to the proportions $H_2/O_2 = 3/1$, i.e.

$$H + 3H_2 + O_2 = 3H + 2H_2O \qquad (6.21)$$

This stoichiometry is a correct representation of 'chemical autocatalysis' in which the H atom regenerates itself. There is also a net gain of two free radicals in the autocatalytic cycle, which is an inevitable consequence of the separation of a pair of electrons in a chemical bond to two individual entities which are then capable of continuing chain propagation.

Equation (6.7) has an activation energy of about $70\,kJ\,kJ\,mol^{-1}$, just slightly greater than its endothermicity. The activation energies of eqns (6.19) and (6.20) are not as high, so the rate constant of eqn (6.7) is appreciably lower than that of eqns (6.19) and (6.20), certainly at temperatures below 1000 K. This causes eqn (6.7) to be rate determining and the OH and O concentrations remain in stationary state with respect to the H atom concentration throughout. This type of analysis is common in branching-chain reactions although it must be used with caution [93]. At higher temperatures the rate constant of eqns (6.7) and (6.19) are much closer, and so the rate-determining role of eqn (6.7) is only marginal.

Although these three reactions are responsible for the exponential increase in oxidation rate, they do not correspond to the stoichiometry for the complete oxidation of hydrogen,

$$2H_2 + O_2 = 2H_2O \qquad (6.22)$$

Table 6.2 Arrhenius parameters for some elementary reactions involving H, O, OH and HO_2 in the form $k = AT^n\exp(-E/RT)$ [14, 90, 91]

Reaction	A ((cm^3 mol^{-1})$^{n-1}$ s^{-1})	n	E (kJ mol^{-1})
$H + O_2 \rightarrow OH + O$	2.24×10^{14}	0	70.3
$O + OH \rightarrow O_2 + H$	1.46×10^{13}	0	2.1
$O + H_2(RH) \rightarrow OH + H(R)$[a]	5.12×10^{4}	2.67	26.2
$OH + H \rightarrow O + H_2$	8.3×10^{9}	1.0	45.7
$OH + H_2 \rightarrow H_2O + H$	1.02×10^{8}	1.6	13.8
$H + H_2O \rightarrow OH + H_2$	1.40×10^{14}	0	42.6
$O + H_2O \rightarrow OH + OH$	5.75×10^{13}	0	75.4
$OH + OH \rightarrow O + H_2O$	2.10×10^{8}	1.4	−1.6
$H + H_2O \rightarrow H_2 + OH$	8.41×10^{13}	0	84.1
$OH + H_2 \rightarrow H + H_2O$	2.16×10^{8}	1.5	14.4
$H + O_2 + M \rightarrow HO_2 + M$[b]	2.30×10^{18}	−0.8	0
$HO_2 + M \rightarrow H + O_2 + M$	2.80×10^{15}	0	191.2
$HO_2 + H \rightarrow O_2 + H_2$	2.50×10^{13}	0	2.9
$HO_2 + H \rightarrow OH + OH$	1.50×10^{14}	0	4.2
$HO_2 + OH \rightarrow H_2O + O_2$	6.02×10^{13}	0	0
$HO_2 + HO_2 \rightarrow H_2O_2 + H$	1.06×10^{11}	0	7.1
$HO_2 + H_2(RH) \rightarrow H_2O_2 + H(R)$[a]	6.00×10^{11}	0	77.3
$H_2O_2 + M \rightarrow 2OH + M$[c]	3.20×10^{17}	0	196.5
$H_2O_2 + H \rightarrow H_2O + OH$	1.00×10^{13}	0	15.0
$H_2O_2 + H \rightarrow H_2 + HO_2$	1.70×10^{12}	0	15.8
$H_2O_2 + O \rightarrow HO_2 + OH$	2.80×10^{13}	0	26.8
$H_2O_2 + OH \rightarrow H_2O + HO_2$	7.00×10^{12}	0	6.0
$H + H + M \rightarrow H_2 + M$	6.40×10^{17}	−1.0	0
$O + O + M \rightarrow O_2 + M$	1.00×10^{17}	−1.0	0
$H + OH + M \rightarrow H_2O + M$	2.20×10^{22}	−2.0	0

Reaction order n; [a] data for H_2; [b] third-order, $M = N_2$; [c] first-order, high-pressure limit, $M = N_2$.

so other reactions must become important as reaction proceeds. In flames, for example, the branching chain reaction ceases when the reverse reactions shown in eqns (6.7), (6.19) and (6.20)

$$OH + O \rightarrow H + O_2 \tag{6.23}$$

$$OH + H \rightarrow O + H_2 \tag{6.24}$$

$$H + H_2O \rightarrow OH + H_2 \tag{6.25}$$

become significant. A quasi-equilibrium is then established in which the reaction has proceeded perhaps three-quarters of the way to completion but a large proportion of the available enthalpy is still contained in the high atom and radical concentrations. This energy is then released by third-order recombination reactions which take place on a longer time scale than the second order processes above. In the post-flame gases of fuel-rich mixtures the dominant reactions are

$$H + OH + M \rightarrow H_2O + M \tag{6.26}$$

$$H + H + M \rightarrow H_2 + M \tag{6.27}$$

In fuel-lean conditions, processes involving the HO_2 radical, e.g.

$$OH + HO_2 \rightarrow H_2O + O_2 \quad (6.28)$$

$$H + HO_2 \rightarrow H_2 + O_2 \quad (6.29)$$

become important. Many of the reactions in hydrogen oxidation that are also relevant to hydrocarbon combustion are summarised in Table 6.2 [90, 91].

6.4.2 The oxidation of carbon monoxide

From a practical point it is difficult to free CO from all traces of hydrogenous material for the purpose of investigating the oxidation of the pure substance. This arises mainly from the production of cylinder-gas CO by fractionation: there are often parts per million (ppm) traces of CH_4 (b. pt. 109 K) in CO (b. pt. 81 K). It is likely that methane is the main impurity but, for historical reasons, the common terminology is to describe impure carbon monoxide (with respect to hydrogenous material) as 'wet CO' [23], alluding to the possible presence of water vapour.

The oxidation of hydrogen (or methane) each involve OH and HO_2 radicals. Thus, if there are small quantities of either H_2 or CH_4 also present, chain propagation and branching in CO oxidation is promoted via the following reactions:

$$CO + HO_2 \rightarrow CO_2 + OH \quad \text{propagation} \quad (6.30)$$

$$CO + OH \rightarrow CO_2 + H \quad \text{propagation} \quad (6.31)$$

followed by

$$H + O_2 \rightarrow OH + O \quad \text{branching} \quad (6.7)$$

or

$$H + O_2 + M \rightarrow HO_2 + M \quad \text{propagation} \quad (6.8)$$

and also by

$$CO + O + M \rightarrow CO_2 + M \quad \text{termination} \quad (6.32)$$

Apart from contributions to chain propagation, there is also a supplementary branching step which involves the impurity, e.g.

$$O + CH_4 \rightarrow CH_3 + OH \quad (6.33)$$

Despite the very small impurity concentrations, these steps are competitive with the termolecular termination of eqn (6.32) because it has such a low probability. Moreover, only exceedingly small amounts of hydrogen-containing compounds need be present because the hydrogen atom content is recycled via the reactions shown in eqns (6.30) and (6.31). H atoms are lost only if they are removed in termination processes in which water is formed, or by adsorption at a surface of a reaction vessel. Reactions involving CHO have also been invoked in more detailed interpretations of the oxidation of carbon monoxide [94], but these do not seem to be essential to a qualitative explanation of the main features of carbon monoxide oxidation.

Equations (6.30)–(6.32) are the main processes by which carbon monoxide is oxidised to carbon dioxide in the combustion of all carbon-containing fuels, eqn (6.31) being by far the most important. In general, carbon dioxide cannot be formed as a final product of hydrocarbon combustion without carbon monoxide being a precursor to it. Thus, any interference with the ability of OH, HO_2 or O to react with CO, must reduce the efficiency of combustion and exacerbate the problem of carbon monoxide pollutant emissions. Equation (6.31) is relatively slow compared with H atom abstraction by OH, and so it cannot compete very successfully with hydrocarbon fuels in reaction with OH radicals. For example (using $k = AT^n \exp(-E/RT)$ [14]), from

$$CH_4 + OH \rightarrow CH_3 + H_2O;$$

$$k = 1.56 \times 10^7 T^{1.83} \exp(-1400/T)\,\mathrm{cm^3\,mol^{-1}\,s^{-1}} \qquad (6.6)$$

$$CO + OH \rightarrow CO_2 + H;\; k = 6.32 \times 10^7 T^{1.5} \exp(250/T)\,\mathrm{cm^3\,mol^{-1}\,s^{-1}} \qquad (6.31)$$

the ratio of the rate constants, $k_{6.6}/k_{6.31}$, is 4.6 at 1000 K, 9.2 at 1500 K and 13.3 at 2000 K. CO may remain as long as there is still unreacted fuel in the system, and the formation of carbon monoxide as a final product of combustion is inevitable under fuel-rich conditions.

6.5 Relationship between high temperature chemistry and the spatial structure of premixed flames

In a premixed flame most of the enthalpy is released rapidly in a narrow reaction zone leading to the production of very high temperatures, which produce steep temperature gradients. The transport of heat or of free radicals causes self-propagation of the flame. Although most of the reaction is associated with the narrow spatial region of the reaction zone, the preheat and post-flame (or recombination) zones may be distinguished from this, and different chemical processes take place in these.

For a stable molecule such as methane, with a first-order decomposition rate constant of only $10^4\,\mathrm{s^{-1}}$ even at 2000 K, little or no pyrolysis can occur within the short residence time in the preheat zone of a premixed laminar flame. For the majority of hydrocarbons, considerable degradation occurs and the fuel fragments leaving this zone comprise mainly lower hydrocarbons, alkenes and hydrogen. Thus the composition in the reaction zone is very similar, irrespective of the nature of the fuel, which accounts, in part, for the similar flame temperatures and burning velocities for a wide range of fuels. Only a small proportion of oxygen is consumed in the preheat zone.

The relative importance of the processes occurring may be altered quite considerably by transport processes (i.e. diffusion of intermediates into the preheat zone) and other supplementary effects. The dominant reactions are

$$H + O_2 \rightarrow OH + O \qquad (6.7)$$

$$CO + OH \rightarrow CO_2 + OH \qquad (6.31)$$

to the extent that the velocities of alkane flames can be modelled within a factor of two using the pure H_2–O_2–CO mechanism with the addition of attack by H, O and OH on the hydrocarbon, the resulting alkyl radicals being assumed to form CO and H_2O directly at a rate which is effectively infinitely fast. A more satisfactory account of the behaviour of premixed flames of higher alkanes is obtained numerically through a much more detailed kinetic mechanism [95]. At the temperatures prevailing in the flame, the alkyl radicals break down very rapidly to alkenes and a simple radical, CH_3 or C_2H_5. The alkenes themselves are also rapidly attacked by H, O and OH; in general, H atoms add to give the corresponding alkyl radical, while attack by O and OH yields acetyl radicals or acetaldehyde together with an alkyl radical. The subsequent steps described in Fig. 6.2 can be used to accurately reproduce the composition profiles through lean and moderately rich alkane flames and also to predict the way flame speed varies with mixture composition [95].

At high temperature, equilibrium considerations favour dissociation, and typical flames in oxygen at 2000–3000 K may contain 10–50% of the fuel in the form of radicals. Furthermore, because the kinetic order of dissociation reactions is lower than that of recombination, radical concentrations will tend to 'overshoot' in the reaction zone (Fig. 3.2), that is to exceed the local equilibrium values. They will return only relatively slowly. Although as a result of the high temperature all reactions will be correspondingly faster, radical–radical reactions become progressively more significant in the high radical concentration environment. In particular, the reaction of hydrogen atoms becomes very important, owing to the increased significance of processes such as the reactions in eqn (6.14), the major sequence usually being closely related to that which occurs in the hydrogen oxidation

$$H + O_2 \rightarrow OH + O \qquad (6.7)$$

$$O + H_2(RH) \rightarrow OH + H(R) \qquad (6.19)/(6.15)$$

$$OH + H_2(RH) \rightarrow H_2O + H(R) \qquad (6.20)/(6.16)$$

The reaction zone proper terminates in a type of quasi-equilibrium region with radical–radical reactions of the type

$$O + OH \rightarrow H + O_2 \qquad (6.23)$$

which balances the reverse branching reaction of eqn (6.7). However, thermodynamic equilibrium has not yet been attained because the radical concentrations are still very high and the oxidation has proceeded mainly to carbon monoxide rather than carbon dioxide. The post-flame or recombination zone is a more extended region in which the slower recombination reactions occur, leading to further enthalpy change and carbon monoxide is oxidised to carbon dioxide if the fuel to oxygen ratio permits it. The dominant reactions in the recombination zone of rich H_2+O_2 flames are [34]

$$H + H + M \rightarrow H_2 + M \tag{6.27}$$

or reaction with oxygen atoms which are certainly present under flame conditions and, in lean flames, they probably include

$$H + OH + M \rightarrow H_2O + M \tag{6.26}$$

$$H + O_2 + M \rightarrow HO_2 + M \tag{6.8}$$

The concentrations of free radicals and atoms are found to be in close accord with the predictions of thermodynamic equilibrium at the prevailing temperature in the post-flame region.

The different nature of the reaction zone and the post-flame regions is revealed also in other ways. In the luminous reaction zone, the high radical concentrations are manifest in non-equilibrium chemiluminescence and by excess ion concentrations, whereas in the post-flame gases heat is still evolved but the radiation corresponds closely to thermal equilibrium.

6.6 Light emission from flames

Almost all combustion phenomena are accompanied by light emission, although most information has been gained from studies on flames. If a flame were to be at thermodynamic equilibrium then the hot gases would emit the continuous radiation predicted by the Planck radiation law for the appropriate flame temperature. For a true black body, the emissivity would have a value of unity at all wavelengths. In flames, the emissivity is closer to zero in the ultraviolet and visible regions, although broad infrared bands with emissivities close to unity are to be found. Once solid particles appear in the flame, the emissivity becomes much closer to that of a black or 'grey' body with only slight variation with wavelength. This is the origin of the orange emission from the post-flame region of diffusion flames and fuel-rich premixed flames.

The main source of non-thermal, light emissions arises from chemical excitation processes. In combustion systems, the high temperatures in the reaction zone lead to high radical and atom concentrations, even exceeding the concentrations predicted at equilibrium (section 6.5). Some reactions of these reactive species, especially those involving radical + radical or radical + atom, are sufficiently exothermic for the products to be formed in excited electronic states, which have quantised energy levels. The excited state then deactivates to the ground state with the emission of light at the frequency appropriate to the energy change. This process is known as *direct chemiluminescence*. The criterion that has to be satisfied is that the exothermicity of reaction exceeds the energy required for electronic excitation of the product species. The excited product species may not be a strong emitter, but it may be able to exchange energy with another species which is able to radiate. This type of process is referred to as *indirect chemiluminescence*, as demonstrated in the diffusion flame of diatomic sodium and chlorine [96].

$$Na_2 + Cl \rightarrow NaCl^* + Na \tag{6.34}$$

The product NaCl is formed in a vibrationally excited state, and its energy is then transferred to sodium atoms

$$NaCl^* + Na \rightarrow NaCl + Na^* \tag{6.35}$$

followed by

$$Na^* \rightarrow Na + h\nu \tag{6.36}$$

The frequency of the light emission in this case corresponds to the orange, sodium 'D' lines at the wave numbers 169.6 and 169.7 cm^{-1}, respectively. A classification of the types of elementary reactions that lead to chemiluminescence has been made by Laidler [97].

The absence of low-lying electronic energy states of the stable products of hydrocarbon combustion (H_2O, CO_2, CO) means that the intensity of radiation in the ultraviolet and visible regions is low. Some of the transient species (OH, CH, CN, C_2, HCO, N, H and NH_2) do have more accessible energy states, from which discrete spectra may be detected [98]. Part of the banded structure of emissions from a methane + air Bunsen burner flame is shown in Fig. 6.5 [98]. Processes that are known to be capable of forming CH^* and C_2^* (the 'Swan bands') in hydrocarbon flames are

$$C_2 + OH \rightarrow CO + CH^* \tag{6.37}$$

$$O + C_2H \rightarrow CH^* + CO \tag{6.38}$$

$$O_2 + C_2H \rightarrow CH^* + CO_2 \tag{6.39}$$

and

$$CH_2 + C \rightarrow C_2^* + H_2 \tag{6.40}$$

The spectrum of fuel-lean flames often contains features of HCO^*, which may be produced by

$$CH + HO_2 \rightarrow HCO^* + OH \tag{6.41}$$

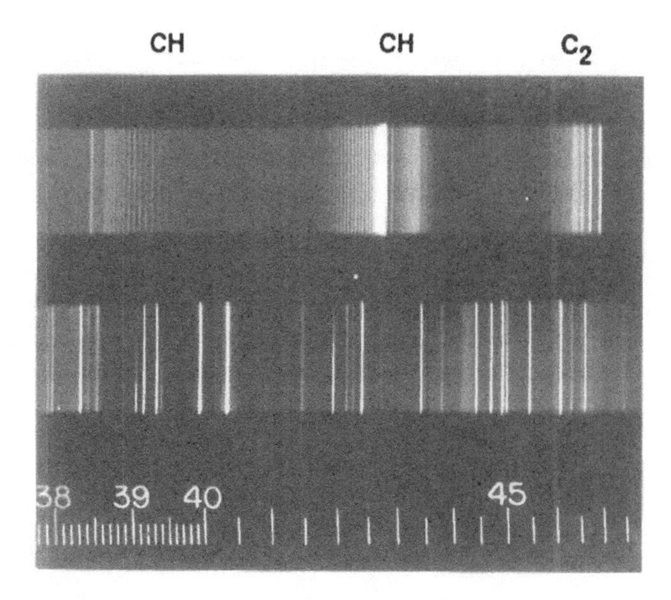

Figure 6.5 Part of the emission spectrum associated with CH and C_2 from a natural gas flame on a Bunsen burner, resolved by optical spectroscopy. The transitions from which these banded structures originate are discussed by Gaydon [98]. The lower scale represents wavelength in μm. The centre spectrum shows calibration lines obtained from a copper arc.

Given that a pseudo-equilibrium exists between the concentrations of radicals in flames, in general the intensity of the blue emissions from premixed hydrocarbon flames gives an indication of the radical and atom concentrations in the reaction zone and post-flame gases. The intensity of emissions from a propane + air flame is appreciably weaker than that from a propane + oxygen flame, in accord with the data in Tables 2.3 and 2.4. The lack of visible emissions from a pure hydrogen + oxygen flame arises not because there are few radicals or atoms, but because of the absence of carbon, the emissions being associated with $OH^{\bullet}$, which occur in the ultraviolet range.

Non-thermal distributions are not restricted to the population of excited electronic levels. Rotational and vibrational excitations also may occur, the probability of which may be characterised as a 'temperature' based on the distribution among the appropriate energy levels of each of these modes of excitation. The measurement of a vibrational or rotational temperature which does not coincide with the adiabatic flame temperature then also provides evidence for a chemiluminescent excitation mechanism. Non-equilibrium distributions in the OH band system arise in rotational and vibrational excitations. The rotational temperature of $OH^{\bullet}(^2\Sigma^+)$ has been measured [99] as 9000 K in an acetylene + oxygen flame at a pressure of 0.2 kPa, although the adiabatic flame temperature is only 3320 K. High rotational temperatures for $OH^{\bullet}$ in hydrocarbon + oxygen flames result from the reaction

$$CH + O_2 \rightarrow CO + OH^{\bullet} \quad (6.42)$$

Vibrational non-equilibrium occurs in the reaction zone of the $H_2 + O_2$ flame [100].

The 'flame bands' of the $CO + O_2$ flame are attributed to

$$O(^3P) + CO(^1\Sigma^+) + M \rightarrow CO_2^{\bullet}(^3\Pi) + M \quad (6.43)$$

$$CO_2^{\bullet}(^3\Pi) \rightarrow CO_2(^1\Sigma_g^+) + h\nu \quad (6.44)$$

Some direct two-body association also occurs, producing a continuum down to 250nm. An appreciable number of the $CO_2^{\bullet}$ molecules transfer energy to O_2

$$CO_2(^3\Pi) + O_2(^3\Sigma_g^+) \rightarrow CO_2(^1\Sigma_g^+) + O_2^{\bullet}(^3\Sigma_u^-) \quad (6.45)$$

which give emission from the Schumann–Runge system of oxygen.

Light emission from hydrocarbon oxidation at low temperatures results from $CH_2O^{\bullet}$, and is termed the 'cool flame' emission (Chapter 9). The main reactions are

$$CH_3O + CH_3O \rightarrow CH_2O^{\bullet} + CH_3OH \quad (6.46)$$

$$CH_3O + OH \rightarrow CH_2O^{\bullet} + H_2O \quad (6.47)$$

6.7 Ionisation processes

The majority of combustion phenomena involve some degree of ionisation and the presence of ions in a flame may be demonstrated by the distortion which occurs when an electric field is applied across it. Ions are detected in flames by introducing a pair of electrodes and measuring the current flowing when a potential difference is applied across them. The measured current depends on the mobilities of the ions and on space charge and boundary layer effects. Ionisation has also been detected by measurements of microwave attenuation and by direct detection in a mass spectrometer. Ionisation produces mainly positive ions and free electrons, although electron capture can create negative ions.

The degree of ionisation depends critically on the ionisation potential. The total ion density is therefore very sensitive to the nature of trace quantities of any easily ionised impurities. If present, 1 part in 10^8 of potassium will create a high ion density, and would dominate the formation of ions in flames. All of the common flame gases and most of the radicals, such as O_2, N_2, H_2, H_2O, CO, CO_2, OH and H, have high ionisation potentials (Table 6.3 [10]) and normally make negligible contribution to the total ionisation. For example, CO is able to form only 10^6 ions cm^{-3} at 3000 K. The most common species with a low ionisation potential found in flames is NO (9.26 eV) and, at 1% concentration, it

Table 6.3 Ionisation potentials of some typical flame components [10]

Species	E_i (eV)
O_2	12.1
H_2	15.4
N_2	15.6
CO	14.0
CO_2	13.8
H_2O	12.6
H	13.5
CH	13.2
NO	9.3
NO_2	9.8
C_2H_2	11.4
CH_4	12.9
CH_3	9.8
CHO	9.8

can yield 10^{11} ions cm^{-3} at 3000 K. Although in the absence of impurities the total ion concentration in hydrocarbon flames, around 10^{11} ions cm^{-3}, far exceeds that expected on equilibrium grounds, the overall concentrations remain sufficiently small that these species have no effect on the reaction rate in the flame.

6.7.1 *Mechanisms of ion formation in flames*

The predominant positive ion in flames is usually H_3O^+ although a large range of polymeric ions $C_nH_n^+$ also exist. Low concentrations of HCO^+, negatively charged species and free electrons are also observed. The extents of ionisation are greatest in the reaction zone and fall off in the burned gases. The origins are chemi-ionisation processes, for which similar energetic criteria as those for chemiluminescence have to be satisfied. Chemi-ionisation occurs close to the reaction zone as a result of

$$O + CH \rightarrow HCO^+ + e \tag{6.48}$$

although ions are also produced by

$$CH + C_2H_2 \rightarrow C_3H_3^+ + e \tag{6.49}$$

particularly in fuel-rich systems. The charge transfer process which follows eqn (6.48) is

$$HCO^+ + H_2O \rightarrow H_3O^+ + CO \tag{6.50}$$

in which a proton is transferred to a molecule such as H_2O with a proton affinity greater than that of CO. Protons are then exchanged with species of higher proton affinity, resulting in the formation of a wide range of

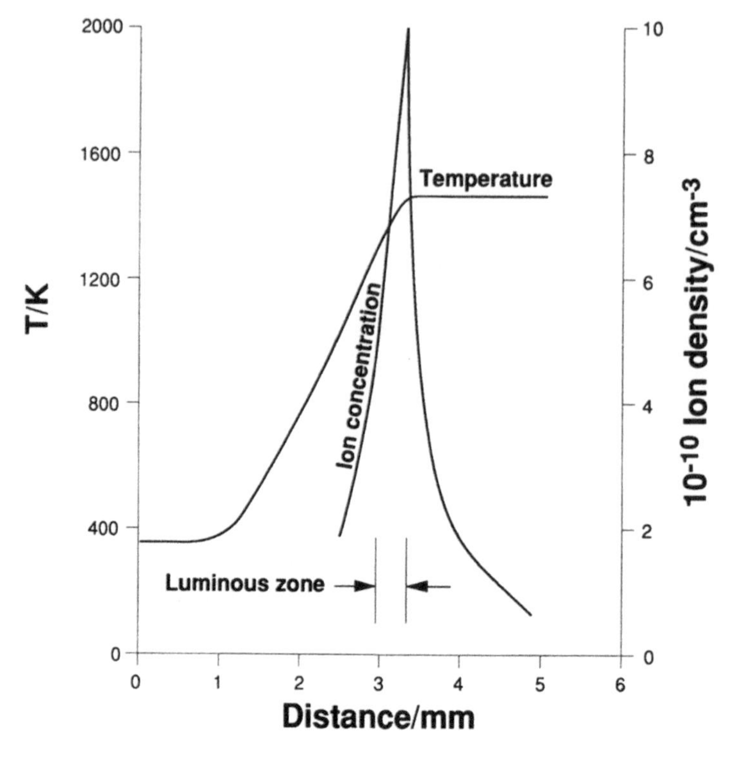

Figure 6.6 An ion density profile through a flame (after Calcote and King [101]).

ions of different structures. Reaction of $C_3H_3^+$ with hydrocarbon molecules leads to the large polymeric ions which are probably associated with incipient soot formation (section 6.8.3).

In the reaction zone radical concentrations frequently exceed the equilibrium values and thus ion concentrations greater than those corresponding to equilibrium are also produced by the reactions of eqns (6.48) and (6.49). The 'radical concentration overshoot' is thereby converted to an 'ionisation overshoot' which persists into the post-flame region (Fig. 6.6 [101]) where the major recombination reaction is

$$H_3O^+ + e \rightarrow H + H + OH \tag{6.51}$$

with some contribution from

$$H_3O^+ + e \rightarrow H_2O + OH \tag{6.52}$$

Negative ions are created by

$$O_2 + e + M \rightarrow O_2^- + M \tag{6.53}$$

followed by electron transfer to species of high electron affinity, such as CH, O, OH, O_3, C_2H, CH_3O_2 and HO_2. Ions such as O^- and OH^- are strong bases and so abstract protons, for example, by

$$CH_3OH + OH^- \rightarrow H_2O + CH_3O^- \tag{6.54}$$

Negative ions are lost by dissociative attachment, for example,

$$OH^- + H \rightarrow H_2O + e \tag{6.55}$$

Ions originate from potassium either by direct ionisation of the alkali metal atom or by charge transfer, such as

$$K + H_3O^+ \rightarrow K^+ + H_2O + H \tag{6.56}$$

and the ionisation tends to persist because of the relative slowness of

$$K^+ + e + M \rightarrow K + M \tag{6.57}$$

which requires the presence of a third body, M. Reaction between alkali metals and halogens can be used to enhance still further the ion concentration, for example, beginning with

$$K + Cl \Leftrightarrow K^+ + Cl^- \tag{6.58}$$

6.7.2 Aerodynamic effects and applications of ion formation in flames

The presence of ions in flames accounts for the aerodynamic effects produced by imposed electric fields. A wide range of effects can be observed, the most general feature being that the flame bends towards a negatively charged body. Limits of stability and flame speeds are also affected by applied electric fields. The free electrons, which constitute the majority of the negative species, have a much greater mobility than any other species and are rapidly collected at the positive electrode. This leaves the flame gas with a net positive charge which then creates movement towards the negative electrode, and is called the Chattock electric wind [102].

There is no chemi-ionisation in a pure hydrogen + oxygen flame, because no carbon is present. This ion free system is the foundation for the flame ionisation detector that is used in gas chromatography. Detection and measurement of each CH-containing substrate eluted from the chromatographic column into the hydrogen flame of the detector is possible from the ion current that is created at the electrodes.

Ion probes are sufficiently robust to survive in the combustion chambers of engines and the detection of ions at a probe may be used to identify when a flame arrives at a particular location in the chamber. This technique has been used to identify when engine knock occurs. The electrical signal from an ion probe can be used as a trigger for electronic adjustments in the engine management system.

Finally, ionisation in flames offers the potential for the direct generation of electricity from combustion between the poles of a magnet and the extraction of a current via electrodes at right angles to both the direction of motion of the flame and the magnetic field. This is termed magnetohydrodynamic generation of electricity (MHD). However, the process has not fulfilled in a practical way the promise that was expected of it some years ago.

6.8 Soot formation in flames

Although the oxidation of organic compounds leads eventually to carbon monoxide and carbon dioxide, fuel-rich flames tend also to produce solid carbon. This is the origin of the intense yellow light emission which is characteristic of such flames and may lead to the formation of carbon particles as soot. The radiation from such flames approaches black body intensity and can therefore be put to advantage in industrial burners because of the increased efficiency of heat transfer. However, the formation of solid particles which can deposit on cool surfaces must be kept to a minimum. Soot is also an undesirable product of combustion in engines where some is deposited as solid carbon in the combustion chamber or it is emitted as solid particles in the exhaust.

6.8.1 The nature of soot

The nature of soot varies both chemically and physically during its life. When first observed, soot has a particle diameter of about 5 nm which then increases, mainly by agglomeration in the later stages, to 200 nm or more. During this process the highly reactive, free radical character of the soot decreases. The concentration of soot particles in post-flame gases lies typically in the range 10^6–$10^9\,cm^{-3}$. The hydrogen content of soot is quite high, about 1% by weight or 12% of the number of atoms present, and corresponds approximately to the empirical formula $(C_8H)_n$. From electron microscopy deposited carbon appears to consist of chain-like aggregates of roughly spherical particles comprising crystallites measuring about 1.3 nm by 2.1 nm. The carbon atoms are shown by X-ray analysis to be a graphitic structure with the planes parallel to each other. By contrast to graphite, the planes are placed randomly one above the other with the hydrogen atoms distributed between them [103].

6.8.2 Conditions for soot formation

The presence of soot is dictated by the balance between the rates of reactions forming soot and those which cause its oxidation, largely by OH radicals (Fig. 6.7 [104]). These depend very much on the fuel to

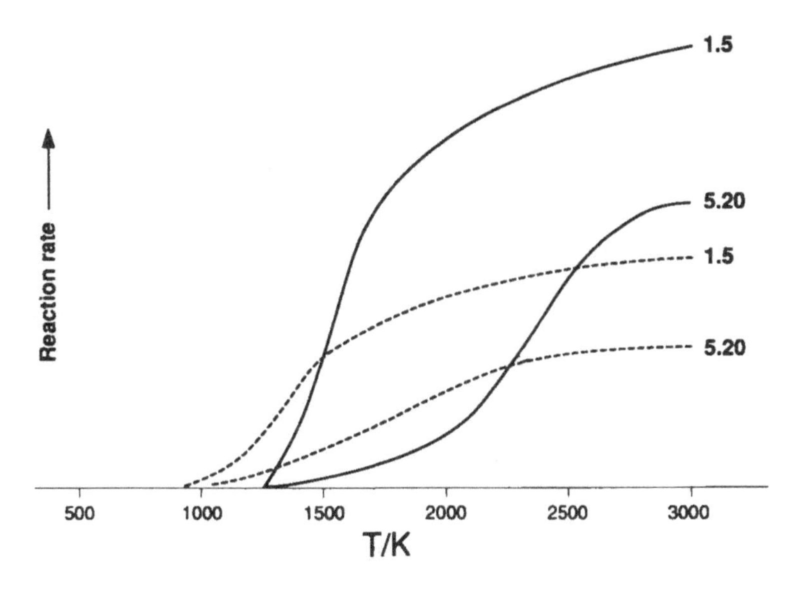

Figure 6.7 A schematic representation of the relative rates of soot formation by pyrolysis (broken line) and its oxidation (solid line) over a range of temperatures at the equivalence ratios $\phi = 1.5$ and 5.2. Soot formation is favoured at low temperatures in all fuel-rich mixtures, but it is likely to persist at higher temperatures as mixtures are enriched (after Homan [104]).

oxygen ratio and the effects observed are different in premixed and diffusion flames. The luminosity of the flame is characteristic of carbon formation even though luminous flames do not necessarily lead to deposition of soot.

Contours of the regions of increasing formation of soot from self-sustained flame fronts with reference to temperature and equivalence ratio are shown schematically in Fig. 6.8 [104]. The highest sooting tendency occurs in the presence of some oxygen, which may be attributed to a sensitising role of oxygen in pyrolytic processes. Figure 6.8 is applicable to either the post-flame gases of premixed flames or the soot yield on the fuel side of diffusion flames. Although not intended to be used in a quantitative sense, this diagram conveys trends for soot formation in hot, fuel-rich zones.

The sooting tendency of different fuels in laminar diffusion flames is

$$\text{aromatics} > \text{alkynes} > \text{alkenes} > \text{alkanes}$$

Sooting tendencies in premixed flames are not so susceptible to fuel structure because there is an increasing probability of soot formation at lower fuel to air ratios as the number of C–C bonds increases within a given class [105]. Thus, the minimum at $\phi > 1.5$ may vary for different fuels to some extent. Much higher temperaturess are required for soot oxidation process to be dominant in richer mixtures (Fig. 6.7).

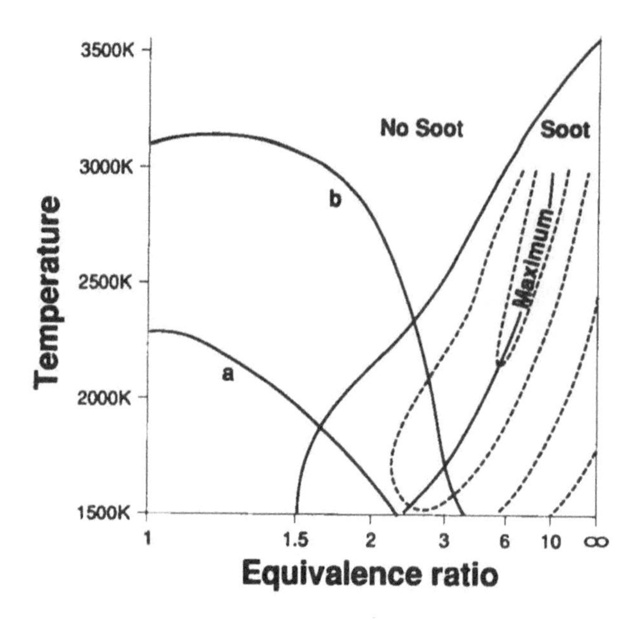

Figure 6.8 A schematic representation of soot yield trends as a function of temperature and equivalence ratio. As a guide to typical post-flame gas temperatures, the adiabatic temperatures for fuel mixtures (a) in air, and (b) in oxygen are also included (after Homan [104]).

The soot formation contours depicted in Fig. 6.8 represent the net balance of pyrolytic, soot forming, and oxidative, soot consuming, processes. Consequently, despite its luminosity, soot may not be a final product of an overventilated (i.e. conical) diffusion flame because sufficient oxygen is able to diffuse in at the tip to reduce the fuel:oxygen equivalence ratio below that at the 'soot/no soot' boundary at the prevailing temperature. There is a minimum sooting height for overventilated diffusion flames, above which soot first appears at the tip of the flame. Soot formation is always restricted to the upper part of the diffusion flame. This sooting tendency may be attributed to the development of an isolating layer of combustion products between the fuel and oxygen, which ensures that the fuel to oxygen equivalence ratio is maintained within the sooting range until the product gases have cooled sufficiently for no soot oxidation to be possible.

6.8.3 Gas-phase mechanisms leading to soot formation

In addition to soot, a range of polycyclic aromatic hydrocarbons (PAH) can also be detected as products of fuel-rich flames [106] and from other combustion systems [107, 108]. These products have carcinogenic properties.

PAH compounds with more than 16 C atoms are tarry substances at normal temperatures (Fig. 6.3).

The C to H ratio diminishes as the relative molecular mass (M_r) of PAH increases, although it has a long way to go to match C:H = 8 for 'young' soot. Although not fully substantiated, one compelling argument that PAH compounds are precursors to soot is that if the general decreasing trend of the concentration of PAH species over the range $150 < M_r < 300$ is extrapolated to $M_r \approx 600$, the species number density is of the same order of magnitude as that for the particles in sooting flames [109].

Since fuel degradation is a prominent feature of flame chemistry, a starting point for PAH formation must be a pyrosynthetic route to aromatic rings from small aliphatic units. There is agreement that acetylene and its derivatives, or similar highly unsaturated units such as butadiene, are amongst the earliest precursors to soot. A possible free radical mechanism to aromatic structures is [105, 110]

$$C_2H_2 + C_2H_3^\bullet \rightarrow n\text{-}C_4H_5^\bullet\ ({}^\bullet CH{=}CH-CH{=}CH_2) \quad (6.59)$$

$$C_4H_5^\bullet + C_2H_2 \rightarrow n\text{-}C_6H_7^\bullet\ ({}^\bullet CH{=}CH-CH{=}CH-CH{=}CH_2) \quad (6.60)$$

$$n\text{-}C_6H_7^\bullet \rightarrow c\text{-}C_6H_7^\bullet \rightarrow C_6H_6 + H^\bullet \quad (6.61)$$

This sequence signifies a cyclisation to a six-membered ring, and then to benzene by the loss of an H atom. Supplementary steps that may lead to the synthesis of higher PAH products from benzyl radicals are shown in Fig. 6.9 [111]. The abstraction of H atoms or their dissociation at various stages is a prerequisite for the dehydrogenation that leads to an increasing C to H ratio. These illustrative mechanisms for pyrosynthetic chemistry in flames represent a simplified view of an exceedingly complex problem. It seems unlikely that the same mechanism applies to all flames,

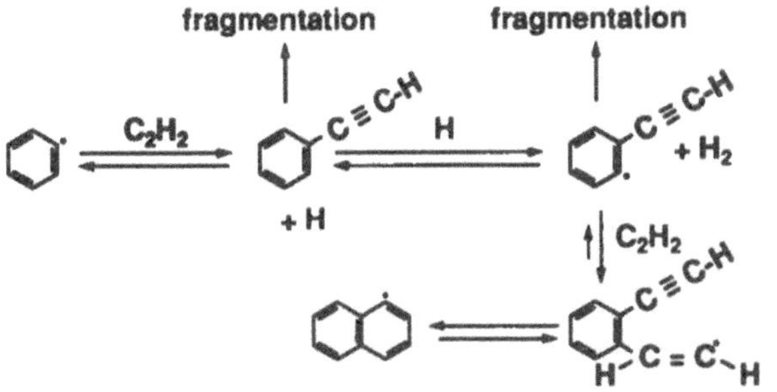

Figure 6.9 A mechanism proposed for the pyrosynthesis of polycyclic aromatic hydrocarbons during soot formation (after Frenklach *et al.* [111]).

Some interpretations of soot formation include ion–molecule reactions [112]. In rich flames, the dominant ion is $C_3H_3^+$ and its concentration falls rapidly at the critical equivalence ratio for soot formation, its place being taken by larger ions. Thus, it is possible that ionic processes are involved even from the earliest stages, for example,

$$C_3H_3^+ + C_2H_2 \rightarrow C_5H_3^+ + H_2 \quad (6.62)$$

$$C_3H_3^+ + C_4H_2 \rightarrow C_7H_5^+ \quad (6.63)$$

These larger ions may grow via

$$C_5H_3^+ + C_2H_2 \rightarrow C_7H_5^+ \quad (6.64)$$

or

$$C_7H_5^+ + C_2H_2 \rightarrow C_9H_7^+ \quad (6.65)$$

Large gaseous ions are known to rearrange very rapidly to their most stable, polynuclear aromatic structures. Three dominant ions in soot-forming flames are $C_{13}H_9^+$ (three rings), $C_{17}H_{11}^+$ (four rings) and $C_{19}H_{11}^+$ (five rings) and their relative concentrations decrease with mass in a similar way to that of the neutral PAH species [109]. Nevertheless, the proportions of ions relative to radicals present in the gas phase are too low to have significant effect on the main reaction processes, so ionic theories to explain the gaseous stages of soot formation must be regarded with some caution.

6.8.4 The condensed phase of soot formation

Just as the production of liquid water droplets from water vapour requires nuclei, so it is assumed that nucleation (the transformation from a molecular to a particulate system) must occur first. This is followed by growth to the spherical particles that are detected by electron microscopy. Finally, aggregation of these units occurs to yield chain structures.

Ion studies have contributed very substantially to a better understanding of the nucleation phase. Structural information is not obtained, but it is possible to measure the overall concentrations of charged species of relative molecular masses in the range $500 < M_r < 5 \times 10^4$, say. This bridges the gap between the limits for mass spectrometric and electron microscopic detection, and represents the equivalent approximate diameters 0.5–5 nm, in which range the transition occurs from a molecule to a small solid body [109]. The earliest stage of the soot nucleation zone in flames is thought to be in the range $650 < M_r < 700$, speculative structures and the empirical formulae, $C_{54}H_{18}$–$C_{54}H_{30}$, for which have been proposed [113].

The ratio of charged to uncharged species is appreciably larger than the corresponding ratio in the gaseous reaction stages. Moreover, soot formation is strongly influenced by applied electric fields [102]. For example, larger particles are produced when an electric field is applied so as to increase the residence time of positive ions in the flame. The addition of alkali metal ions to the flame gases also has a profound effect on soot growth. These features may testify to a role of ionic effects in soot growth, but there is no concensus on this as yet [109].

Once particles appear, a range of species can condense on their surface and eventually react to increase the size of the spherical particles to between 10 and 50 nm. Acetylene is the most likely growth species. The surface is very active in catalysing these growth processes owing to the presence of free radical centres. The particles then grow by aggregation to form chains.

6.8.5 Fullerenes

Fullerene has a spherical cage structure, comprising 60 carbon atoms (C_{60}), which resemble the geodesic dome designed by Buckminster Fuller. The commonplace example is to be found in the combination of hexagonal and pentagonal segments that are sewn together to make up a soccer ball. A more ellipsoid shape, somewhat like a rugby ball, comprises 70 carbon atoms. Although first prepared by evaporation of graphite [114], fullerenes may be formed in premixed, sooting flames at low pressures [115]. Under the most favourable conditions, samples of condensable compounds drawn by a quartz microprobe from a laminar, flat flame of benzene, and extracted for analysis by HPLC contained typical C_{60} and C_{70} units up to about 20% of the soot mass [115].

By contrast to PAH and soot, fullerenes are free from hydrogen and have curved structures. Fullerene precursors appear to be a subset of the (planar) PAH growth reactions, perhaps identifiable at the onset of curvature (Fig. 6.10) when a five-membered ring is surrounded completely

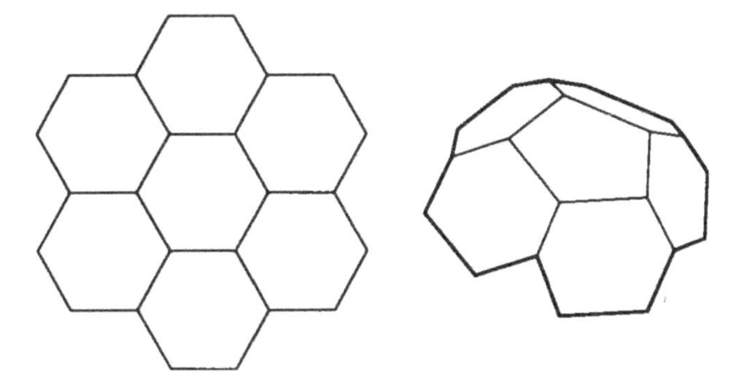

Figure 6.10 Representations of the planar structure of coronene ($C_{24}H_{12}$) and the curved structure of corannulene ($C_{20}H_{10}$).

by six-membered rings (corannulene, $C_{20}H_{10}$). Growth is thought to occur through a sequence of events involving species containing hydrogen up to $C_{50}H_{10}$, with the final stages being completed by C_2 addition and H_2 elimination [115].

6.9 Diamond synthesis in flames

A novel development in flame technology has emerged in recent years which is the synthesis of diamond in atmospheric pressure flames by chemical vapour deposition of carbon onto a silicon or a molybdenum surface. Generally C_1 and C_2 hydrocarbons are used in fuel-rich mixtures, and the flame is directed onto a water-cooled holder, on which the deposition substrate (Si or Mo) is itself deposited. The growth rate of the deposited film of carbon can be appreciable (100–150 $\mu m\,h^{-1}$). Raman spectroscopic studies of the deposited carbon show the characteristic signal of pure diamond at the Raman shift wavelength of 1333 cm^{-1}.

Acetylene is a favoured fuel, and a conventional oxy-acetylene welding torch can be used as a burner. However, a more uniform growth of diamond is possible in the vicinity of the stagnation point of a flat-flame just below a flat, water-cooled substrate surface because there is a spatially uniform temperature and concentration field. The presence of hydrogen in a fuel-rich $C_2H_2/H_2/O_2$ flame tends to reduce the burning velocity from that of acetylene alone. With $C_2H_2/O_2 \approx 1$, $H_2/O_2 \approx 0.5$, a 40 μm layer of diamond can be built up in 1 h on a surface controlled at about 1180 K [116].

Further reading

Calcote, H.F. (1981). *Combust. Flame*, **42**, 215.

D'Alessio, A. (1994). Smoke, diamonds and stardust. *Twenty-Fifth Symposium (International) on Combustion*. The Combustion Institute, Pittsburg, PA, USA (in press)

Dixon-Lewis, G. and Williams, D.J. (1977). The oxidation of hydrogen and carbon monoxide. In *Comprehensive Chemical Kinetics* (Vol. 17) (ed. C.H. Bamford and C.F.H. Tipper). Elsevier, Amsterdam, The Netherlands, p. 1.

Fenimore, C.P. (1964). *Chemistry in Pre-Mixed Flames*. Pergamon, Oxford, UK.

Gaydon A.G. (1974). *The Spectroscopy of Flames* (2nd edn). Chapman and Hall, London, UK.

Gaydon, A.G. and Wolfhard, H G. (1979). *Flames: Their Structure, Radiation, and Temperature* (4th edn). Chapman and Hall, London, UK.

Gardner, W.C. (ed.) (1984). *Combustion Chemistry*. Springer Verlag, Heidelberg, Germany,

Glassman, I. (1987). *Combustion*, (2nd edn). Academic Press, New York, USA.

Glassman, I. (1988). Soot formation in combustion processes. In *Twenty-Second Symposium (International) on Combustion*. The Combustion Institute, Pittsburgh, PA, USA, p. 295.
Harris, S.J. and Weiner, A.M. (1985). *Ann. Rev. Phys. Chem.*, **36**, 31.
Homann, K.H. (1984). Formation of large molecules, particulates and ions in premixed hydrocarbon flames. In *Twentieth Symposium (International) on Combustion*. The Combustion Institute, Pittsburgh, PA, USA, p. 857.
Howard, J.B. (1990). Carbon addition and oxidation reactions in heterogeneous combustion and soot formation. In *Twenty-Third Symposium (International) on Combustion*. The Combustion Institute, Pittsburgh, PA, USA, p. 1107.
Howard, J.B. (1992). Fullerenes formation in flames. In *Twenty-Fourth Symposium (International) on Combustion*. The Combustion Institute, Pittsburgh, PA, USA, p. 933.
Hucknall, D.J. (1985). *Chemistry of Hydrocarbon Combustion*. Chapman and Hall, London, UK.
Lawton, J. and Weinberg, F.J. (1969). *Electrical Aspects of Combustion*. Clarendon Press, Oxford, UK.
Lewis, B. and Von Elbe, G. (1987). *Combustion, Flames and Explosions of Gases* (3rd edn). Academic Press, New York, USA.
Minkoff, G.J. and Tipper, C.F.H. (1962). *Chemistry of Combustion Reactions*. Butterworths, London, UK.
Mulcahy, M.F.R. (1973). *Gas Kinetics*. Nelson, London, UK.
Palmer, H.B. and Cullis, C.F. (1965). Formation of carbon from gases. *Chemistry and Physics of Carbon* (Vol. 1) (ed. P.L. Walker). Marcel Dekker, New York USA, p. 200.
Wagner, H. Gg. (1978). Soot formation in combustion. In *Seventeenth Symposium (International) on Combustion*. The Combustion Institute, Pittsburgh, PA, USA, p. 3.

Problems

(1) When CO is burned at 1 bar (100 kPa) in a stoichiometric mixture with oxygen, the adiabatic flame temperature is 3000 K. Calculate the equilibrium partial pressure of CO_2, CO, O_2 and O in the post-flame gas. The relevant equilibrium constants at 3000 K, $K^{\ominus}(=K_{p\,p}{}^{\ominus})$ are

(a) $0.5O_2 \Leftrightarrow O \qquad K_1^{\ominus}=0.1125$

(b) $CO+0.5O_2 \Leftrightarrow CO_2 \qquad K_2^{\ominus}=3.0$

Suggested method. There are four unknowns, so four independent equations are required, in terms of partial pressures. These can be represented by the two equilibria, Dalton's law of partial pressures and the element balance ($n_C/n_O = 1/2$. From Avogadro's law, the element balance ratio can be written in terms of the sum of the partial pressures of the components containing each species. The final expression is obtained in terms of one component, which then has to be solved numerically. The method is described by Gaydon and Wolfhard [10].

(2) When CO is burned at 1 bar (100 kPa) in a stoichiometric mixture with air, the adiabatic flame temperature is 2400 K. Calculate the equilibrium partial pressure of CO_2, CO, O_2 and O in the post-flame gas, given that $K_1^\ominus = 8.61 \times 10^{-3}$ and $K_2^\ominus = 47.4$ at 2400 K. Hence estimate the equilibrium partial pressure of NO that might be expected to be present in the post-flame gas. For the process

$$0.5N_2 + 0.5O_2 \Leftrightarrow NO$$

the equilibrium constant at 2400 K, $K^\ominus (= K_{p/p^\ominus}) = 0.05$. Assume that air comprises 21% O_2 and 79% N_2 by volume.

7 Low temperature chemistry ($T < 1000\,K$)

7.1 Introduction and background

Whereas an understanding of the kinetics and mechanism of methane combustion at $T > 1000\,K$ is relevant to the detailed interpretation of the high temperature combustion of higher alkanes and other hydrocarbons, there is only a limited relationship between methane oxidation at $T < 1000\,K$ and that of higher alkanes. This distinction arises because so much of the low temperature chemistry of hydrocarbons is governed by the size and structure of the carbon backbone. The kinetics and mechanism of higher alkanes have yielded to quantitative interpretations in recent years. The common ground does not begin to emerge until there are four or more carbon atoms in the fuel molecule. The seeds of modern mechanistic interpretations of alkane oxidation were sown in the 1960s.

By virtue of its position as the simplest hydrocarbon and also its importance as a heating and power generation fuel, the oxidation of methane has been studied extensively. The oxidation is extremely slow at temperatures below about 700 K, at normal pressures. Only at exceptionally high pressures (> 3 MPa), is there a reasonably significant reaction rate of oxidation at lower temperatures. There are some similarities with respect to ethane and ethene. There is useful common ground with the chemistry of other alkanes at temperatures in the range from about 750 K. Some aspects of the oxidation of hydrogen and of carbon monoxide and related phenomonology also stand alone (Chapter 9). Features of their chemistry, especially in the temperature range 750–1000 K, that relate to hydrocarbon combustion will be discussed as appropriate.

7.2 The oxidation of methane

Initiation in the gas phase is by

$$CH_4 + O_2 \rightarrow CH_3 + HO_2 \qquad (7.1)$$

The HO_2 radicals may either attack methane

$$CH_4 + HO_2 \rightarrow CH_3 + H_2O_2 \qquad (7.2)$$

or undergo a radical recombination reaction such as

$$HO_2 + HO_2 \rightarrow H_2O_2 + O_2 \quad (7.3)$$

or

$$CH_3O_2 + HO_2 \rightarrow CH_3OOH + O_2 \quad (7.4)$$

The oxidation of methyl radicals plays an important part in the overall process, and occurs at relatively low temperatures by

$$CH_3 + O_2 + M \Leftrightarrow CH_3O_2 + M \quad (7.5)$$

Both the forward and reverse processes are relevant at temperatures below 1000 K. The equilibrium constant is given by [117]

$$K(\mathrm{cm^3\,mol^{-1}}) = RT\exp(15520/T) - 17.33 - 1.961((\ln(298/T) + (T - (298/T)/T) \quad (7.6)$$

The equilibrium is displaced to the right hand side at 750 K or below, so that reactions of CH_3O_2 are then dominant.

Propagation through H atom abstraction by CH_3O_2 is possible,

$$CH_3O_2 + CH_4 \rightarrow CH_3OOH + CH_3 \quad (7.7)$$

but this has a relatively high activation energy ($E > 60\,\mathrm{kJ\,mol^{-1}}$) because the primary C–H bond of methane is amongst the strongest of the C–H single bonds. Consequently, in the low temperature range the favoured alternatives are radical–radical interactions of the following kind:

$$CH_3O_2 + CH_3O_2 \rightarrow CH_3O + CH_3O + O_2 \quad (7.8)$$

$$CH_3O_2 + HO_2 \rightarrow CH_3O + OH + O_2 \quad (7.9)$$

$$\rightarrow CH_3OOH + O_2 \quad (7.4)$$

$$CH_3O_2 + CH_3 \rightarrow CH_3O + CH_3O \quad (7.10)$$

The main molecular products of the methyl + oxygen reaction are known to be methanol and formaldehyde, the proportions of which depend on the reaction conditions. The reaction

$$CH_3O + CH_3O \rightarrow CH_3OH + CH_2O \quad (7.11)$$

yields equal proportions of methanol and formaldehyde. An excess of methanol would suggest the reaction

$$CH_3O + CH_4 \rightarrow CH_3OH + CH_3 \quad (7.12)$$

whereas an excess of formaldehyde in the products would suggest the reactions

$$CH_3O + M \rightarrow CH_2O + H + M \quad (7.13)$$

or

$$CH_3O + O_2 \rightarrow CH_2O + HO_2 \quad (7.14)$$

The equilibrium shown in eqn (7.5) is displaced to the left hand side at $T > 850$ K, say, which favours methyl radical reactions, either as the recombination

$$CH_3 + CH_3 + M \rightarrow C_2H_6 + M \quad (7.15)$$

or oxidation,

$$CH_3 + O_2 \rightarrow CH_2O + OH \quad (7.16)$$

Equation (7.16) is not a simple bimolecular process but probably involves the methylperoxy radical which undergoes an intramolecular hydrogen atom transfer

$$CH_3O_2 \rightarrow [CH_2OOH] \rightarrow CH_2O + OH \quad (7.17)$$

Since a primary C–H bond and a four-membered transition state ring are involved, this path has a high energy barrier and so it can become important only at higher temperatures. Chain propagation by hydroxyl radicals produced in eqn (7.16) occurs by

$$CH_4 + OH \rightarrow H_2O + CH_3 \quad (7.18)$$

leading to formaldehyde and water as the molecular products of eqns (7.16) and (7.18). Equation (7.15) is the principal chain terminating process in methane oxidation. During slow oxidation in closed vessels, reaction chains may also terminate at the walls probably by removal of species such as HO_2.

This overall sequence of reactions probably gives a fair representation of the main elementary reactions that occur in methane at $T < 850$ K, but there are further possibilities. The more important features are governed by the reactivity of the molecular intermediates. Formaldehyde is able to offer a very labile H atom ($D_{C\text{-}H} = 368\,kJ\,mol^{-1}$ compared with $D_{C\text{-}H} = 435\,kJ\,mol^{-1}$ in methane). Thus, not only is a secondary initiation of reaction possible by

$$CH_2O + O_2 \rightarrow CHO + HO_2 \quad (7.19)$$

but also radical attack by OH and HO_2 is favoured

$$OH + CH_2O \rightarrow H_2O + CHO \quad (7.20)$$

$$HO_2 + CH_2O \rightarrow H_2O_2 + CHO \quad (7.21)$$

Although the difference between the reactivities of eqns (7.18) and (7.20) is not very significant, the rate constant of eqn (7.21), for which $E \approx 42\,kJ\,mol^{-1}$, is considerably lower than that for eqn (7.2), $E > 60\,kJ\,mol^{-1}$. Subsequent reactions of the formyl radical, CHO, are

$$CHO + O_2 \rightarrow CO + HO_2 \quad (7.22)$$

and

$$CHO + M \rightarrow CO + H + M \quad (7.23)$$

and the oxidation of CO to CO_2 and H to HO_2 may also occur (see Chapter 6). The products CH_2O, H_2O_2, H_2O and CO predicted by this mechanism can all be detected during the low temperature oxidation of methane.

The additional formation of HO_2 in eqn (7.22) sustains the hydroperoxy radical as a propagating species and the formation of hydrogen peroxide either in the abstraction reactions of eqns (7.2) and (7.21) or the radical recombination of eqn (7.3). Hydrogen peroxide decomposes quite readily, as follows:

$$H_2O_2 + M \rightarrow 2OH + M \quad (7.24)$$

Relatively unreactive HO_2 radicals are converted to highly reactive OH radicals, in consequence, and, if the major route to H_2O_2 is via the abstraction reactions of eqns (7.2) and (7.21), this also constitutes a chain branching sequence. The overall reaction is not a chain branching sequence if eqn (7.3) is the sole source of H_2O_2, because two HO_2 radicals are involved in the formation of two OH radicals.

Taking the rate constant for the unimolecular decomposition to be in the first-order (high pressure) limit, for which $k = 3.0 \times 10^{14} \exp(-24400/T)\,s^{-1}$, the half-life for H_2O_2 decomposition is 3.2 s at 700 K, 40 ms at 800 K and 1.4 ms at 900 K. Thus, the rate of acceleration of reaction as a result of this chain branching process is very strongly temperature dependent. Development on such a slow timescale as that at low temperatures relative to the normal propagation rate was called a *degenerate chain branching* reaction by Semenov [118]. In this type of reaction the rate accelerates rather slowly to a maximum and then decays. The controlled development, as opposed to a chain branching explosion (see Chapter 9), arises because there is a significant depletion of the reactant concentration owing to concurrent propagating reactions.

There is another possibility of degenerate branching during methane oxidation, but it is likely to be prevalent only at conditions in which reaction is promoted at temperatures below 700 K. These include reaction in very fuel-rich conditions at high pressure or when there is a sensitiser (such as HBr) present, and results from the formation and decomposition of methyl hydroperoxide, as

$$CH_3O_2 + CH_4 \rightarrow CH_3OOH + CH_3 \quad (7.7)$$

or, more favourably

$$CH_3O_2 + CH_2O \rightarrow CH_3OOH + CHO \quad (7.25)$$

followed by

$$CH_3OOH \rightarrow CH_3O + OH \qquad (7.26)$$

The (first-order) rate constant for methyl hydroperoxide decomposition is believed to be about $8 \times 10^{14}\exp(-21500/T)\,s^{-1}$, which gives a half-life of 3.2 s at 600 K and 19 ms at 700 K. However, as discussed above, the primary H atom abstraction of eqn (7.7) is not strongly favoured so this mode of chain branching is more strongly affected by the formation of formaldehyde as an molecular intermediate if the reaction is not catalysed by additives.

Hydrogen bromide is an effective catalyst because it is able to offer a labile H atom:

$$CH_3O_2 + HBr \rightarrow CH_3OOH + Br \qquad (7.27)$$

The activation energy for eqn (7.27) is approximately $30\,kJ\,mol^{-1}$. The catalytic nature arises because hydrogen bromide may be regenerated in the reaction

$$Br + CH_4 \rightarrow HBr + CH_3 \qquad (7.28)$$

The supplementary reactions that can occur during methane oxidation depend on the reactant temperature. The mechanism that is required to represent the reaction over very wide temperature ranges may be very complicated indeed.

7.3 An overview of the oxidation of higher alkanes and other organic compounds

The low temperature oxidation of organic compounds involves a considerable variety of reactive intermediates and there is a correspondingly large number of molecular products. There is a coherent structure to the elementary reactions involved, which can be coordinated in a formal structure for the kinetics. 'Higher alkanes' refers essentially to the series starting with normal butane and isobutane. There is a limited overlap with the behaviour of propane. The main, gaseous reaction pathways spanning the temperature range of approximately 500–850 K are encapsulated in Fig. 7.1, taking butane as a representative case. A number of simplifications are made to maintain clarity.

Oxidation can also take place at temperatures below 500 K, and the qualitative kinetic features are characterised by the very lowest temperature branch that is displayed in Fig. 7.1. The reaction rate becomes exceedingly slow and it relates mainly to the liquid-phase oxidation of much higher hydrocarbons, especially with regard to oxidative degradation when they are used as lubricants or the long-term stability of animal or vegetable oils and fats. Oxidation also progresses smoothly at temperatures above 850 K, but there is an increasing intervention of the alkyl radical decomposition reactions of eqn (6.13), which are not shown in Fig. 7.1 but which lead eventually to 'high temperature' mechanisms

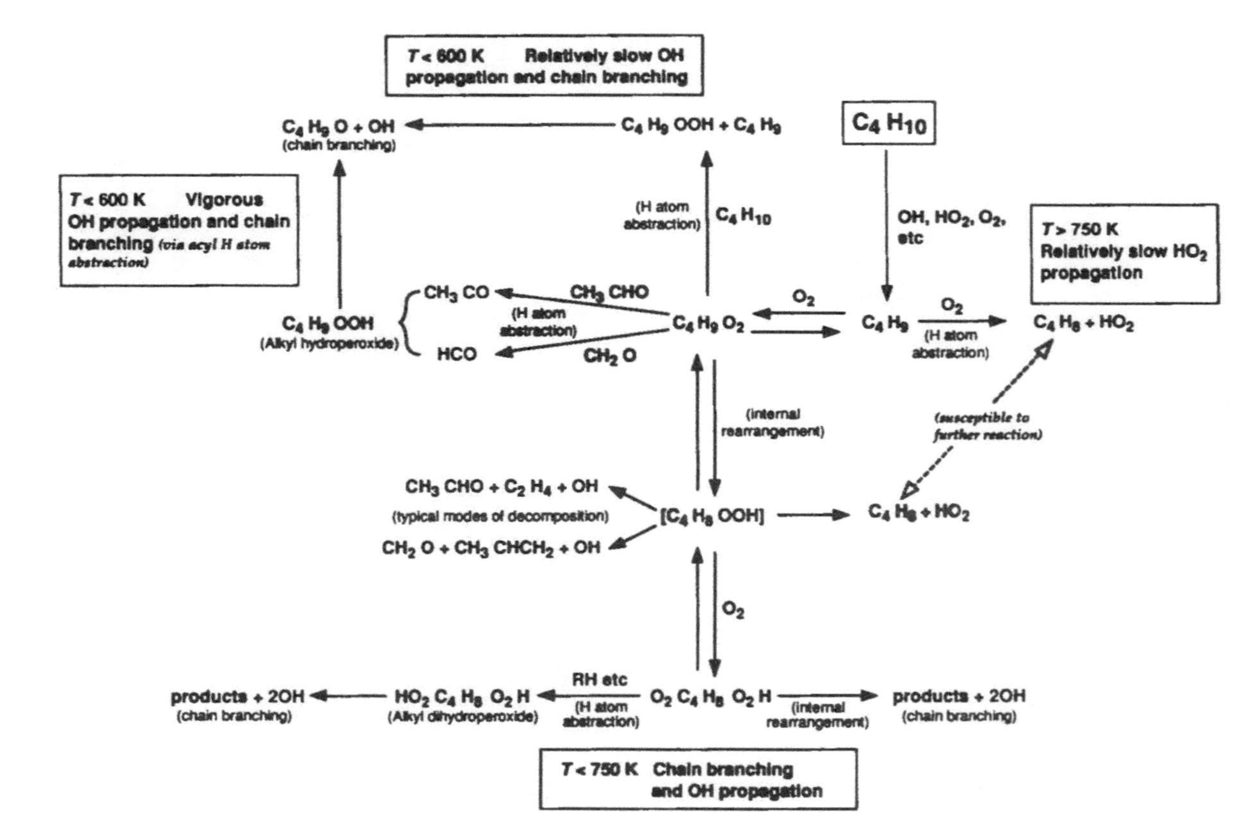

Figure 7.1 A schematic representation of normal butane oxidation in the temperature range 500–850 K. This is typical of the mechanisms involved in the oxidation of alkanes with four or more carbon atoms.

(Chapter 6). If the temperature is sufficiently high for rapid alkyl radical decomposition, the overall process may be characterised by the oxidation of a lower molecular mass alkene.

Figure 7.1 is divided into three temperature regimes, which signify the shift in the predominant reaction modes as the temperature is changed. Each alkane, or its alkyl radical derivative, responds in ways which correspond qualitatively to the reactions shown. Quantitative differences in the kinetics are reflected in different transitional temperatures, as discussed in section 7.4.

At the centre of Fig. 7.1 is a simplified construction of the $R+O_2/RO_2$ equilibrium and its competition with the overall abstraction process yielding an alkene (see section 7.5). As long as the oxidation to alkylperoxy radicals is favoured in the $R+O_2/RO_2$ equilibrium, reaction is propagated by OH radicals. Not only is this vigorous because the associated activation energies are negligible, but also it tends to be quite strongly exothermic:

$$RH+OH \rightarrow R+H_2O; \quad \Delta H^{\ominus}_{298} \approx -50\,kJ\,mol^{-1} \tag{7.29}$$

for an alkane. The subsequent oxidation of the alkyl radical gives considerably more heat, especially when carbon oxides are eventually formed. Degenerate chain branching modes also follow from the alkylperoxy radical reaction route.

By contrast, if the equilibrium is displaced towards the dissociation of alkylperoxy, the way is left open for an H atom abstraction process of the kind

$$R+O_2 \rightarrow \text{alkene}+HO_2 \tag{7.30}$$

which gives rise to a predominance of HO_2 radical propagation of the form

$$RH+HO_2 \rightarrow R+H_2O_2 \tag{7.31}$$

Not only is this reaction relatively slow because of the associated activation energies ($E>60\,kJ\,mol^{-1}$ at a primary C–H bond, $>50\,kJ\,mol^{-1}$ at a secondary C–H bond and $>40\,kJ\,mol^{-1}$ at a tertiary C–H bond), but also there is little contribution of branching from hydrogen peroxide at temperatures below 750 K. Moreover, the stoichiometry from eqns (7.30) + (7.31),

$$RH+O_2 = \text{alkene}+H_2O_2 \tag{7.32}$$

is virtually thermoneutral.

The overall feature of alkane oxidation and (for reasons connected with similarities of the reaction mechanism) of other organic compounds, is a transition from fairly vigorous, chain branching, exothermic oxidation to an essentially non-branching reaction of rather low exothermicity as the temperature is raised through a certain range.

The main processes at the lowest temperatures (typically $T<650\,K$) are attributed to intermolecular H atom abstraction by alkylperoxy radicals of the form

$$RH + RO_2 \rightarrow R + ROOH \qquad (7.33)$$

where ROOH represents a molecular, alkyl hydroperoxide. The activation energies of these types of processes are believed to be rather similar to the corresponding reactions for HO_2 radicals, which means that chain propagation is relatively slow.

As for the decomposition of CH_3OOH (eqn (7.26)), the alkyl hydroperoxides are appreciably less stable than hydrogen peroxide ($E \sim 170$–$180\,kJ\,mol^{-1}$). Degenerate chain branching is possible, but the acceleration of the reaction rate will not be fast since, typically, the half-life for the decomposition is 3 s at 600 K falling to about 20 ms at 700 K. These times may seem short, but combustion processes are noted for their rapidity, and there are applications, such as in diesel engine combustion (section 13.3), for which this is too slow.

The decomposition of the alkyl hydroperoxide is promoted at the weak peroxide bond

$$ROOH \rightarrow RO + OH \qquad (7.34)$$

so that the subsequent propagation occurs via the hydroxyl and alkoxyl radicals. The alkoxyl radical could be a source of an alcohol as an intermediate molecular product, by

$$RO + RH \rightarrow ROH + R \qquad (7.35)$$

and this is the simplest means of reverting to alkyl and alkylperoxy radical propagation. A relatively high activation energy for eqn (7.35) allows the occurrence of alternative reactions (section 7.7.1).

The most complicated kinetic region displayed in Fig. 7.1 is associated with the intermediate temperature range, loosely assigned as 650–750 K. Additional reactions are possible because the alkylperoxy radical is able to undergo an internal rearrangement or *isomerisation*, which involves an intramolecular H atom transfer,

$$RO_2 \Leftrightarrow QOOH \qquad (7.36)$$

The product of this reaction (commonly assigned as QOOH in general terms) is represented as C_4H_8OOH, which signifies that a hydroperoxide linkage has been created and that the free radical site is then associated with a carbon atom somewhere further along the structure. The importance of these types of reactions was identified in the early 1960s [119, 120]. The current view is that the isomerisation reaction of eqn (7.36) is regarded to be reversible, as discussed further in section 7.4. QOOH is a free radical which also includes the weak O–O bond and so it is very reactive.

Two possible modes of decomposition of C_4H_8OOH are shown in Fig. 7.1, but there may be a number of alternatives which are governed by the relative locations of the free radical site and the -OOH moiety (see section 7.4), represented generally as

$$QOOH \rightarrow AO + OH \quad (7.37)$$

The dissociation of the peroxide group is inevitable, yielding a partially oxygenated molecular intermediate (AO) and a hydroxyl radical. These products are represented as aldehydes in Fig. 7.1 because they may be contributory to one of the important, supplementary, kinetic features in this particular temperature regime, namely

$$RCHO/CH_2O + RO_2 \rightarrow RCO/CHO + ROOH \quad (7.38)$$

Equation (7.38) is distinguished from eqn (7.33) by the low activation energy ($E \approx 40\,kJ\,mol^{-1}$) which results from the weak acyl C–H bond in the aldehydes. The ability of the alkylperoxy radical to isomerise and generate molecular intermediates that are more reactive than the primary fuel greatly facilitates a route to the alkyl hydroperoxide, and chain branching.

Further oxidation of QOOH may also yield alkyl dihydroperoxides, first identified in 1961, in the low temperature oxidation of heptane [121]. Their formation is believed to include the reversible addition of molecular oxygen

$$QOOH + O_2 \Leftrightarrow O_2QOOH \quad (7.39)$$

followed by either a further, reversible, intramolecular rearrangement (eqn (7.40)) or intermolecular H atom abstraction (eqn (7.41)). The former would lead to another free radical (designated as HOOQ′OOH below), which is then likely to decompose to give a chain branching reaction (eqn (7.40)). The latter also gives degenerate chain branching as a result of the decomposition of the dihydroperoxide itself (eqn (7.42)).

$$O_2QOOH \Leftrightarrow HOOQ'OOH \rightarrow 2OH + \textit{products of lower molecular mass} \quad (7.40)$$

$$O_2QOOH + RH \rightarrow HOOQOOH + R \quad (7.41)$$

$$HOOQOOH \rightarrow 2OH + \textit{diketone or products of lower molecular mass} \quad (7.42)$$

Diketone and hydroxyketo derivatives have been identified from the low temperature combustion of heptane and other alkanes [122, 123].

The decomposition of the isomerised free radical HOOQ′OOH (eqn (7.40)) is likely to be similar to that for the hydroalkylperoxy radical, QOOH (eqn (7.37)), so that degenerate chain branching via eqn (7.40) is able to occur much more rapidly than through the molecular, organic peroxide intermediates.

The activation energies associated with the alkylperoxy/hydroalkylperoxy isomerisations fall in the range 90–150 kJ mol^{-1} [124], according to the mode of internal rearrangement (section 7.4). It is this energy barrier that prohibits the isomerisation pathways from being the dominant chain branching route throughout the entire range at temperature below about 750 K. Although complicated by the various equilibria involved, this is another example of the gradual transition of one predominant reaction route to another as a consequence of competition between a bimolecular interaction (eqn (7.33)) and a unimolecular reaction (eqn (7.37)) which has a higher activation energy. The main steps of Fig. 7.1. may be summarised in alternative form as a generalised terminology, as follows:

RH (↓ O_2, X) non-branching + OH (↑ from QO_2H)

$$\text{alkene} + HO_2 \leftarrow R \overset{O_2}{\Leftrightarrow} RO_2 \Leftrightarrow QO_2H \overset{O_2}{\Leftrightarrow} O_2QO_2H \Leftrightarrow HO_2Q'O_2H \rightarrow \text{branching} \quad (7.43)$$

alkene + R′ (← R) RH (↓ from RO_2) RH (↓ from O_2QO_2H)

branching ← ROOH + R alkene + HO_2 HO_2QO_2H → branching

7.4 Alkylperoxy radical formation and isomerisation

The formation of alkylperoxy radicals occurs through the fast, reversible, exothermic process

$$R + O_2 \Leftrightarrow RO_2 \quad (7.44/7.45)$$

The exothermicity of the forward reaction is approximately 40 kJ mol^{-1}, with relatively little variation over a range of alkyl radicals. Equilibrium constants for eqn (7.44/7.45) may be reliably estimated from enthalpy and entropy data [8, 88]. They may be regarded to be the same for all primary (R_p), secondary (R_s) or tertiary (R_t) alkyl radicals (Table 7.1 [124]). The equilibrium value of $[RO_2]/[R]$ is likely to be greater than unity at the temperatures for slow combustion.

Another way of presenting the difference, is as the 'ceiling temperature' [125], that is the temperature at which $[RO_2]/[R] = 1$. The partial pressure of oxygen is also required in this calculation, and the values given in Table 7.1 pertain to normal atmospheric conditions. The primary rate parameters for the forward or reverse reactions would normally be derived from the equilibrium constant by assuming a temperature independent rate constant for the association step

$$R + O_2 \rightarrow RO_2 \quad (7.44)$$

Typically, $k_{7.44} = 5 \times 10^{12} \text{cm}^3 \text{mol}^{-1}$.

For all but the simplest alkylperoxy radical several alternative internal hydrogen abstractions may be possible in the isomerization to the

Table 7.1 Equilibrium constants for $R + O_2 \Leftrightarrow RO_2$, where R represents an alkyl radical [124]

Equilibrium	K_c (mol^{-1} cm^3) = $[RO_2]/[R][O_2]$	T(K) at $[RO_2] = [R]^a$
$R_p + O_2 \Leftrightarrow R_pO_2$	$7.1 \times 10^{-5}\exp(16358/T)$	750
$R_s + O_2 \Leftrightarrow R_sO_2$	$9.0 \times 10^{-4}\exp(17080/T)$	886
$R_t + O_2 \Leftrightarrow R_tO_2$	$1.9 \times 10^{-3}\exp(15625/T)$	843

[a] $[O_2] = 4.715 \times 10^{-6}$ mol cm^{-3} (≡ 29.4kPa at 750 K).

alkylhydroperoxy radical QOOH (eqn (7.36)). Their relative importance will depend on the steric and energetic factors involved. These arise from the strength of the C–H bond that is being broken and the 'strain energy' involved in the ring structure that exists in the transition state, exemplified as follows:

$$\underset{\displaystyle CH_3CHCH_2CH_3}{\overset{O-O^\bullet}{|}} \rightarrow \underset{\displaystyle CH_3CHCH_2CH_2}{\overset{O-O\cdots H}{|\quad\;\;\vdots}} \rightarrow \underset{\displaystyle CH_3CHCH_2CH_2^\bullet}{\overset{O-OH}{|}} \tag{7.46}$$

The six-membered ring structure shown in this example is a transition state of intermediate strain energy, the eight-membered rings being regarded to be those of lowest energy and therefore leading to the lowest isomerisation activation energies. A small highly strained ring in the transition state implies a large activation energy. The strength of the C–H bond from which internal abstraction occurs is also important, for which D_{C-H}(tertiary) < D_{C-H}(secondary) < D_{C-H}(primary).

Although specifically representing a secondary butylperoxy radical, eqn (7.46) exemplifies a much more general classification for the isomerisation reactions. It involves an H atom transfer from a primary C–H bond to a position five atoms removed. Thus, it is termed a 1:5p transition, and can be represented in a generalised way as

$$R_sO_2 \rightarrow Q(1{:}5p)OOH \tag{7.47}$$

Implicit in this generalisation is that there is no difference in the kinetic parameters for the corresponding reaction in a different alkyl radical. For an *n*-heptyl radical, this would be

$$\underset{\displaystyle CH_3CH_2CH_2CH_2CHCH_2CH_3}{\overset{O-O^\bullet}{|}} \rightarrow \underset{\displaystyle CH_3CH_2CH_2CH_2CHCH_2CH_2^\bullet}{\overset{O-OH}{|}} \tag{7.48}$$

The pre-exponential factors (per C–H bond) and activation energies for various internal hydrogen abstraction reactions have been derived by

Table 7.2 Rate parameters for $RO_2 \rightarrow QOOH$ [124]

Reaction	A (per C–H) (s^{-1})	E/R(K)	Example
$R_pO_2 \rightarrow Q(1.4p)O_2H$	1.41×10^{12}	18535	$CH_3CH_2O_2^{\bullet} \rightarrow {}^{\bullet}CH_2CH_2O_2H$
$R_pO_2 \rightarrow Q(1.5p)O_2H$	1.76×10^{11}	14758	$(CH_3)_3CCH_2O_2^{\bullet} \rightarrow {}^{\bullet}CH_2C(CH_3)_2CH_2O_2H$
$R_pO_2 \rightarrow Q(1.6p)O_2H$	2.20×10^{10}	12617	$CH_3(CH_2)_2CH_2O_2^{\bullet} \rightarrow {}^{\bullet}CH_2(CH_2)_2CH_2O_2H$
$R_pO_2 \rightarrow Q(1.7p)O_2H$	2.75×10^{9}	11126	$CH_3(CH_2)_3CH_2O_2^{\bullet} \rightarrow {}^{\bullet}CH_2(CH_2)_2CH_2O_2H$
$R_pO_2 \rightarrow Q(1.6s)O_2H$	2.20×10^{10}	10909	$CH_3CH_2(CH_2)_2CH_2O_2^{\bullet} \rightarrow CH_3CH^{\bullet}(CH_2)_2CH_2O_2H$
$R_pO_2 \rightarrow Q(1.4t)O_2H$	1.41×10^{12}	14181	$(CH_3)_2CHCH_2O_2^{\bullet} \rightarrow (CH_3)_2{}^{\bullet}CCH_2O_2H$
$R_sO_2 \rightarrow Q(1.4p)O_2H$	1.41×10^{12}	18655	$CH_3CH(O_2^{\bullet})CH_2CH_3 \rightarrow {}^{\bullet}CH_2CH_2(O_2H)CH_2CH_3$
$R_sO_2 \rightarrow Q(1.4s)O_2H$	1.41×10^{12}	16538	$CH_3CH(O_2^{\bullet})CH_2CH_3 \rightarrow CH_3CH_2(O_2H)CHCH_3$
$R_sO_2 \rightarrow Q(1.5s)O_2H$	1.76×10^{11}	13074	$CH_3CH(O_2^{\bullet})CH_2CH_2CH_3 \rightarrow CH_3CH_2(O_2H)CH_2CHCH_3$

reference to the direct experimental measurement in carefully selected cases (Table 7.2 [124]).

The variations in the magnitudes of the rate parameters have a considerable bearing on differences in the reactivity of alkanes and their isomers. Whilst it is not appropriate to discuss these aspects in very great detail, the consequences of changes of reactivity are important to the combustion of alkanes and their mixtures, which is reflected in practical problems such as spontaneous ignition hazards or 'knock' in spark ignition engines (Chapters 12 and 13). Consider, as a simple example, the alkyl radicals formed by abstraction of an H atom during the oxidation of normal butane, namely 1-butyl (${}^{\bullet}CH_2CH_2CH_2CH_3$) or 2-butyl ($CH_3CH^{\bullet}CH_2CH_3$), and of isobutane, namely 2-methyl propyl ($(CH_3)_2CHCH_2^{\bullet}$) or t-butyl ($(CH_3)_3C^{\bullet}$). From the data given in Table 7.3 for these radicals in the abbreviated form of eqn (7.47), it seems unlikely that the oxidation of isobutane is as reactive as that of normal butane because none of the associated isomerisations have activation anergies as low as $105\,kJ\,mol^{-1}$. A complete analysis, which would normally require numerical simulation, must take into account not only the relative ease of formation of the alkyl radicals throughout the temperature range but also reactions of the products formed from the decomposition of the

Table 7.3 Alkylperoxy isomerisation reactions associated with butyl radicals

		E(kJ mol^{-1})		E(kJ mol^{-1})
From n-butane				
	1-Butylperoxy		2-Butylperoxy	
	$R_pO_2 \rightarrow Q(1{:}4s)OOH$	133	$R_sO_2 \rightarrow Q(1{:}4s)OOH$	133
	$R_pO_2 \rightarrow Q(1{:}5s)OOH$	109	$R_sO_2 \rightarrow Q(1{:}4p)OOH$	155
	$R_pO_2 \rightarrow Q(1{:}6p)OOH$	105	$R_sO_2 \rightarrow Q(1{:}5p)OOH$	125
From i-butane				
	2-Methyl propylperoxy		t-Butylperoxy	
	$R_pO_2 \rightarrow Q(1{:}5p)OOH$	125	$R_tO_2 \rightarrow Q(1{:}4p)OOH$	155
	$R_pO_2 \rightarrow Q(1{:}4t)OOH$	118		

Although the decomposition of 'QOOH' is illustrated in Fig. 7.1 with respect to aldehyde formation, *O*-heterocyclic compounds (cyclic ethers such as oxirans, oxetans and furans) are detected in the products of the low temperature oxidation of alkanes or other hydrocarbons. The range may be quite varied even from the butanes [126] and may be exceedingly complicated when higher alkanes are oxidised. The formation of the cyclic ethers also results in chain propagation by OH radicals, as follows:

$$\mathrm{H(R)C(OOH)CH_2CH_2\dot{C}HR'} \rightarrow \mathrm{H(R)C(O^{\bullet})CH_2CH_2\dot{C}HR'} + \mathrm{OH} \rightarrow \mathrm{H(R)C(-O-)CH_2} + \mathrm{CH_2CHR'} + \mathrm{OH} \quad (7.49)$$

If in a general kinetic interpretation a specific selectivity is not being sought, it is common to represent the rate of attack by free radicals on a particular alkane from an additivity rule. The contribution per C–H bond to the total rate constant is assumed to be the same for all primary, all secondary, or all tertiary bonds in each hydrocarbon. The overall rate constant is then given by eqn (7.50):

$$k = n_p A_p \exp(-E_p/RT) + n_s A_s \exp(-E_p/RT) + n_p A_t \exp(-E_t/RT) \quad (7.50)$$

where n is the number of bonds of a specific type, A is the Arrhenius pre-exponential factor per C–H bond, and E is the corresponding activation energy. The subscripts p, s, t, refer to attack at primary, secondary and tertiary C–H bonds, respectively. Values of the Arrhenius parameters for some radical–alkane reactions are given in Table 7.4 [127, 128].

Laser resonance fluorescence of hydroxyl radicals has been used extremely successfully to determine elementary rate parameters associated with $RO_2 \Leftrightarrow QOOH$ isomerisation reactions [129], and a laser perturbation of hydroxyl radical concentrations in stabilised cool flames has been used to obtain 'global' kinetic data for chain branching rates in the low temperature region [130].

Table 7.4 Arrhenius parameters for H atom abstraction from hydrocarbons at primary, secondary and tertiary C–H sites (A (per C–H bond) ($cm^3\,mol^{-1}\,s^{-1}$) and E ($kJ\,mol^{-1}$)) [127, 128]

Reaction	Primary		Secondary		Tertiary	
	A	E	A	E	A	E
H + RH	2.2×10^{13}	39.2	4.9×10^{13}	33.3	5.1×10^{13}	25.2
O + RH	5.0×10^{12}	24.2	1.3×10^{12}	18.8	1.6×10^{13}	13.8
OH + RH	6.1×10^{11}	6.9	1.4×10^{12}	3.6	1.25×10^{12}	−0.8
HO_2 + RH	4.9×10^{10}	62.5	4.9×10^{10}	52.6	4.9×10^{10}	41.5
RO_2 + RH	4.9×10^{10}	62.5	4.9×10^{10}	52.6	4.9×10^{10}	41.5
CH_3 + RH	4.9×10^{11}	49.0	3.3×10^{11}	42.3	2.4×10^{11}	33.6
CH_3O + RH	5.3×10^{10}	29.5	3.6×10^{10}	18.7	1.9×10^{10}	11.6

Although limited to laboratories in Armenia [131] and France [132], electron spin resonance is used for the detection of alkylperoxy and hydroperoxy radicals. The procedure involves the freeze-trapping of free radicals on a liquid nitrogen cooled trap (77 K) located within the cavity of the spectrometer. The reaction products, including the relatively long-lived RO_2 radicals, are pumped continuously at low pressure from a flow reactor to the collection point. The accumulated signal is scanned repeatedly across the appropriate microwave frequency range, during which time the concentration of radicals builds up at the cold trap.

7.5 Extent of reaction and the negative temperature dependence of reaction rate

Representative analyses of the products of isothermal oxidation of *n*-heptane and of 2.2.4-trimethylpentane ('iso-octane') oxidation at 1.0 MPa in a jet-stirred flow reactor at a mean residence time of 1 s over the temperature range 550–1200 K are given in Table 7.5 [133]. These data may be broadly classified from the mole fractions of components in three temperature regimes, and features that are compatible with the mechanism outlined for butane in Fig. 7.1 may be identified in this overview. A number of different partially oxygenated products of heptane oxidation are detected at lower temperatures. Water and hydrogen peroxide may also be expected. The carbon oxides and formaldehyde exhibit peaks in their concentrations at temperatures both below and above 850 K in *n*-heptane oxidation, and alkyl hydroperoxides are known to be formed at low temperatures, but were not identified in this work. Alkenes of relatively low molecular mass were prominent amongst the reaction products in the higher temperature ranges from both fuels, which indicates that the oxidation route via alkylperoxy radicals is not favoured and that the heptyl and octyl radicals decompose readily at temperatures above about 850 K.

There is a considerable difference of reactivity of these two alkanes, as expected from the relative difficulty for 2.2.4-trimethylpentylperoxy radicals to undergo isomerisation reactions compared with the reactivity of *n*-heptylperoxy radicals. Whereas the oxidation of *n*-heptane is quite extensive from about 550 K, there is very little reaction of iso-octane below 850 K.

The reaction rate in a stirred flow reactor relates directly to the difference between the inflow and outflow reactant concentrations (see section 10.1). In view of a high dilution by nitrogen in these particular experiments [133], a good approximation of the dependence of reaction rate on temperature is demonstrated with respect to the outflow mole fraction of each fuel over a wide temperature range. The mole fraction of *n*-heptane decreases considerably as the vessel temperature is raised from 550 K, to a point at 630 K where more than half of the fuel has reacted (Fig. 7.2 [133]). This represents a maximum in the reaction rate since, as the temperature is raised further to 750 K, the mole fraction of *n*-heptane

Table 7.5 Typical molecular products of *n*-heptane and 2,2,4-trimethylpentane (*i*-octane) oxidation over the temperature range 550–1200 K listed according to their maximum mole fractions in three vessel temperature ranges in a CSTR operated at 1.0 MPa [133].

	n-Heptane			*i*-Octane		
Mole fraction	$T < 800\,K$	$T = 800–850\,K$	$T > 850\,K$	$T < 800\,K$	$T = 800–850\,K$	$T > 850\,K$
10^{-2}			Carbon monoxide Carbon dioxide			Carbon monoxide Carbon dioxide
10^{-3}			Hydrogen Ethene			Hydrogen
	Carbon monoxide Carbon dioxide Formaldehyde	Methane Propene Formaldehyde			*i*-Butene Methane Ethene Propene	
10^{-4}	Methanol Acetaldehyde Propionaldehyde				Ethane Formaldehyde	
		1-Butene 1-Pentene				
10^{-5}	Heptenes Ethylene oxide Acetone	Acetylene	Carbon monoxide Propane	1-Butene Carbon dioxide	Acetylene Butadiene	Pentenes
Trace	2-Methyltetrahydrofuran 2-Methyl-5-ethyltetrahydrofuran					

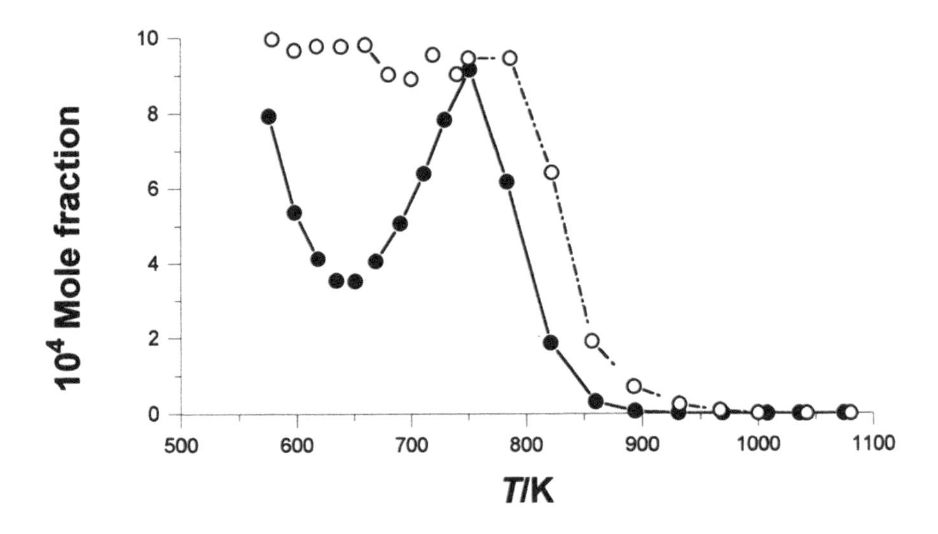

Figure 7.2 Experimental measurements of the residual fuel mole fraction in a CSTR at a mean residence time of 1 s and a total pressure of 1 MPa over a range of vessel temperatures. ●, $nC_7H_{16} + O_2 + N_2$; $iC_8H_{18} + O_2 + N_2$. The fuel + O_2 are in a stoichiometric proportion ($\phi = 1$), but there is an excess of N_2 in order to maintain isothermal conditions. The extent of reactant consumption is proportional to the reaction rate (after Dagaut *et al.* [133], primary data courtesy of M. Cathonnet).

at the outflow increases. Very little conversion occurs at this temperature, which represents a minimum in the reaction rate. Thus, there is a negative temperature coefficient of the overall reaction rate over the vessel temperature range 630–750 K under the chosen conditions. The reaction rate increases considerably at higher vessel temperatures. The appreciably lower reactivity of 2.2.4-trimethylpentane throughout the lower range of temperatures makes it difficult to distinguish similar features. Nevertheless, there is some indication of the existence of a weak negative temperature dependence of the reaction rate over a limited range of temperatures below 800 K.

The observation of a negative temperature dependent region of the overall reaction rate during hydrocarbon oxidation is not new; it was first identified during propane oxidation by Pease in a flow reactor in 1929 [134] and in a closed vessel in 1938 [135]. However, the kinetic foundation for its existence is now clear, and may be summarised as follows.

(i) At low temperatures the $R+O_2/RO_2$ equilibrium is displaced towards the formation of RO_2. The oxidation proceeds via alkyl hydroperoxide formation and, involving RO_2 isomerisation, the formation of dihydroperoxy species.
Autocatalysis occurs through degenerate chain branching and the reaction is propagated mainly by OH radicals. These modes predominate up to about 630 K in the present example.

(ii) As the temperature is raised, a decreasing fraction of alkyl radicals are converted to alkylperoxy radicals. The alternative mode is the formation of alkenes with propagation occurring via HO_2 radicals. This displacement of the predominant mode is virtually complete by 750 K. The system is then at its minimum reactivity because the prevailing radical concentration is exceedingly low.

(iii) As the temperature is raised beyond 750 K, an increased reactivity becomes possible through further reaction of the molecular intermediates, such as CH_2O and an increasing rate of decomposition of H_2O_2 which may promote autocatalysis.

The part played by the existence of a negative temperature coefficient of reaction rate to the occurrence of oscillatory cool flames in closed vessels under non-isothermal conditions is discussed in section 10.4.

7.6 Alkene formation from alkylperoxy radicals

There is clear evidence from experimental and theoretical studies of ethyl radical reactions [136] that ethene formation via an activated complex ($C_2H_5O_2^*$) competes with a direct abstraction mode.

$$C_2H_5 + O_2 \Leftrightarrow C_2H_5O_2^* \rightarrow C_2H_4 + HO_2 \qquad (7.51)$$

$$C_2H_5 + O_2 \rightarrow C_2H_4 + HO_2 \qquad (7.52)$$

The significance is that complex calculations are required in order to predict the rate of alkene formation over a range of temperatures. Hitherto, this underlying structure has been taken into account in numerical work only in part, the various modes of alkene formation being attributed to the reactions summarised as eqns (7.30) and (7.44/7.45).

7.7 Reactions of other species

7.7.1 Alkoxyl radicals

Alkoxyl radicals, RO, are formed in the decomposition of hydroperoxides and in radical–radical reactions involving RO_2. Like alkyl radicals they may decompose, react with oxygen or abstract hydrogen. Several modes of decomposition the RO radical are possible but the primary fission always appears to take place at the carbon atom adjacent to the oxygen, leading to an aldehyde and an alkyl radical, for example,

$$R'CH(O)R \rightarrow R + R'CHO \tag{7.53}$$

This provides an alternative route for the formation of lower aldehydes. Reaction with oxygen can lead to HO_2 radicals and aldehydes, thus

$$RCH_2O + O_2 \rightarrow RCHO + HO_2 \tag{7.54}$$

Secondary alkoxy radicals yield ketones by a similar process. Alkoxyl radicals can also abstract hydrogen atoms from alkanes, in this case producing alcohols and regenerating an alkyl radical.

$$RO + R'H \rightarrow ROH + R' \tag{7.55}$$

7.7.2 Aldehydes and their derivatives

Aldehydes may be formed from alkanes at low temperatures as illustrated in Fig. 7.1 or eqns (7.53) and (7.54), and the major process removing them always leads to acyl radicals (RCO) by H atom abstraction. At temperatures above 550 K decomposition of acyl radicals can occur to form alkyl radicals and carbon monoxide. In the low temperature range, reaction with oxygen may also be important

$$RCO + O_2 \rightarrow RCO_3 \tag{7.56}$$

and this leads to supplementary degenerate branching via

$$RCO_3 + RH \rightarrow RCO_3H + R \tag{7.57}$$

$$RCO_3H \rightarrow R + CO_2 + H \tag{7.58}$$

If RH is replaced by RCHO in eqn (7.57) the subset of eqns (7.56)–(7.58) becomes part of the mechanism associated with aldehyde oxidation [137].

7.7.3 *Toluene and other aromatics*

As discussed in section 6.3 the reactivity of aromatic compounds is expected not to be significant in the low temperature region. Abstraction of an H atom from toluene, for example, yields the relatively inert methylbenzyl radical, which is unable to sustain the degenerate chain-branching cycle. Although methylbenzyl radicals may combine with oxygen to form a benzylperoxy radical [138], there is no opportunity for an RO_2 isomerisation (eqn (7.36)) that confers a reactivity that is similar to that of the long-chain alkanes.

The extent to which an induced oxidation of toluene is possible at low temperatures in the presence of other hydrocarbons is determined by the relative rate for abstraction by propagating free radicals. The main propagation reactions involving OH are

$$RH + OH \rightarrow R + H_2O;\ k = 1.0\times 10^{13}\exp(-433/T)\,\mathrm{cm^3\,mol^{-1}\,s^{-1}} \quad (7.59)$$

$$C_6H_5CH_3 + OH \rightarrow C_6H_5CH_2 + H_2O;$$

$$k = 5.2\times 10^{9}\exp(-1440/T)\,\mathrm{cm^3\,mol^{-1}\,s^{-1}} \quad (7.60)$$

where reaction (7.59) represents a typical secondary H atom abstraction rates from an alkane to form an alkyl radical. The relative rate of these reactions is given by

$$v_{7.59}/v_{7.60} = 1920\exp(1007/T)([RH]/[C_6H_5CH_3]) \quad (7.61)$$

and is only weakly temperature dependent. This suggests that only limited extents of reaction of toluene are likely to occur even if reaction can be initiated by alkanes. Competitive reactions with other propagating species, such as HO_2, may also be important.

Further reading

Ashmore, P.G. (1963). *Catalysis and Inhibition of Chemical Reactions*, Butterworths, London, UK.

Ashmore, P.G., Dainton, F.S. and Sugden, T.M. (eds) (1967). *Photochemistry and Reaction Kinetics*. Cambridge University Press, Cambridge, UK.

Barnard, J.A., (1977). *Comprehensive Chemical Kinetics* (Vol. 17) (eds C.H. Bamford and C.F.H. Tipper). Elsevier, Amsterdam, The Netherlands, p. 441.

Benson, S.W. (1981). *Prog. Energy Combust. Sci.*, **7**, 125.

Griffiths, J.F. (1995). *Prog. Energy Combust. Sci.*, **21**, 25.

Hancock, G. (ed.) (1995). Oxidation kinetics and autoignition of hydrocarbons. *Comprehensive Chemical Kinetics*. Elsevier, Amsterdam, The Netherlands.

Hucknall, D.J. (1985). *The Chemistry of Hydrocarbon Combustion*. Chapman and Hall, London, UK.

Minkoff, G.J. and Tipper, C.F.H. (1962). *Chemistry of Combustion Reactions*. Butterworths, London, UK.
Pollard, R.T. (1977). *Comprehensive Chemical Kinetics* (Vol. 17) (eds C.H. Bamford and C.F.H. Tipper). Elsevier, Amsterdam, The Netherlands, p. 249.
Semenov, N.N. (1958). *Some Problems in Chemical Kinetics and Reactivity* (Vol. 2). Pergamon, Oxford, UK.
Shtern, V.Ya. (1964). *Gas Phase Oxidation of Hydrocarbons*, Pergamon, Oxford, UK.
Walker, R.W. (1975). A critical survey of rate constants for gas-phase hydrocarbon oxidation. In *Specialist Periodical Report, Reaction Kinetics* (Vol. 1). The Chemical Society, London, UK, p. 161.
Walker, R.W. (1977). Rate constants for reactions in gas-phase hydrocarbon oxidation. In *Specialist Periodical Report, Gas Kinetics and Energy Transfer* (Vol. 2). The Chemical Society, London, UK, p. 296.

Problems

(1) Calculate the relative rates of the two following reactions at 500 K and 750 K. Assume that the concentration of $O_2 = 10^{-4}\,\text{mol}\,\text{dm}^{-3}$.

(a) $CH_3CO + O_2 \rightarrow CH_3CO_3$; $A = 10^{10}\,\text{mol}^{-1}\,\text{dm}^3\,\text{s}^{-1}$, $E = 0$

(b) $CH_3CO \rightarrow CH_3 + CO$; $A = 10^{13}\,\text{s}^{-1}$, $E = 75\,\text{kJ}\,\text{mol}^{-1}$

(2) On the basis of the outline mechanism

(a) $R + O_2 \rightarrow AB + HO_2$

(b) $R + O_2 \rightarrow RO_2$

(-b) $RO_2 \rightarrow R + O_2$

(c) $RO_2 + RH \rightarrow ROOH + R$

in which RH, AB, ROOH and R represent an alkane, alkene, alkyl hydroperoxide and alkyl radical respectively, by use of the stationary state approximation with respect to R show that the relative rates of formation of the products may be expressed in the approximate form

$$\frac{d[ROOH]}{d[AB]} \approx \frac{k_c K_b [RH]}{k_a}$$

K_b is an equilibrium constant. Under what conditions would this expression be valid? By use of the following data, derive an expression for the relative yields of hydroperoxide to alkene. Hence, determine the temperature at which the yields of hydroperoxide and alkene are equal when the partial pressure of the RH is 13 kPa (100 torr).

$$k_a = 3.5 \times 10^{12} \exp(-3850/T)\,\text{cm}^3\,\text{mol}^{-1}\,\text{s}^{-1}$$

$$K_b = 4.3 \times 10^{-9} \exp(13600/T)\,\text{cm}^3\,\text{mol}^{-1}$$

$$k_c = 1 \times 10^{12} \exp(-7253/T)\,\text{cm}^3\,\text{mol}^{-1}\,\text{s}^{-1}$$

(3) After an initiating step producing alkyl radicals R, the main linear chain leading to products in combustion reactions can be written

(a) $R+O_2 \rightarrow A+HO_2$

(b,-b) $R+O_2 \Leftrightarrow RO_2$

(c) $RO_2+RH \rightarrow ROOH+R$

(d,-d) $RO_2 \Leftrightarrow QOOH$

(e_1) $QOOH \rightarrow B+OH$

(e_2) $\rightarrow C+OH$

(e_3) $QOOH+O_2 \rightarrow D+OH$

(f) $RH+OH \rightarrow R+H_2O$

where A is the conjugate alkene, and B, C and D represent stable molecular products. Write the differential equations for the formation of the radical species RO_2 and QOOH. Then using the stationary state approximation for radical concentration show that

$$\frac{d[B]}{dt}=\frac{k_b k_d[R][O_2]}{\{(k_{-b}+k_c[RH])(k_{-d}+\Sigma k_e)+k_d\Sigma k_e\}}$$

where $\Sigma k_e = k_{e1}+k_{e2}+k_{e3}[O_2]$. Assuming that reaction b is equilibrated (i.e. $k_{-b} \gg k_c[RH]$ or k_d) and that k_{-d} is negligible, show that the expression can be reduced to

$$\frac{d[B]}{dt}=k_d k_b F\,[R][O_2]$$

where F is the fraction of QOOH forming B.

(4) The initial products of neopentane ($(CH_3)_4C$) combustion at 753 K include *i*-butene, dimethyl ketone and 3,3-dimethyloxetan (DMO). A mechanism proposed for the formation of these products from the neopentyl radical (C_5H_{11}) is

(a) $C_5H_{11} \rightarrow iC_4H_8+CH_3$

(b, -b) $C_5H_{11}+O_2 \Leftrightarrow C_5H_{11}O_2$

(c) $C_5H_{11}O_2 \rightarrow C_5H_{10}OOH$

(d) $C_5H_{10}O_2H \rightarrow DMO+H$

(e) $C_5H_{10}O_2H+O_2 \rightarrow OOC_5H_{10}O_2H$

(f) $OOC_5H_{10}O_2H \rightarrow CH_3COCH_3+2CH_2O+OH$

(g) $C_5H_{12}+OH \rightarrow C_5H_{11}+H_2O$

Show that both $[CH_3COCH_3]/[DMO]$ and $[DMO]+[CH_3COCH_3]/[iC_4H_8]$ should be proportional to the concentration of O_2. Assume that $k_{-b} \gg k_c$.

Thermal ignition 8

8.1 Introduction

Spontaneous, thermal ignition in a closed system implies that the whole volume is simultaneously involved in the phenomenon and we can disregard propagating combustion waves. The analysis is applicable to all phases. In liquids and solids diffusion coefficients are several orders of magnitude lower than in gases, such that diffusion of active species is virtually impossible in solids until melting occurs. This means that chain branching reactions do not usually control ignition in condensed phases. By contrast, thermal conductivities in liquids or solids are typically higher than in gases by an order of magnitude. Consequently, explosions in condensed phases generally have their origin in thermal effects.

8.2 Thermal ignition theory

At the end of the last century it was recognised by van't Hoff [139] that the autocatalytic action required for an explosion can arise from the self-heating produced by an exothermic reaction because rates of chemical reactions increase dramatically with rising temperature. The criterion for occurrence of an explosion must therefore be related to the net rate of heat gain or loss in a volume element of the reacting system. If the rate of heat loss owing to conduction, convection and radiation remains equal to the rate of heat generation by reaction, then a stable temperature distribution will be established, while if the rate of heat loss cannot keep pace with that of heat generation then 'thermal runaway' to ignition will follow.

Some 30 years later, Semenov [140] developed an elementary model for thermal explosions which serves to demonstrate the principal quantitative features of the phenomenon. In this model, the temperature of the reacting system, T, is assumed to be uniform over the whole volume but to differ from that of the walls of the container which are at a temperature T_a (Fig. 8.1a). This condition pertains to well-mixed fluids; the conductivity of solids is usually too low to match this behaviour in a quantitative way. Although the fully quantitative interpretation of thermal ignition in solids really requires both temporal and spatial variations to be taken into account (Fig. 8.1b), there is hardly any qualitative distinction from the predictions of the spatially uniform model.

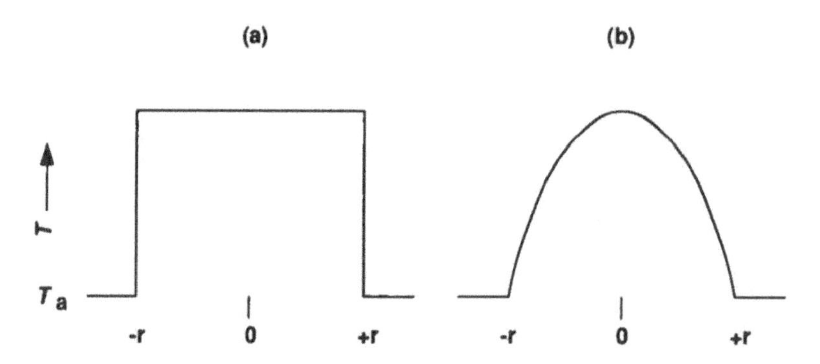

Figure 8.1 Spatial temperature profiles that are assumed in thermal ignition theory. (a) Semenov conditions, all reactant at T, heat transfer at the surface ($\pm r$). (b) Frank-Kamenetskii conditions, heat transport by internal thermal conduction, $T = T_a$ at the surface ($\pm r$).

8.2.1 Heat release rate dependences as criteria for spontaneous ignition

As long as the temperature difference is not too large, the rate of heat loss $\boldsymbol{L}$ from reactants at a spatially uniform temperature, T, per unit volume may be written as

$$L = h(S/V)(T - T_a) \tag{8.1}$$

where h is a heat transfer coefficient, S is the surface area of the reaction vessel and V is its volume. The vessel itself is thermostatted at T_a. Equation (8.1) is a representation of Newtonian heat loss applied to a spatially uniform temperature environment with all resistance to heat loss at the walls. The rate of heat production per unit volume, $\boldsymbol{R}$, is given by

$$R = vq \tag{8.2}$$

where $v(\text{conc}^{-1}\,\text{time}^{-1})$ is the reaction rate per unit volume and $q(\text{J mol}^{-1})$ is the molar exothermicity. For a process taking place at constant volume, $q = -\Delta U$, while for a constant pressure process, $q = -\Delta H$. If it is assumed that the rate of reaction displays the normal Arrhenius temperature dependence then

$$R = -kf(c)q = A\exp(-E/RT)f(c)q \tag{8.3}$$

where $f(c)$ is the appropriate function of reactant concentration, as applied to gases. A density term is also required if $\boldsymbol{R}$ is developed in terms of the reactant mass. The relationship between these heat flow terms against temperature is shown in Fig. 8.2. The rate of heat loss is a straight line passing through T_a and the rates of heat production form a family of curves related to the concentrations of reactants. This diagram

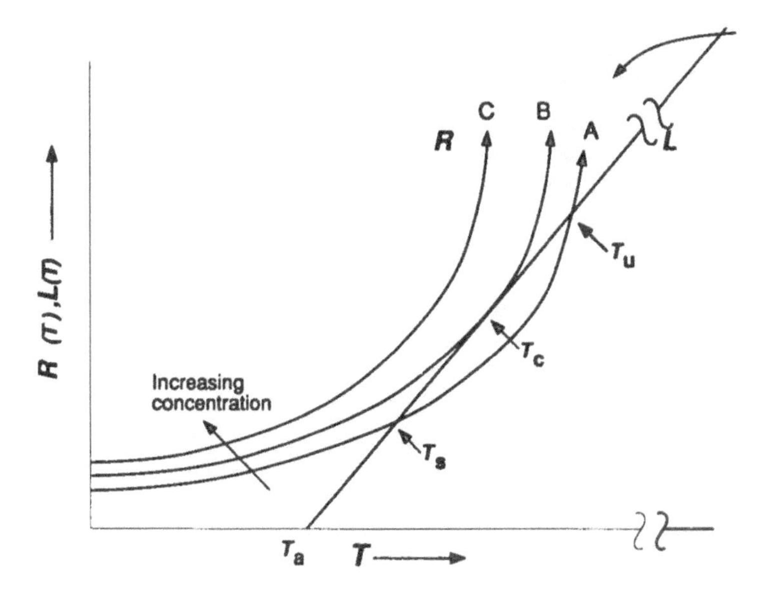

Figure 8.2 Relationships between the heat release rate $\boldsymbol{R}(T)$, governed by the Arrhenius reaction rate dependence, at different reactant concentrations in a closed vessel and $\boldsymbol{L}(T)$ at a fixed surface temperature (T_a). T_s, T_u and T_c signify stable, unstable and critical stationary states, respectively. The upper branch shows the form of the heat release curve at extremely high temperature ($T > E/R$).

was first constructed by Taffanel and Le Floch in 1912 [141], pre-dating the analytical approach by Semenov.

Consider, first, that a gaseous reactant is admitted to a container, surface temperature T_a, under concentration conditions corresponding to the heat release curve A. The reactant itself is also at T_a and, because the heat production curve lies above the heat loss curve, the system will heat up until a temperature T_s is attained. At this point, the rates of heat generation and heat loss are equal and the reaction will proceed steadily without further acceleration. This intersection of the two curves corresponds to a stable, stationary state, and would be termed *slow reaction* or *subcritical* reaction. If a small temperature excursion is induced in either direction the system will return to T_s($\boldsymbol{R} > \boldsymbol{L}$ as a result of a negative excursion; $\boldsymbol{L} > \boldsymbol{R}$ as a result of a positive excursion).

A stable condition cannot be maintained at the second intersection, T_u, in the way that it could at T_s. If the temperature rises above T_u the reaction accelerates ($\boldsymbol{R} > \boldsymbol{L}$), whereas if it falls below T_u, the system will drop back to the stable reaction condition at T_s($\boldsymbol{L} > \boldsymbol{R}$). T_u is, therefore, an unstable stationary state and it represents a watershed for the behaviour of the system, as follows. If the *initial temperature* of the reactants, as opposed to the *ambient temperature* of the reaction vessel

T_a (the boundary condition), is above the second intersection, T_u, the system becomes unstable and ignition occurs. This can occur in practice, and is known to have been the cause of industrial fires. It is recognised as the 'hot stacking problem', which signifies circumstances in which a product is taken from a hot process line and stockpiled before it has been allowed to cool appreciably.

In constructing the line to represent $\boldsymbol{R}$ it has been assumed that there is no reactant consumption as a result of reaction taking place (i.e. $f(c)$ relates to the initial reactant concentration, c_0, throughout). This assumption is sometimes called the *pool chemical approximation*, and if it is invoked it enables the behaviour in a closed system to be defined in terms of stationary states. Without it there is only one true stationary state, the final position of equilibrium, and the different modes of interesting behaviour that may be possible in the system cannot be easily explored. It is necessary to justify this assumption of course, but it is presumed to be satisfactory for most circumstances to be considered here.

If the temperature scale in Fig. 8.2 were to be extended, eqn (8.3) would lead to an inflexion in $\boldsymbol{R}$ and the asymptotic approach to a constant value given by $\boldsymbol{R}=Af(c)q$. This corresponds to $\exp(-E/RT)\rightarrow 1$, or $E/R \ll T$, and is an inevitability of the form of the Arrhenius expression. Taking a 'reasonable' activation energy for combustion systems, 150 kJ mol^{-1} say, T must exceed 18000 K when $E/R \approx T$, which does not seem to be physically very accessible. Nevertheless, in principle, there is a third intersection between $\boldsymbol{R}$ and $\boldsymbol{L}$ in Fig. 8.2, which also represents a stable stationary state, in the sense defined for T_s. We may choose to recognise this as representing an 'ignited state' of the system, although the physical realism has been upset because, in practice, the pool chemical approximation is no longer valid in these circumstances.

The conditions for reaction in open systems make the 'ignited state' physically accessible, though not at 18 000 K (see Chapter 10). The above properties exemplify a feature that is common in combustion, namely that of *reaction multiplicity*, which means that the system can exist in one of several states (three in this case, T_s, T_u and the 'ignited state') for a given set of boundary conditions. Which of these states is achieved is determined by the starting point, or at least the history of events.

Curve C in Fig. 8.2 represents an experiment with a higher initial concentration. In this case the reaction is immediately explosive since the rate of heat production always exceeds that of heat loss at reasonable temperatures, and the event would be termed *ignition* or *supercritical reaction*. The only intersection between $\boldsymbol{R}$ and $\boldsymbol{L}$ is the hypothetical, ignited state. The limiting condition, which marks the boundary between the two types of behaviour, is illustrated by curve B to which the heat loss curve forms a tangent, and at which the intersection points T_s and T_u have coalesced to give a single point intersection at T_c. The system

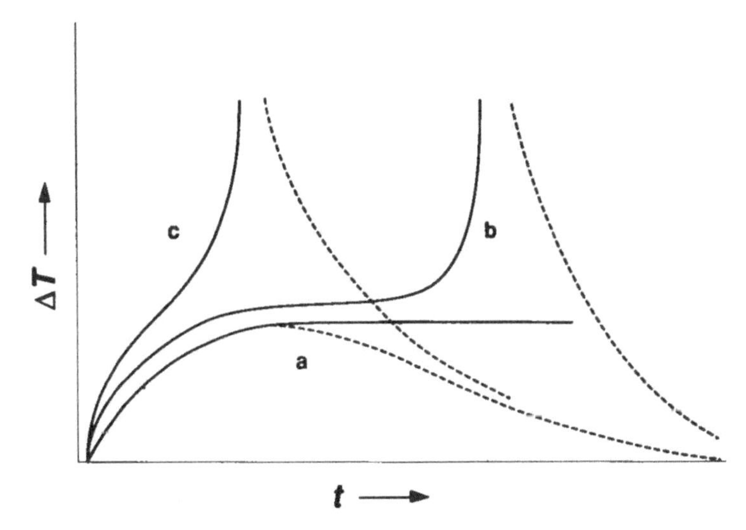

Figure 8.3 Typical ΔT versus time profiles in an exothermic system. The solid lines represent solutions to the energy conservation eqn (8.4) with no reactant consumption (a) in subcritical non-isothermal reaction, (b) ignition at $T_a \approx T_{a,c}$, and (c) ignition at $T_a > T_{a,c}$. The broken lines represent the effect of reactant consumption.

will heat up slowly from T_a to T_c after which rapid self-heating and acceleration to ignition occur. The conditions of a simultaneous equality of and tangency between R and L describe the minimum thermal requirements for spontaneous ignition, or autoignition. They yield the minimum reactant concentration at a given ambient temperature T_a, termed the *critical conditions*, at which spontaneous ignition is possible.

A thermocouple located within the reaction volume (with a reference junction at the external surface of the vessel) would have yielded, in the three separate experiments, the temperature excess (ΔT)-time records shown in Fig. 8.3. The solid lines represent ΔT assuming no reactant consumption. The broken lines represent the response that would be expected when the reactants are consumed. In reality, in closed systems, both ignition and slow reaction are transitory states of the system *en route* from its initial to its final state [142]. The profiles in Fig. 8.3 are the form of the solutions to the energy conservation equation, which links the heat release and loss rate terms:

$$C_v \sigma \mathrm{d}T/\mathrm{d}t = R - L$$

or

$$C_v \sigma \mathrm{d}T/\mathrm{d}t = A\exp(-E/RT)f(c)q - (hS/V)(T - T_a) \qquad (8.4)$$

The molar heat capacity is given by C_v and σ is a molar density. The equation must be integrated numerically to obtain the solution for $T(t)$, regardless of whether or not reactant consumption is assumed. The following qualitative features emerge:

(i) $dT/dt > 0$ initially, at $R > L$;
(ii) $dT/dt \rightarrow 0$ as $R \rightarrow L$, and for a stationary state $dT/dt = 0$, at which $R = L$;
(iii) $dT/dt \gg 0$ in circumstances when $R \gg L$;
(iv) $dT/dt < 0$ at $R = 0$ in the post-ignition period, after complete reactant consumption.

8.2.2 Vessel temperature and heat loss rate as criteria for spontaneous ignition

If the ambient temperature of the vessel (T_a) is changed, then the line representing the heat loss rate, L, will be displaced either to the left or to the right (Fig. 8.4). For a given initial reactant concentration, an increase in the value of T_a to T_a' takes L beyond the 'tangency condition' which marks the point of criticality at the vessel temperature $T_{a,c}$.

Alternatively, since the gradient of L is determined by the term (hS/V), if the rate of heat transfer changes owing to an alteration in vessel surface to volume ratio or in the heat transfer coefficient itself, the slope of the heat loss curve is modified, as shown in Fig. 8.5. There is a

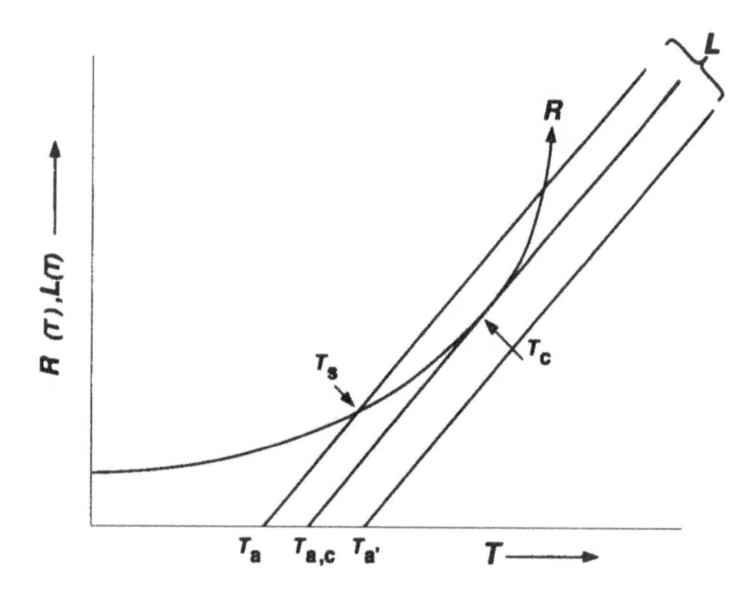

Figure 8.4 Relationships between the heat release rate $R(T)$ in a closed vessel and $L(T)$ at increasing surface temperature (T_a). Criticality occurs at $T_{a,c}$.

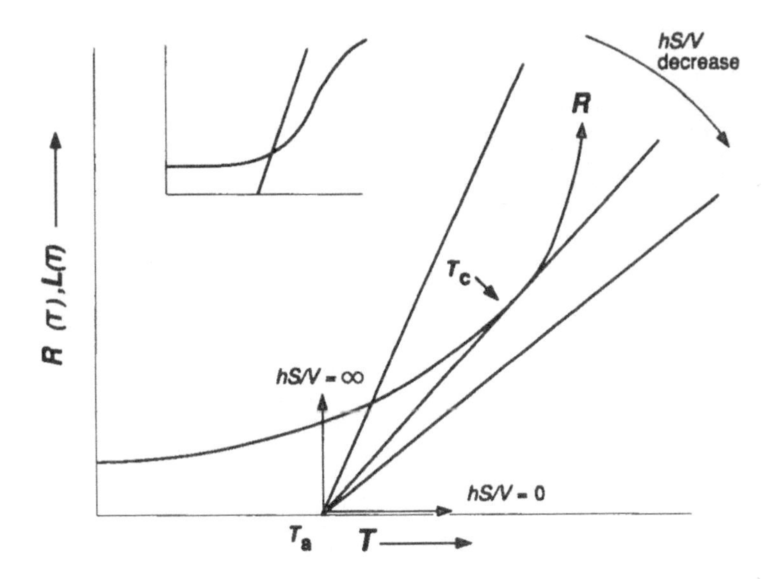

Figure 8.5 Relationships between the heat release rate $R(T)$ in a closed vessel and $L(T)$ at decreasing surface heat transfer rates, governed by S/V. The limiting conditions of $hS/V = \infty$ (isothermal) and $hS/V = 0$ (adiabatic) are also shown. The inset represents the relationship $dL/dT > dR/dT$ throughout the temperature range.

transition from subcritical reaction to supercritical reaction if the experiment is performed in a vessel of lower S/V, even though neither the reactant concentration nor the vessel temperature are varied. In this case the amount of material is seen to be important because it governs the surface to volume ratio, the simplest example (though not a very practical one) being that of spherical geometry, for which $S/V = 3/\text{radius}$. Since the problem of spontaneous ignition may be addressed in the context of stockpiled materials, such as grain, animal foods, fertiliser, coal, explosives (section 8.5), the answer to the question: 'If I store material X at a temperature T_a will it be safe?' must be, 'It all depends how much there is.'

At an exceptionally high heat loss rate, $(hS/V) \rightarrow \infty$, the reactants always remain at T_a. That is, the reaction is isothermal and can never show a critical jump to ignition. There can also be only one intersection between R and L for a given value of T_a, regardless of its value. By contrast, if the system is adiabatic, $(hS/V) \rightarrow 0$, R and L never intersect and ignition is inevitable, regardless of the ambient temperature.

Although specified here strictly with respect to the magnitude of hS/V, the condition for near-isothermal behaviour in an exothermic reaction is satisfied at $L \gg R$, so that in a stationary state $(T - T_a) \rightarrow 0$ (eqn (8.4)). The corollary of this in the context of chemical kinetics is that if the rate constant of an exothermic reaction is being measured in an apparatus at a controlled temperature T_a, it is essential to ensure the condition $L \gg R$ otherwise the prevailing temperature within the reactants is different from T_a, and the reaction occurs at a rate corresponding to the higher temperature.

R could be drawn in such a way that even at the steepest part the gradient is never steeper than the gradient of L (see inset, Fig. 8.5). Under these circumstances there can never be a 'critical jump', i.e. an ignition, and there is a specific criterion which determines the condition at which this change in behaviour occurs (see section 8.3).

8.2.3 Summary of conditions for thermal ignition to be possible

The principal features may be summarised as follows:

- exothermic reaction with a significant activation energy,
- only simple kinetics are required,
- an internal temperature rise occurs,
- only in adiabatic conditions is ignition an inevitability,
- critical conditions for ignition exist in a non-adiabatic system,
- ignition criteria are governed by an interplay between R and L,
- the ambient temperature and the size or shape of the reactant system are important.

8.3 Analytical interpretation of criteria for thermal ignition

Consider the stationary state condition of the energy conservation eqn (8.4) applied specifically to a first-order reaction [143]

$$A\exp(-E/RT)c_0q-(hS/V)(T-T_a)=0 \tag{8.5a}$$

or

$$A\exp(-E/RT)c_0q=(hS/V)(T-T_a), \qquad R=L \tag{8.5b}$$

To obtain an analytical solution it is necessary to assume that no reactant consumption occurs ($c \to c_0$), otherwise the changing reactant concentration has to be considered. The 'tangency condition' between R and L applies at the critical point, T_c (Fig. 8.2). Thus

$$\frac{dR}{dT}=\frac{dL}{dT} \tag{8.6}$$

Assuming the identities in eqn (8.5), differentiating eqns (8.1) and (8.3) with respect to T and equating the terms according to eqn (8.6), we obtain

$$A\exp(-E/RT_c)(E/RT_c^2c_0q=(hS/V) \tag{8.7}$$

Dividing eqn (8.5b) by eqn (8.7), and identifying $T = T_c$, leads to the equation

$$(T_c - T_a) = RT_c^2/E \qquad (8.8)$$

This is a quadratic in T_c, which can be solved to give

$$T_c = (E/2R)[1 \pm (4RT_a/E)^{1/2}] \qquad (8.9)$$

Only the root with the negative sign is of interest, since the other solution corresponds to the physically unrealistic state at exceedingly high temperature. Expanding the square root term using the binomial theorem leads to the conclusion that, to a good approximation ($E/R \gg T_a$),

$$\Delta T_c = (T_c - T_a) = RT_a^2/E \qquad (8.10)$$

This temperature change represents the highest temperature excess that can be sustained within the reactants without ignition occurring. Its value will normally be quite small. For example, with $E = 150\,\text{kJ mol}^{-1}$ and $T_a = 500\,\text{K}$, the temperature rise is about 15 K. It follows also from eqns (8.10) and (8.3) that at criticality, for which $T = T_c$, the rate of heat release $\boldsymbol{R}$ increases by a factor of $e(=2.718)$ from its initial value at T_a. That is,

$$\exp(-E/RT_c) = (e)\exp(-E/RT_a) \qquad (8.11)$$

Spontaneous ignition limits in gases are usually described in terms of a $(p - T_a)$ relationship [144]. The concentration corresponding to the critical condition may be obtained by substituting the expression for T_c in eqn (8.5). Subsequent manipulation, including eqn (8.11), leads to the expression

$$c_o = \frac{hSRT_a^2}{VEAqe}\exp(E/RT_a) \qquad (8.12)$$

or

$$\ln(c_o) = (E/RT_a) + \ln(hSR/VEAqe) \qquad (8.13)$$

Assuming ideal gas behaviour, from which $c_o = p/RT$, the critical reactant pressure at a given ambient temperature T_a is

$$p_c = \frac{hSR^2T_a^3}{VEAqe}\exp(E/RT_a) \qquad (8.14)$$

or

$$\ln(p_c/T_a^3) = (E/RT_a) + \ln(hSR^2/VEAqe) \qquad (8.15)$$

as shown in Fig. 8.6.

Fully dimensionalised terms have been retained throughout this discussion, but most of thermal ignition theory has been developed by grouping

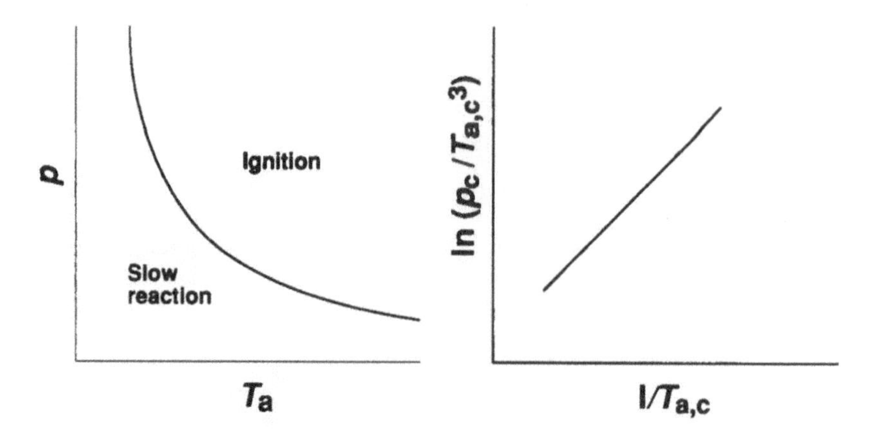

Figure 8.6 A representation of the ($p - T_a$) ignition boundary in an exothermic gaseous reaction with thermal feedback in a closed vessel. The linear relationship ln ($p/T^3_{a,c}$) versus $1/T_{a,c}$ for a first order reaction is also shown. The gradient of the line is E/R.

and non-dimensionalising parameters. Certain benefits accrue, as discussed in Chapter 1, although sometimes the simplification is illusory. For example, for understandable reasons a dimensionless temperature excess, θ, was defined by Frank-Kamenetskii [145] in the form

$$\theta = (T - T_a)E/RT_a^2 \tag{8.16}$$

which may cause difficulties in separating the dependences on E and T_a [146].

Nevertheless, the application of a dimensionless temperature excess of this form has been exploited extensively throughout thermal ignition theory [147] and chemical reactor theory [148]. The importance of θ is drawn mainly from the extremely useful 'exponential approximation' which (although not unique in this respect) allows solution of equations from which analytical solutions would not otherwise be possible. Thus, the Arrhenius dependence of the rate constant may be written as

$$\exp(-E/RT) = \exp(-E/RT_a)\exp(\theta/(1 + \varepsilon\theta)) \tag{8.17}$$

In this expression $\varepsilon = RT_a/E$, and T has been interpreted as $T = T_a + \Delta T$. The exponential approximation is applicable under conditions for which the product $\varepsilon\theta \ll 1$ or, if $\theta \sim 1$, $\varepsilon \ll 1$ which is certainly true in the examples cited above. Thus

$$\exp(-E/RT_a)\exp(\theta/(1 + \varepsilon\theta)) \rightarrow \exp(-E/RT_a)\exp\theta \tag{8.18}$$

The approximation is equivalent to that made in connection with eqn (8.10), and has the additional consequence that the upper 'ignited state' no longer exists because the inflexion of the Arrhenius term no longer exists. Equation (8.18) will be encountered also in Chapter 9. For the present, invoking the approximation, with some rearrangement, the stationary state energy conservation eqn (8.5) can be written

$$\frac{VqEc_oA}{hSRT_a^2}\exp(-E/RT_a)\exp(\theta)-\theta=0 \tag{8.19}$$

In this form, as developed by Semenov, the control parameters have been separated from the independent variable T. He abbreviated the first term as the Semenov parameter, ψ, or the 'Semenov dimensionless heat release rate', giving

$$\psi\exp(\theta)-\theta=0 \tag{8.20}$$

The elegantly simple solution for thermal criticality (i.e. the simultaneous stationary state and tangency condition) is then

$$\theta=1 \tag{8.21}$$

The stationary state solutions to eqn (8.20) and the association of criticality with the loss of solutions to $\psi(\theta)$ at $\psi=\exp^{-1}(=0.368)$ are shown in Fig. 8.7. Numerical solutions may be obtained for the exact dimensionless energy equation [149]

$$\psi\exp(\theta/(1+\varepsilon\theta))-\theta=0 \tag{8.22}$$

the values of which depend on the magnitude of ε. Either 1 or 3 solutions are recovered for $\varepsilon>0$ (Fig. 8.7), as opposed to 0 or 2 for the approximate equation, and only 1 solution is found throughout the entire range of ψ at $\varepsilon>0.25$. The latter is in accord with the single real solution to the quadratic equation of eqn (8.9) which is obtained when $4RT_a/E>1$. The condition $\varepsilon>0.25$ signifies that criticality, i.e. a transition from one state to another, is no longer possible. It matches the inset in Fig. 8.5. Thermal ignition is unlikely to be a feature of a reaction of low activation energy, especially if relatively high temperatures are required.

The Semenov parameter ψ includes terms to represent both chemical and physical parameters which are influential in determining the conditions under which spontaneous ignition may be possible. These features can be extracted by writing ψ in another form, to give eqn (8.20) as follows:

$$\frac{Bt_N}{t_{ch}}\exp(\theta)-\theta=0 \tag{8.23}$$

B represents the dimensionless adiabatic temperature rise associated with complete combustion, and is given by $Eqc_o/\sigma C_vRT_a^2$. The term (E/RT_a^2) is used also in this case to non-dimensionalise the adiabatic temperature excess. The characteristic times are the Newtonian cooling time, $t_N=\sigma C_vV/hS$, and the chemical time at the ambient temperature, $t_{ch}=$

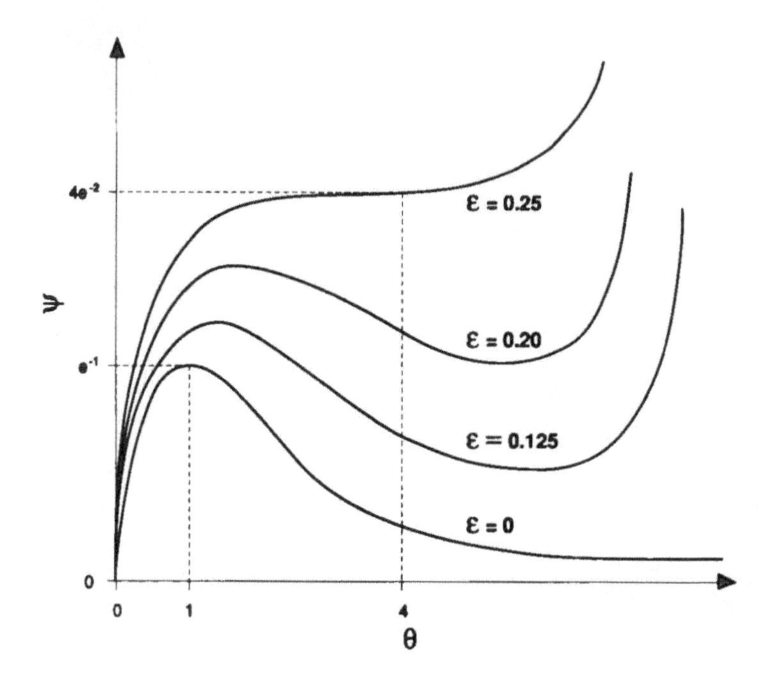

Figure 8.7 $\psi(\theta)$ derived from the dimensionless Semenov equation for thermal ignition (eqns (8.20) and (8.22)). There are 0 or 2 solutions when the exponential approximation is made (eqn (8.20)) and 1 or 3 solutions when no approximation is made (eqn (8.22)). There is only one solution at $\varepsilon > 0.25$.

$1/A\exp(-E/RT_a)$, which is the reciprocal of the rate constant evaluated at T_a. For reactions of order greater than unity this would have to be defined in terms of a suitable pseudo first order rate constant. The observed time-dependent behaviour is reflected strongly in the physical properties of the system that are incorporated in the Semenov parameter $\psi(=Bt_N/t_{ch})$. Writing the energy conservation equation (eqn (8.4)) as

$$t_N \mathrm{d}\theta/\mathrm{d}t = (Bt_N/t_{ch})\exp(\theta) - \theta \qquad (8.24)$$

on integration gives a representation of θ as $f(t)$, with a maximum subcritical rise at $\theta = 1$. The qualitative shape is the same as the $\Delta T - t$ profiles shown in Fig. 8.3, However, the timescale on which reaction develops is governed by t_N in particular, which is proportional to the reactant density. Since the product Bt_N, and therefore the right-hand side of eqn (8.24) is independent of density, the temperature gradient, $\mathrm{d}\theta/\mathrm{d}t$, is affected by the density dependence of t_N. This leads to quite different timescales for reaction in gases when compared with those in solids since $t_N(\mathrm{s}) \gg t_N(\mathrm{g})$. Laboratory experiments of thermal ignition in gases ($V \approx 0.5\,\mathrm{dm}^3$) last no more than a few seconds. Packed powders or fibres at similar scale and oven temperature require several hours to fully react.

Large stockpiles of materials, such as haystacks, can survive for many weeks or months before evidence of spontaneous ignition emerges. A greater density than that adopted in laboratory experiments may be contributory but the major control is exercised by the much lower ambient temperatures, which increase t_{ch} considerably.

8.4 Frank-Kamenetskii theory for internal distributed temperatures

The Semenov theory provides a qualitative understanding of the nature of critical conditions for explosion. Subject to the constraints of the exponential approximation and the assumption of no reactant consumption, it leads to solutions for the maximum subcritical temperature rise and for the way that the critical concentration varies under conditions of a spatially uniform internal temperature. This is satisfactory if all resistance to heat transport is at the walls. At another extreme, heat is transported with relatively low thermal conductivity within the reactant. That is, all resistance to heat transport exists within the reactants. This is illustrated in Fig. 8.1b, and has been treated theoretically by Frank-Kamenetskii [39]. In the stationary state, the differential heat balance may be written

$$\kappa\nabla^2 T + qf(c)A\exp(-E/RT) = 0 \qquad (8.25)$$

with the boundary conditions $T = T_a$ at the wall and, for a symmetrical vessel, $dT/dr = 0$ at the centre. ∇^2 is the Laplacian operator $(\partial^2/\partial x^2 + \partial^2/\partial y^2 + \partial^2/\partial z^2)$ in three dimensions and q is the exothermicity of the reaction. Just as there is a limiting value for $\psi(\varepsilon \to 0)$, solutions to this equation exist only when the Frank- Kamenetskii (dimensionless) parameter, δ, is less than a certain value. Solutions in the form of stable temperature profiles $T = f(r)$ are then possible and, under these circumstances, steady reaction takes place with a certain temperature distribution defined by $f(r)$ in the reaction medium. Above the critical value of δ, no stable temperature profile is possible and thermal ignition occurs.

The Frank-Kamenetskii parameter, δ, is also a 'dimensionless heat release rate' appropriate to the temperature of the surroundings and its value is given by

$$\delta = \frac{qEr^2A}{\kappa RT^2}\exp(-E/RT_a)f(c) \qquad (8.26)$$

There is a similarity to the Semenov parameter, ψ, but whereas ψ depends directly on the characteristic dimension (given by V/S), δ depends on the square of the characteristic dimension (r). The corresponding dimensionless energy conservation equation, expressed in terms of θ is

$$\delta\exp(\theta/(1+\varepsilon\theta)) - \theta = 0 \qquad (8.27)$$

Solutions to this equation have very similar properties to that under Semenov conditions (eqn (8.22)), or its counterpart in terms of $\exp^{\theta}$. It is easiest to focus attention on the behaviour at the centre of the reacting system; thus Frank-Kamenetskii theory leads to an expression for the maximum subcritical temperature rise, ΔT_c, at the centre of the vessel, which differs from ΔT predicted by Semenov theory only by a numerical factor of order unity. The temperature profile in the subcritical case is close to, but not exactly, parabolic. The value of δ at the spontaneous ignition limit (δ_c) and the corresponding values of $(\Delta T_c)_{r=0}(=(T_c - T_a)_{r=0})$ depend on the geometry of the system. The results for three simple cases are given in Table 8.1. The two treatments become identical if

$$\psi_c = \delta_c \kappa / hr \tag{8.28}$$

which can be rearranged to

$$h = b(\kappa / r) \tag{8.29}$$

where the Biot number, b, is 2.39, 2.72 and 3.01 for a slab, cylinder and sphere, respectively [147]. Each of these shapes have only one characteristic dimension (half-thickness or radius), so the Laplacian in eqn (8.25) can be written in one dimension [143, 147].

The Frank-Kamenetskii analysis retains a number of features implicit in the Semenov theory. No allowance is made for consumption of reactants [150] or for dealing with vessels of complex geometry, although an adaptation to many different shapes was developed by Boddington *et al.* [151] and has been tested experimentally [152]. Heat transfer is taken to occur only by conduction whereas, in gases at least, convective heat transfer is important under some conditions. In experiments with gases in closed vessels, convection begins when the Rayleigh number (*Ra*) exceeds 600. Convection is negligible at pressures below about 10 kPa, but above this pressure heat transfer is enhanced (and explosion correspondingly hindered) by convection [153]. The Rayleigh number is a dimensionless parameter defined as follows

$$Ra = \frac{g \alpha r^3 \sigma^2 C_p \Delta T}{\kappa \eta} \tag{8.30}$$

Table 8.1 Values of δ_c, ΔT_c and the characteristic dimension for various vessel geometries according to Frank–Kamenetskii theory for thermal ignition [39,143]

Shape of vessel	δ_c	ΔT_c (at centre)	Characteristic dimension, r
Infinite slab	0.88	1.20 RT_a^2/E	Half-width of slab
Infinite cylinder	2.00	1.39 RT_a^2/E	Radius of cylinder
Sphere	3.32	1.60 RT_a^2/E	Radius of sphere

where g is the gravitational acceleration, α is the volume coefficient of thermal expansion and η is the dynamic viscosity.

8.5 Combustion of particulate and fibrous materials in storage or transport

Whichever treatment is appropriate for any particular system, governed by the temperature, density and geometry, there is a critical dimension below which ignition cannot occur, but there is a potential combustion hazard (i.e. 'thermal runaway') when materials that are capable of undergoing exothermic oxidation are stored or transported in bulk. The chemical industry has its own protocols and procedures for assessing hazards associated with 'runaway reactions' during industrial processing [154].

Problems for materials stored at ambient temperature are extremely common and each year enormous losses are sustained world-wide. The common casualties are coal or peat, cellulosic materials (wood chippings, grain, hay, bagasse, animal feedstocks), but inert materials that are contaminated with organic solvents or oily substances are also vulnerable and can be the source of fires in industrial environments. Each of these requires the intervention of oxygen, so they must be porous. Thermal runaway can also occur in substances that can undergo exothermic decomposition, such as fertilisers. Bowes has reviewed this field extensively [147].

Stockpiles of military explosives and propellants present particular difficulties because they are kept for such a long time (50 years is not uncommon) and in enormous quantities. Nitrocellulose and nitroglycerine are the main constituents. Their decomposition cannot be avoided and the nitrogen oxides released can promote exothermic reaction quite readily within the propellant structure. Stabilisers, such as diphenylamine, are added during manufacture in order to inhibit further reaction by selective removal of the nitrogen oxides. The build-up over a period of years of nitrated derivatives of diphenylamine can be measured, and modern methods of surveillance of the safe 'shelf-life' of propellants involve the quantitative determination of these derivatives by chromatographic methods. Once full nitration of the stabiliser is approached, the stores must be destroyed in a controlled manner.

The more general problem arises from the failure to recognise that stockpiles of material which are capable of undergoing exothermic oxidation cannot be increased in size indefinitely without promoting 'thermal runaway'. That the seat of ignition is deep within the stockpile is part of the difficulty, but the preignition stages, i.e. self-heating during the induction time (or ignition delay), seem to be so innocuous. Since the critical temperature excess above ambient is typically only 20 K (eqn (8.1) and Table 8.1), even a stockpile that appears to be 'getting a bit warm' may already be in a supercritical state. It is then only a matter of time before fire breaks out.

One way of predicting the potential combustion hazard of materials when stored in large amounts, which obviates the need to have fundamental knowledge of kinetic and thermochemical data, is to establish the critical temperature of different amounts of the material in laboratory experiments and then to extrapolate the data to the industrial scale. Since in most circumstances conductive heat transport would be most likely, a scaling relationship may be derived from the Frank-Kamenentskii equation (eqn (8.26)). This gives the scaling as a function of the square of the characteristic dimension r in the form

$$\ln\frac{(\delta_c T_{a,c}^2)}{(\sigma r^2)} = \ln\frac{(qAE)}{(\kappa R)} - \frac{E}{RT_{a,c}} \tag{8.31}$$

The concentration term in eqn (8.26) has been replaced in eqn (8.31) by the density, σ, of the solid material. The subscript c signifies the values at criticality. The quotient in the first term on the right hand side may be regarded to be a constant if there is no change of kinetics and thermochemistry over the temperature range of interest. The terms δ_c and σ could also be included in this term if the large-scale shape and density are the same as those of the laboratory experiments. The extrapolation to large scale may be deduced by the linear relationship $\ln(\delta_c T_{a,c}^2/\sigma r^2)$ versus $1/T_{a,c}$. Using the geometric relationships [152], the critical size may then be obtained in terms of the characteristic dimension r at the appropriate value of δ_c for a particular stockpiled shape (conical, cubic, etc.), given a known storage temperature T_a.

A qualification is that the chemistry associated with many of the materials that may undergo spontaneous ignition is unlikely to be represented by a single reaction, although (to be consistent with thermal ignition theory) it may be possible to specify an overall activation energy and pre-exponential factor to represent the kinetics over a certain temperature range. However, the kinetics and mechanism may change throughout the entire temperature range of interest, so quite detailed information may be required to be able to predict the potential hazard. Microcalorimetry is a technique which may be used to obtain the appropriate data within specified temperature ranges. Also, we should be mindful that heterogeneous reactions may not be represented satisfactorily by the conventional kinetic laws [155].

Finally, water that is present within a stockpile of material may contribute to stability. Heat is generated in lawn-grass cuttings, for example, by microbial activity. Evaporation of water from the hottest regions constitutes a 'heat sink' through the enthalpy of vaporisation, and the movement of the vapour through the pile to colder regions or as total loss to the environment enhances the heat transfer properties considerably. Fires do occur in haystacks, which are not only much larger in size but also very much drier. Water also plays a part in the development of events but it is a much more complicated involvement [156].

8.6 Asymmetric heating and initiation by hot spots

The theories discussed above relate to uniformly heated reactants. There are many circumstances in which asymmetric heating occurs, such as in a reactant adjacent to a hot surface. In analytical interpretations of these circumstances, derived from Frank-Kamenetskii theory, the distinction is made between criticality at a constant surface temperature or with a constant heat flux at the surface [147]. Although related, the spontaneous ignition of gases at a hot surface is much more complicated because convection effects that move reactants away from the hot zone mean that the residence time of the gas in the vicinity of the surface has some bearing on whether or not initiation of ignition occurs. This is not amenable to a stationary state treatment.

Combustion can be initiated also in a large mass of material by the formation of a local hot spot. This may be brought about in a variety of ways, often mechanical in nature. For example, frictional heating, adiabatic compression of trapped gas bubbles and shock compression can all transform the mechanical disturbance into a small zone of increased temperature. Local heating from an electrical source is also possible. Provided that the size of the region and the temperature achieved are together sufficient to sustain self-heating, the hot spot necessary for ignition will have been created. The size of the heated volume and temperature within the hot spot that is necessary to cause ignition depends mainly on the reactivity of the material. There is an important distinction between a constant heat flux at the source and a momentary or short duration energy input [147].

Further reading

Bowes, P.C. (1984). *Self-Heating: Evaluating and Controlling the Hazard.* HMSO Books, London, UK.

Frank-Kamenetskii, D.A. (1969). *Diffusion and Heat Transfer in Chemical Kinetics* (2nd edn) (translated by J.P. Appleton). Plenum Press, New York, USA.

Gray, P. and Lee P.R. (1968). Thermal explosion theory. In *Oxidation and Combustion Reviews* (vol. 2) (ed. C.F.H. Tipper). Elsevier, Amsterdam, The Netherlands.

Merzhanov, A.G. and Averson, A.E. (1971). *Combust. Flame*, **16**, 89.

Mulcahy, M.F.R., (1973). *Gas Kinetics.* Nelson, London, UK.

Problems

(1) Calculate the critical temperature excess of methyl nitrate vapour in a well-stirred reaction vessel maintained at 520 K. Compare this with the critical temperature excess at the centre of a spherical reaction vessel at 520K when heat transfer occurs by conduction. The activation energy for CH_3ONO_2 decomposition is 151 kJ mol^{-1}.

(2) Show that

$$(1-4RT_a/E)^{1/2} = 1-2(RT_a/E)-2(RT_a/E)^2-2(RT_a/E)^3 \ldots$$

Hence, by neglecting terms of order >2, derive eqn (8.10) from eqn (8.9). Is the approximation valid for methyl nitrate decomposition at 520 K?

(3) Show that

$$(1+RT_a/E)^{-1}=1-(RT_a/E)+(RT_a/E)^2-(RT_a/E)^3\ldots.$$

Hence, by neglecting terms of order > 1, show that $\exp(-E/RT_c)=\exp(-E/RT_a)\times e$. The substitution $T_c=T_a+RT_a^2/E$ is required. This is the derivation of eqn (8.11).

(4) Obtain an expression for T from $\theta=(T-T_a)/RT_a^2$. Hence, making the substitution $\varepsilon=RT_a/E$, show that

$$\exp(-E/RT)=\exp(-E/RT_a)\times\exp(\theta/1+\varepsilon\theta)$$

(5) If the critical vapour pressure for the first order decomposition of di-*t*-butyl peroxide vapour at 500 K in a well-stirred spherical vessel (0.5 dm^3) is 0.67 kPa, deduce the critical vapour pressure in (a) a 1.0 dm^3, sphere, and (b) a cylindrical vessel (10 cm × 4 cm dia.). Assume that the heat transfer coefficient is the same in each vessel.

(6) The Texas City disaster occurred because bags filled with hot, damp ammonium nitrate fertiliser were packed tightly together in a ship's hold. Given a rate constant for the decomposition of ammonium nitrate $k=6\times10^{13}\exp(-20450/T)\,s^{-1}$ and $\Delta H=-378\,kJ\,mol^{-1}$, $\sigma=1750\,kg\,m^{-3}$ and $\kappa=0.126\,W\,m^{-1}\,K^{-1}$, assuming a spherical reacting mass use the Frank–Kamenetskii theory to estimate the critical radius for self-ignition at 25°C and 100°C, respectively.

(7) Calculate the critical internal temperature excess at the centre of the stockpile of ammonium nitrate, assuming that it is stored at 25°C.

(8) The decomposition of diethylperoxide has been studied by Fine *et al.* [157]. In a spherical reactor at 501.5 K the explosion limit of pure diethylperoxide vapour was 6.1 torr, and in experiments in which slow reaction, but not explosion, occurred the temperature rise at the centre of the vessel never exceeded 20 K. Are these results consistent with explosions occurring by a purely thermal mechanism?

$k=10^{14.2}\exp(-17200/T)\,s^{-1}$, $\kappa=0.027\;W\,m^{-1}\,K^{-1}$,

$\Delta H=-197\,kJ\,mol^{-1}$, $r=60.6\,mm$

Isothermal chain branching and chain–thermal interactions 9

9.1 Introduction

The heat release from most combustion processes originates in a complex chemical mechanism, which includes chain branching. Thus an understanding of how free radical chain branching systems can behave underpins the interpretation of much combustion behaviour.

9.2 Kinetics and mechanisms of chain branching reactions

The main feature of a chain branching (or autocatalytic) reaction is an exponential growth in reaction rate as a result of the multiplication of the primary propagating species. These species are very reactive and are usually atoms or free radicals. Four different types of events can be identified in the overall process, as follows.

Initiation. Atoms or radicals are often produced by the dissociation of either a reactant molecule or some substance (an initiator) added specifically to promote reaction. The rate of initiation is slow because the activation energy of a unimolecular dissociation reaction is equal to, or greater than, the bond dissociation energy, say 200–500 kJ mol^{-1}. Bimolecular initiation reactions (between fuel and oxygen, for example) tend to have lower activation energies but also lower pre-exponential factors, and hence are not necessarily significantly faster. Initiation may take place homogeneously, or heterogeneously at the walls of a reaction vessel. Formally, we may write

$$A \rightarrow 2X \qquad \text{rate, } v_i = k_i[A] \tag{9.1}$$

where A denotes an intial reactant and X an active species (or free radical).

Propagation. The propagation reaction is important because it governs the rate at which the chain continues. The requirement is for a free radical X to react with a molecule producing a second free radical Y, for example,

$$X + A \rightarrow Y + P \qquad \text{rate, } v_p = k_p[X][A] \tag{9.2}$$

where A is a molecule (normally a reactant) and P is a stable product. It is unlikely that X and Y will be identical so the chain normally propagates by a 'shuttle' between two or more types of free radicals

$$Y + A \rightarrow X + P \tag{9.3}$$

The reactant and product molecules in eqns (9.2) and (9.3) may not be the same in practice. For most propagation reactions of importance in combustion, activation energies lie between zero and 60 kJ mol^{-1}.

Branching. In a chain branching reaction three free radicals are formed from the reaction of one radical (i.e. a net gain of two).

$$X + A \rightarrow 3Y + \ldots. \qquad \text{rate, } v_b = k_b[X][A] \tag{9.4}$$

This reaction may be responsible for a ***branching chain explosion*** and is termed ***quadratic branching*** because there is an overall second order dependence of the reaction rate on species concentration. The activation energy for the branching process may be appreciably higher than that of the ordinary propagation reactions. ***Linear branching*** involves decomposition of a stable intermediate product to give free radicals

$$P \rightarrow 2X + \ldots. \qquad v_b = k_b[P] \tag{9.5}$$

This is a relatively slow process, as discussed in Chapter 7, and it is termed ***degenerate branching*** (or secondary initiation).

Termination. A combination of the above processes would cause the overall reaction rate to accelerate without limit were it not for competition for the free radicals by reactions which terminate the chains. Gas-phase termination occurs either by recombination of two radicals to give a stable molecule, or by reaction of a radical with a molecule to give either a molecular species or a radical of lower reactivity which is unable to readily propagate the chain. Since each of these processes is exothermic, a ***third body*** is often required to take up the energy released and prevent redissociation, as discussed in Chapters 2 and 6,

$$X + X + M \rightarrow X_2 + M \qquad \text{rate, } v_{t1} = k_{t1}[X]^2[M] \tag{9.6}$$

$$X + S + M \rightarrow XS + M \qquad \text{rate, } v_{t2} = k_{t2}[X][S][M] \tag{9.7}$$

Removal of radicals at the wall can take place by a variety of processes, the details of which are usually unimportant since the rate-controlling step is normally diffusion through the gas

$$X \rightarrow 1/2X_2 \qquad \text{rate, } v_{t3} = k_{t3}[X] \tag{9.8}$$

Equations (9.7) and (9.8) are referred to as linear termination (with respect to [X]) and eqn (9.6) as quadratic termination. The rate constant k_{t3} is rather different from the previous rate coefficients and should take the form $D\nabla^2[X]$ where D is the diffusion coefficient and ∇^2 is the Laplacian operator introduced in Chapter 8. The significance is that k_{t3} contains quantities to allow for the vessel geometry. As is normal in isothermal kinetics in general, spatial uniformity of species concentrations are assumed, even though the loss of species at the vessel wall would imply the existence of concentration gradients.

9.2.1 A formal analysis of isothermal chain branching

Consider the skeleton scheme:

Scheme I

Initiation	$A \rightarrow X$	rate, $v_i = k_i[A]$	(9.9)
Branching	$A + X \rightarrow mX + \text{product}$	rate, $v_b = k_b[A][X]$	(9.10)
Termination	$X \rightarrow \text{product}$	rate, $v_{t3} = k_{t3}[X]$	(9.8)

The conservation of species X is given by the equation

$$d[X]/dt = k_i[A] + (m-1)k_b[A][X] - k_{t3}[X] \tag{9.11}$$

To proceed analytically the 'pool chemical' approximation, $[A] = [A]_o$, must be made.

$$\int_o^{[X]} \frac{d[X]}{k_i[A]_o + ((m-1)k_b[A]_o - k_{t3})[X]} = \int_o^{[X]} \frac{d[X]}{k_i[A]_o + \phi[X]} = \int_o^t dt \tag{9.12}$$

where ϕ is termed the net branching factor and is defined as

$$\phi = (m-1)k_b[A]_o - k_{t3} \tag{9.13}$$

Integration of eqn (9.12), $\phi \neq 0$, gives

$$[X] = \frac{k_i[A]_o}{\phi}(\exp^{\phi t} - 1) \tag{9.14}$$

There are three possibilities which arise with respect to the branching factor, as follows:

- Case (i): $\phi < 0$, $(k_{t3} > (m-1)k_b[A]_o)$

  $$[X] \rightarrow k_i[A]_o/\phi \text{ at } t \rightarrow \infty \tag{9.15}$$

  and, in the limit, is the stationary state solution to eqn (9.11).

- Case (ii): $\phi > 0$, $((m-1)k_b[A]_0 > k_{t3})$

$$[X] \rightarrow \infty \tag{9.16}$$

i.e. the concentration grows without limit in finite time.

- Case (iii): $\phi = 0$, $((m-1)k_b[A]_0 = k_{t3})$

Integration of eqn (9.12), in the form

$$d[X]/dt = k_i[A] \tag{9.17}$$

yields the linear relationship

$$[X] = k_i[A]t \tag{9.18}$$

The condition $\phi = 0$ (eqn (9.13)) represents the critical boundary between the attainment of a stationary state and an explosion (Fig. 9.1), such that the conditions for an isothermal branched chain explosion boundary, expressed in terms of reactant pressure (p) and the vessel temperature (T_a), can be obtained from it.

Assuming the rate parameters $k_b = A_b\exp(-E_b/RT)$ and $k_{t3} = A_{t3}$, and ideal gas behaviour such that $[A]_0 = p/RT$, the critical pressure at the vessel temperature T_a is given by

$$p_c = \frac{A_{t3}RT_a}{(m-1)A_b}\exp(E_b/RT_a) \tag{9.19}$$

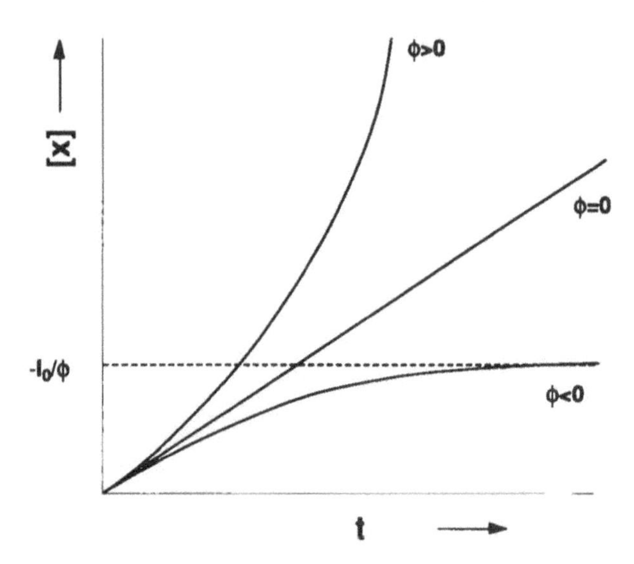

Figure 9.1 The growth of chain carrier concentration, [X], as a function of time that is predicted by an isothermal chain branching scheme with a linear termination reaction at different values of the net branching factor ϕ.

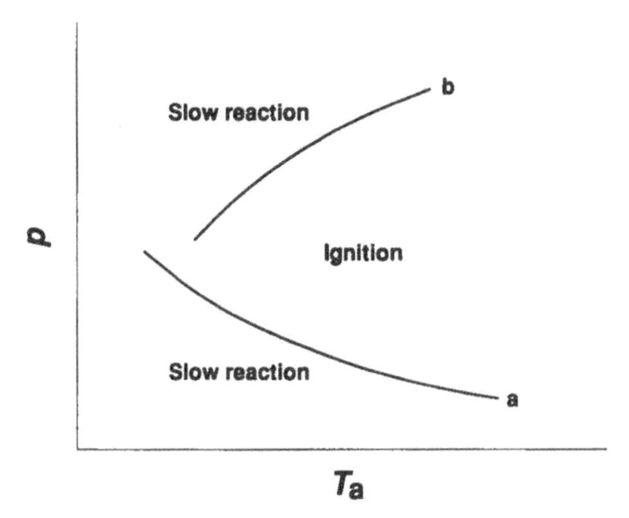

Figure 9.2 The $(p\text{–}T_a)$ ignition boundaries for a gaseous reaction in a closed vessel that are predicted analytically from isothermal chain branching schemes, (a) Scheme I, and (b) Scheme II.

and is illustrated in Fig. 9.2 (curve a). The transition to ignition is brought about by an increase in pressure because there is a higher concentration dependence associated with the branching term than the termination term in the net branching factor.

Suppose that, rather than eqn (9.8) the termination reaction corresponds in form to that in eqn (9.7), in which the molecular species S and the third body M are assumed to be the same species, so that

$$X+M+M \rightarrow XM+M \qquad \text{rate, } v_{t2}=k_{t2}[X][M]^2 \qquad (9.20)$$

Scheme II

Initiation	$A \rightarrow X$	rate, $v_i=k_i[A]$	(9.9)
Branching	$A+X \rightarrow mX+\text{product}$	rate, $v_b=k_b[A][X]$	(9.10)
Termination	$X+M+M \rightarrow \text{product}+M$	rate, $v_{t2}=k_{t2}[X][M]^2$	(9.20)

The net branching factor given by Scheme II is

$$\phi=((m-1)k_b[A]_0-k_{t2}[M]^2) \qquad (9.21)$$

and is equal to zero at criticality, as previously. In order to interpret the form of the $(p\text{–}T_a)$ ignition boundary, suppose that the initial mole fraction of A is x_A, so that its concentration is $x_A[M]$. Thus, the ignition criterion is given by

$$(m-1)k_b x_A[M]-k_{t2}[M]^2=0 \qquad (9.22)$$

assuming ideal gas behaviour $p/RT_a = [M]$, and that $k_{t2} = A_{t2}$, p_c is given by

$$p_c = (m-1)x_A RT_a(A_b/A_{t2})\exp(-E_b/RT_a) \qquad (9.23)$$

Provided that $E_b \gg 0$, the form of the boundary associated with this formal structure corresponds to that shown in Fig. 9.2 (curve b). Only a very weak temperature dependence would be exhibited at the boundary for low values of E_b. The transition to ignition is brought about by a *decrease* in pressure, which is a consequence of the higher order concentration dependence of the termination rate compared with that of chain branching. These are the characteristics of the *second ignition limit* in the reactions of H_2+O_2 and 'wet' $CO+O_2$, which are understood in almost every kinetic detail (sections 9.5 and 9.6). They are also exhibited by other combustion systems, such as SiH_4+O_2, PH_3+O_2, CS_2+O_2 or $P+O_2$, but the kinetics of these systems are far less well characterised. Nevertheless, the foregoing generalisations must form the underlying kinetic structure. On a historical note, van't Hoff [139] recognised that the 'second limit characteristics' of $P+O_2$ could not originate in a thermal instability, though of course the basis of an understanding in branching chain theory did not emerge until 30 years later.

A number of simplifying assumptions and approximations have been made in order to illustrate the principles of isothermal chain branching ignition. Perhaps the most severe kinetic constraint is that the free radical concentration is permitted to grow without limit. As discussed in Chapters 6 and 7, the quadratic recombination of X is inevitable which, disregarding third body effects, takes the form

$$X+X \rightarrow \text{product} \qquad \text{rate, } v_{t1} = k_{t1}[X]^2 \qquad (9.6)$$

Thus, Scheme III may be taken as a more realistic skeleton approach to combustion kinetics.

Scheme III

Initiation	$A \rightarrow X$	rate, $v_i = k_i[A]$	(9.9)
Branching	$A+X \rightarrow mX + \text{product}$	rate, $v_b = k_b[A][X]$	(9.10)
Termination	$X \rightarrow \text{product}$	rate, $v_{t3} = k_{t1}[X]$	(9.8)
Termination	$X+X \rightarrow \text{product}$	rate, $v_{t1} = k_{t1}[X]^2$	(9.6)

Invoking the 'pool chemical' approximation, the mass conservation equation for X is

$$d[X]/dt = k_i[A]_0 + (m-1)k_b[A]_0[X] - k_{t3}[X] - k_{t1}[X]^2$$

or

$$d[X]/dt = k_i[A]_0 + \phi[X] - k_{t1}[X]^2 \qquad (9.24)$$

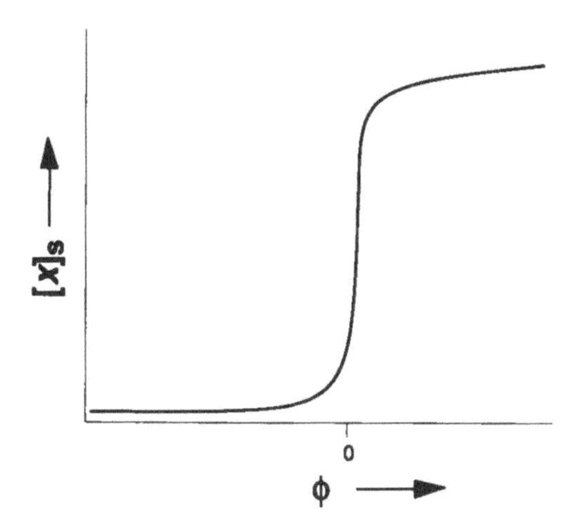

Figure 9.3 The dependence of the stationary state chain carrier concentration, $[X]_s$, on the net branching factor, ϕ, that is predicted by Scheme III.

This equation cannot be integrated analytically, but the stationary state solution to eqn (9.24) is a quadratic, which always has a positive root, given by

$$2k_{t1}[X]_s = \phi + (\phi^2 + 4k_i k_{t1}[A]_0)^{1/2} \tag{9.25}$$

The dependence of $[X]_s$ on ϕ is shown in Fig. 9.3. Although the ignition boundary is no longer marked by a failure to obtain stationary state solutions to the mass balance equation, there is a very high parametric sensitivity of the concentration of X in the vicinity of $\phi = 0$, which may still be interpreted as a critical boundary between slow reaction and ignition.

9.3 Quadratic chain branching with reactant consumption

The analytical derivation of the criteria for ignition has been developed on the assumption that the extent of consumption of the reactant is negligible during 'slow reaction'. There are often circumstances, especially with regard to the degenerate chain branching mechanisms associated with the low temperature oxidation of hydrocarbons, in which this assumption is not valid because the exponential growth rate of the reaction is relatively slow and other reactant consuming processes occur in parallel. It is possible to investigate such behaviour in a closed system in a formal way with chemical autocatalysis described in the form

$$A + X \rightarrow 2X \qquad \text{rate, } v_b = k_b[A][X] \tag{9.10}$$

For reaction to begin, either eqn (9.11) is preceded by an initiation step of the form

$$A \rightarrow X \tag{9.9}$$

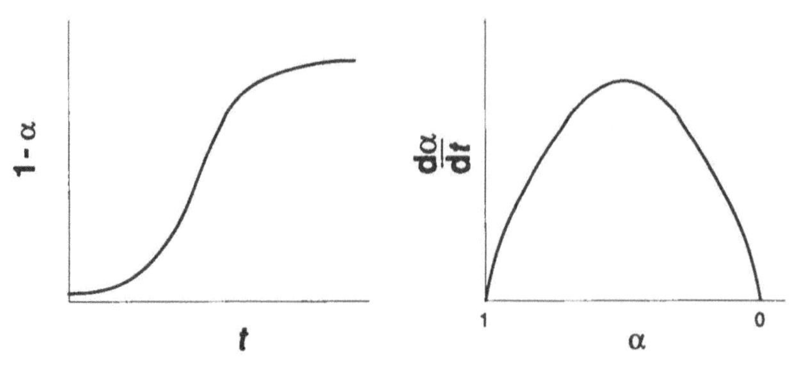

Figure 9.4 The extent of reaction as a function of time and reaction rate as a function of the fraction of remaining reactant in an isothermal quadratic branching reaction.

or a small amount of the catalyst X is added to initiate the reaction. Mass conservation at any time yields the relationship

$$[A]_o + [X]_o = [A] + [X] \tag{9.26}$$

The rate expression then can be expressed in the form

$$v = -d[A]/dt = k_b[A](([A]_o - [A]) + [X]_o) \tag{9.27}$$

and to a good approximation, when $[A]_o \gg [X]_o$

$$-d[A]/dt = k_b[A]([A]_o - [A]) \tag{9.28}$$

According to this rate law the isothermal reaction reaches its maximum rate at 50% consumption of the reactant A, as has been shown experimentally [158, 159] during the oxidation of alkanes in very fuel-rich conditions (Fig. 9.4).

9.4 Relationships between thermal and isothermal branching chain ignition theories

Although isothermal chain branching may seem to be remote from spontaneous combustion processes in which heat release is inevitable, there may be a negligible temperature change during the course of slow reaction region. If this is maintained up to the conditions at which $\phi = 0$ then the isothermal branching criteria are reasonably satisfied. The branching explosion would be strongly augmented by thermal feedback from the heat release rate in supercritical conditions, and this thermal contribution may resolve any doubt about the existence of a critical ignition boundary between slow reaction and ignition under the more realistic kinetic conditions set out in Scheme III.

The simple formal structures of Semenov thermal theory and the isothermal branching chain kinetics of Scheme I have much in common. Foremost is that the predicted (p–T_a) ignition boundaries are similar in form, which causes a particular difficulty in distinguishing thermal ignition from chain branched ignition by investigation of the ignition limits alone. The similarity arises from a non-linear feedback, manifest either as the exponential temperature dependence of the rate constant in eqn (8.3) or as the species multiplication in eqn (9.10). There is also a competing linear loss term, manifest either as the Newtonian heat transport in eqn (8.1) or as the radical termination in eqn (9.8) of Scheme I.

The two may be distinguished experimentally from temperature measurements within the reactants. If little or no temperature change is detected (by a thermocouple) during slow reaction, especially at conditions close to the ignition boundary, then the ignition probably originates predominantly in chain branching criticality. They may also be distinguished by numerical computation. It is comparatively easy to set up thermokinetic models and to integrate the appropriate differential equations for mass and energy. The distinction between isothermal chain branching and chain branching augmented by thermal feedback can then be investigated by 'switching off' the thermal contributions in the model [160].

If the (p–T_a) boundary takes the form exhibited by Scheme II, then the occurrence of a complex chain branching structure is beyond doubt since these are not the features of thermal feedback alone. This does not rule out the intervention of non-isothermal effects supplementing the chemical autocatalysis, but thermal feedback cannot be a primary cause.

Gray [161] derived analytical criteria which could be used to represent the two limiting possibilities, a chain branching reaction perturbed by self-heating, or a thermal reaction perturbed by chain branching. The relationships between thermal, chain branching and chain–thermal criticality are summarised in Fig. 9.5 [161], from which it is seen that both the thermal and the isothermal chain branching criteria are supercritical

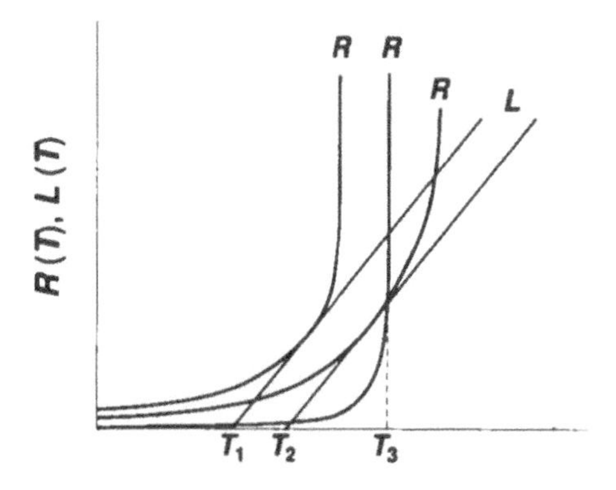

Figure 9.5 A qualitative relationship between the critical vessel temperatures (T_1, T_2 and T_3) determined by the heat release and heat loss rates (*R* and *L*), that are predicted by chain–thermal, thermal and isothermal theories, respectively (after Gray [161]).

with respect to the conditions for chain–thermal criticality. As an approximate prediction for stable operation neither err on the safe side. Quadratic chain branching with reactant consumption (section 9.3) can also be regarded in a non-isothermal context, the mathematical treatment of which is a two-variable problem with respect to [A] and T [159].

9.5 Spontaneous ignition of hydrogen + oxygen

9.5.1 $(p-T_a)$ *ignition boundaries and their kinetic origins*

The observed form of the $(p–T_a)$ limits are represented in Fig. 9.6 for a hydrogen + oxygen mixture. The qualitative features also apply to different compositions and to the same composition in a different size of vessel. At a given vessel temperature within the range approximately 675 to 825 K there are three critical pressures. These are termed the *first* (*or lower*) *limit*, above which ignition takes place, the *second* (*or upper*) *limit*, below which explosion takes place, and also the *third limit* which exists at still higher pressures, and above which explosion is again observed. Identification of the third limit is sometimes difficult because quite vigorous exothermic reaction occurs even in the slow reaction region. These characteristic features and their kinetic origins are very well

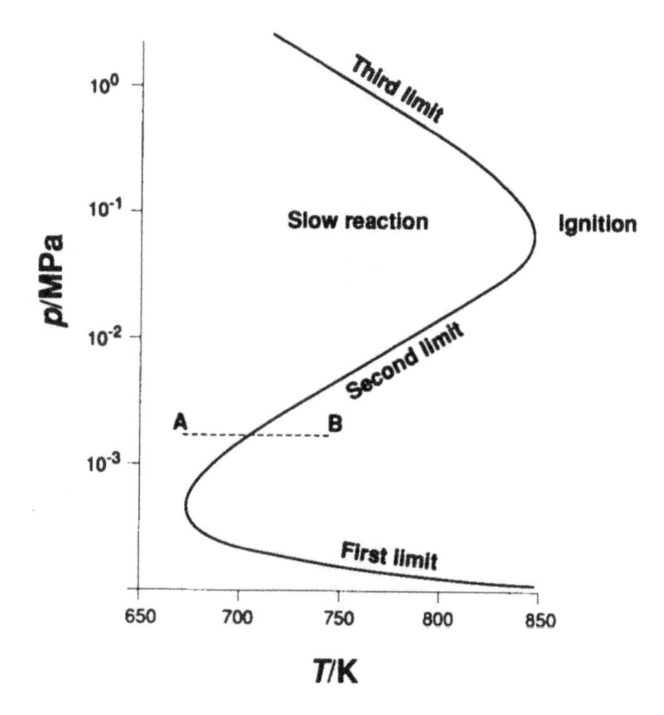

Figure 9.6 $(p–T_a)$ ignition boundaries for the $H_2 + O_2$ reaction in a closed vessel. The reactant pressure is plotted logarithmically. The line AB relates to behaviour in a CSTR.

established [2], and are summarised here as illustration of the foregoing discussion.

Kinetic interpretations of the slow oxidation and the attainment of criticality at the first and second limits have been the subject of experimental study since 1928 [162–164], and the initial understandings of chain branching reactions emerged from such work. Much of the present quantitative understanding of the oxidation at low pressures and temperatures is due to Baldwin and co-workers [165].

Experimental investigations of the spontaneous oxidation of hydrogen at low pressures are susceptible to the nature of the vessel surface, which is normally silica treated by a variety of methods. In the vicinity of the first limit, the rate-determining step probably involves principally the adsorption of H atoms followed by recombination to give stable molecules which are then desorbed. Treatment of the walls by surface coating or by 'ageing' (the performance of successive experiments in the same vessel) modifies the efficiency of the chain breaking step, which can cause up to a 100-fold variation in the pressure at the first explosion limit. KOH, CsCl, Al_2O_3, and K_2HPO_4 are highly efficient at removing chain carriers. By contrast, B_2O_3 coated silica and acid-washed Pyrex surfaces are very inefficient, whereas those coated with KCl, NaCl, $BaCl_2$ and $Na_2B_4O_7$ are of intermediate efficiency [23, 166].

At higher pressures, the kinetics of reaction depend on the fates of the HO_2 radical and of H_2O_2. HO_2 is not destroyed at some surfaces, whilst on others it reacts to give oxygen and water. A corresponding situation occurs for H_2O_2 formed by gas-phase reactions of HO_2. The situation is further complicated by the possibility that certain surfaces favour the decomposition of H_2O_2 to free radicals, which are then released into the gas-phase [167].

The first limit. For a given surface activity, the first limit moves to lower pressures if the size of the vessel is increased or inert gases are added. This is characteristic of a homogeneous chain-branching reaction in competition with a termination which takes place on the vessel walls. The limit is also strongly dependent on the molecular oxygen concentration. The primary kinetic control at the first limit originates in competitive reactions involving H atoms, since not only do they react with oxygen but also they diffuse more rapidly than any other species present. That is,

$$H + O_2 \rightarrow OH + O \tag{9.29}$$

$$H \rightarrow \text{inert (by diffusion to wall)} \tag{9.30}$$

When the supplementary reactions

$$O + H_2 \rightarrow OH + H \tag{9.31}$$

and

$$OH + H_2 \rightarrow H_2O + H \tag{9.32}$$

are included, to give the following stoichiometric equation for autocatalysis:

$$H + 3H_2 + O_2 = 3H + 2H_2O \tag{9.33}$$

the kinetic structure matches that expressed in a formal skeleton way in Scheme I. The branching factor is represented by

$$\phi = (2k_{9.29}[O_2] - k_{9.30}) \tag{9.34}$$

in which the factor 2 arises from the overall net gain of two H atoms in the autocatalytic cycle. The temperature dependence of the critical reactant pressure at the first limit is given by

$$x_{ox}p_c = 0.5RT_a(A_{9.30}/A_{9.29})\exp(8455/T_a) \tag{9.35}$$

where x_{ox} represents the initial mole fraction of O_2. Equation (9.35) signifies that the critical pressure (p_c) decreases as the vessel temperature is raised.

The second limit. Whereas the first limit exists typically at pressures of a few Pa (or <20 torr, say), the second limit is observed at higher pressures. At the second limit, an increase in pressure inhibits the explosive reaction. The dimensions of the vessel and the nature of its surface are not particularly influential, which suggests that the competition involves gas-phase processes in which the a termination process is more strongly pressure dependent than the chain branching process. Since chain branching occurs via

$$H + O_2 \rightarrow OH + O \tag{9.29}$$

the competing reaction, and rate determining step in the termination chain, is

$$H + O_2 + M \rightarrow HO_2 + M \tag{9.36}$$

The HO_2 radical is less reactive than the other free radicals present (H, O and OH) and hence has a greater probability of diffusing to and being destroyed at the vessel surface.

The competition between eqns (9.29) and (9.36) resembles that derived from Scheme II in section 9.2.1 (see also Fig. 6.1). In eqn (9.36), the species M signifies any molecular species present in the system and so [M] represents the total molar concentration. Thus, the net branching factor is

$$\phi = (2k_{9.29} - k_{9.36}[M]) \tag{9.37}$$

Assuming ideal gas behaviour, the temperature dependence of the critical pressure at the second limit is given by

$$p_c = 2RT_a(A_{9.29}/A_{9.36})\exp(-(E_{9.29} - E_{9.36})/RT_a) \quad (9.38)$$

Equation (9.36) exhibits a very slight negative dependence on temperature (see Table 6.2), but to a reasonable approximation $E_{9.36}=0$ (see section 6.1.1). Thus, eqn (9.38) may be simplified to

$$p_c = 2RT_a(A_{9.29}/A_{9.36})\exp(-8455/RT_a) \quad (9.39)$$

which signifies that the critical pressure *increases* as the vessel temperature is raised. The features of Fig. 6.1 also resemble this result.

Different species do not have the same efficiency as chaperone molecules, and the differences can be quite significant. Thus, $k_{9.36}$ has been represented in eqn (9.37) as an average value derived from

$$k'_{9.36,A}\{x_A + \beta_B x_B + \beta_C x_C + \ldots.\} \quad (9.40)$$

where β_B, β_C.... are the collision efficiencies of species B, C, etc. as third bodies relative to that of A, x_i represents the respective mole fraction and $k'_{9.36,A}$ represents $k_{9.36}$ with respect to species A exclusively as the third body. This expression has been verified experimentally for a number of gases (Table 9.1). The relative efficiencies β_B, β_C... are found to be close to the ratios of the collision numbers calculated from momentum transfer cross-sections. Since hydrogen is rather more efficient than oxygen (Table 9.1 [94, 168]), increasing the proportion of hydrogen in an H_2+O_2 mixture will *lower* the critical pressure at the second limit, at a given vessel temperature. That is the termination reaction is more strongly favoured by the presence of a higher proportion of hydrogen.

Whilst in some cases it may be sufficiently precise to derive the average value for $k_{9.35}$ from the initial composition, there are circumstances in which it is necessary to take into account variations of the third body efficiency as reaction proceeds. Water acts as an exceedingly efficient third body in eqn (9.36), so its formation can have an inhibitory effect on the kinetics, with the following consequences.

Table 9.1 Third body efficiencies of some molecular species (M) in the reaction $H+O_2+M \rightarrow HO_2+M$ relative to that of H_2 [94, 168]

Component	Efficiency
H_2	1.0
H_2O	6.4
CO_2	1.50
O_2	0.40
N_2	0.45
He	0.35
Ar	0.35

The ignition diagram shown in Fig. 9.6 is equally appropriate to experiments in a well-stirred reactor (see Chapter 10) as in a closed vessel, but the quantitative features of the diagram are then governed by the mean residence time of the reactants in the vessel. There is an additional closed region of oscillatory ignition at marginally supercritical conditions adjacent to the second limit up to about 150 kPa [169].

Although the underlying kinetic interactions are more complicated [168, 169], there is a simple, qualitative explanation of the oscillatory ignitions of hydrogen [160, 170], as follows. There is a continuous flow of hydrogen and oxygen into the well-stirred reactor which gives a characteristic residence time (t_{res}) at the operating temperature and pressure, typically a few seconds. Residual reactants and products are swept out at a steady rate governed by t_{res}. If the vessel temperature is raised from that at point A in Fig. 9.6, at which the reaction is negligibly slow, towards that at B, the second limit is crossed and oscillatory ignitions begin.

At the onset of an ignition, chain branching via eqn (9.29) dominates the competition with eqn (9.36). The overall stoichiometry of the branching cycle

$$H + 3H_2 + O_2 = 3H + 2H_2O \tag{9.33}$$

ensures the extremely rapid generation of enormous quantities of water and heat ($\Delta H^\circ_{298} = -265\,\text{kJ mol}^{-1}$ of H atoms). The associated temperature rise promotes the 'flame chemistry' reactions discussed in Chapter 6 and reaction effectively ceases at a temperature which approaches the adiabatic temperature. The overall behaviour corresponds to

$$2H_2 + O_2 = 2H_2O \tag{9.41}$$

which has also involved radical recombination reactions. Residual hydrogen or oxygen is left if the reactants are not in stoichiometric proportions at the inflow. The temperature begins to fall at a rate governed by the heat loss rate to the vessel walls, which is independent of and also much faster than the heat carried out by the outflowing gases.

After ignition the predominant component in the vessel is water, although this is gradually replaced by inflowing hydrogen and oxygen. Since the efficiency of water as a third body in eqn (9.36) is six times that of hydrogen and 16 times that of oxygen (Table 9.1), the rate constant (9.40) is enhanced by about 10 times its value just before the onset of ignition. There is no chance of the branching reaction controlling events, provided that the water is the main component in the vessel. Only when virtually all of the water has been replaced by the inflowing hydrogen + oxygen mixture can the branching reaction again predominate, starting the cycle again.

The relationship between $k_{9.29}$ and $k_{9.36}$ $(=k'_{9.35,A}(x_A + \beta_B x_B + \beta_C x_C + \ldots))$ is the major controlling factor, but $k_{9.36}$ varies according to the prevailing composition in the vessel. The period of oscillations (τ) may be expressed, for a reactant mixture comprising H_2+yO_2+zN, as [160, 170]

$$\tau = t_{res} \ln \frac{\beta_{H_2O} - 0.5\beta_{O_2} - 1}{(2k_{9.29}/k_{9.36}[H_2]_o) - 1 - y\beta_{O_2} - z\beta_N} \tag{9.42}$$

N represents any diluting gas, and $k_{9.36}$ refers to the rate constant with respect to hydrogen as the third body.

The third limit. In the hydrogen + oxygen system, an isothermal third limit is possible in principle. If the pressure is increased above the second limit, the HO_2 radical is prevented from diffusing to the walls and can then propagate the chain via [171]

$$HO_2 + H_2 \rightarrow H_2O_2 + H \tag{9.43}$$

$$HO_2 + H_2 \rightarrow H_2O + OH \tag{9.44}$$

$$HO_2 + HO_2 \rightarrow H_2O_2 + O_2 \tag{9.45}$$

Since hydrogen peroxide can dissociate relatively easily by

$$H_2O_2 + M \rightarrow 2OH + M \tag{9.46}$$

the HO_2 radical can also lead to additional degenerate branching under circumstances in which eqn (9.36) is favoured.

The first and second limits can be regarded as virtually isothermal and have been shown to be so at conditions remote from the third limit [172, 173], but a full understanding of the third limit can be reached only when thermal feedback is also taken into account. Points of principle for a unified chain-thermal approach [174, 175] have been raised in section 9.4. Methods of treatment of multivariable problems are discussed in Chapter 10.

9.5.2 Kinetic models for the oxidation in the vicinity of the ignition limits

The main elements of the simplest kinetic model to represent the first and second limit behaviour have been incorporated into the equations for the respective net branching factors, eqns (9.34) and (9.37). Their application is restricted to conditions of very high efficiency of free radical loss at the walls so that none of the elementary rate expression is of order greater than unity with respect to any free radical concentration. Thus an algebraic analysis is possible, but which would suggest also that the reaction rate is negligible (i.e. virtually unmeasurable) within the slow reaction zone. This is not the case in many circumstances, as is confirmed experimentally [165, 171–173].

A representative reaction scheme, appropriate to surfaces of low efficiency especially with respect to the destruction of HO_2 and H_2O_2, must include quadratic interactions of free radicals (see Table 6.2). Subject to the inclusion of an initiation step and a surface termination of H atoms (eqn (9.30)), the reactions given in Table 6.2 could be used as a basis for numerical computation of hydrogen combustion at low temperatures and pressures. Foo and Yang [174] obtained excellent accord with experimental measurements by application of a more limited kinetic scheme than that shown in Table 6.2. If every kinetically reasonable reaction were to be included, the total number of forward and reverse reactions would certainly exceed 100.

Although numerical computation can be very instructive, the kinetics of hydrogen oxidation in aged boric acid-coated vessels was determined experimentally, before computational methods were widely adopted [165]. The slow reaction mechanism is so well understood that a reacting mixture of H_2+O_2 can be used as a controlled source of H, OH and HO_2. Rate constants for the reactions of these species with hydrocarbons have been measured in reacting H_2+O_2 mixtures including hydrocarbon additives [176].

9.6 Carbon monoxide oxidation and ignition

The principal steps involved in the oxidation of pure carbon monoxide are very simple indeed, namely

$$O_2+M \rightarrow 2O+M \quad \text{initiation} \tag{9.47}$$

$$CO+O+M \rightarrow CO_2^{\bullet}+M \quad \text{termination} \tag{9.48}$$

CO_2 may act as an 'energy branching species' since it may be formed in eqn (9.47) in an electronically excited, and therefore highly energetic state [23], such that

$$CO_2^{\bullet}+O_2 \rightarrow 2O+CO_2 \quad \text{branching} \tag{9.49}$$

Whilst an interpretation based on this scheme leads to rewarding academic insights into the theory of isothermal autocatalytic processes [177], it is doubtful that the characteristics of CO oxidation that is completely free from hydrogenous impurities have been observed.

The kinetic association between CO and H_2 oxidation cannot be better exemplified than by the dependence of the location of the second (p–T_a) critical limit on the proportion of hydrogen present in a $CO+O_2$ mixture (Fig. 9.7). It requires less than 1% of H_2 added to the $CO+O_2$ mixture in order to recover a limit which is virtually identical to that of an H_2+O_2 mixture.

The limits that occur at higher temperatures relate to mixtures with diminishing proportions of hydrogen-containing components (Fig. 9.7 [178–182]). At the highest temperatures, the $CO+O_2$ mixture has been freed

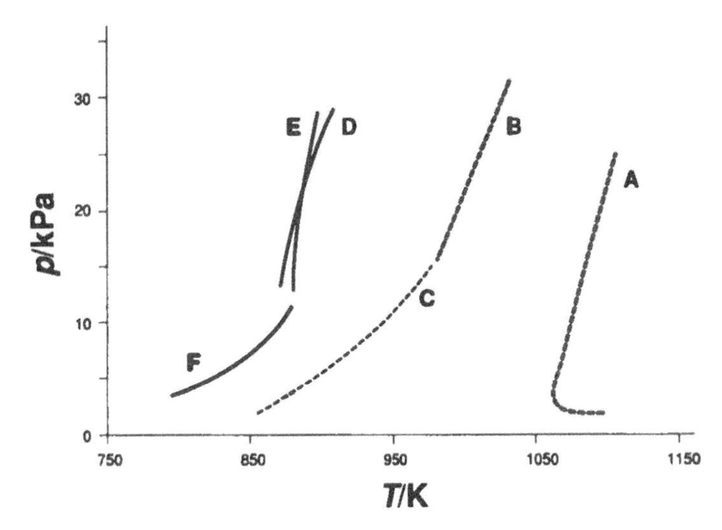

Figure 9.7 Experimental (p–T_a) second ignition limit (solid lines) and glow boundaries (broken lines) for $CO_2 + O_2$ in a closed vessel. The limits shift to higher temperatures with diminishing proportions of hydrogenous material present. A, Dickens *et al.* [178]; B, Gordon and Knipe [179]; C and F (with 0.02 mol% added H_2), Gray *et al.* [180]; D, Lewis *et al.* [181]; E, Hadman *et al.* [182].

as far as possible from hydrogenous material. Following historical convention, this is termed a 'dry' mixture and its limits are termed 'glow' rather than 'ignition' limits. They correspond to a transition from a slow to a rapid reaction rate in much the same manner as the explosion limit but, although there is light emission from excited CO_2 (eqn (9.48) and section 6.4.2), there is very little temperature change and high proportions of carbon monoxide and oxygen remain even when reaction appears to have stopped.

The glow can be manifest as a sharp pulse of short duration which, although weak in intensity relative to normal ignition, can be seen by eye with subdued background light in the laboratory. With reduced 'hydrogen content', the emission is less easy to discern by eye, but the signal from a photomultiplier shows that the glow intensity may continue for many seconds, decaying monotonically [180, 183]. These features have been known for a very long time, and studies of carbon monoxide oxidation pre-date those of hydrogen [184].

With careful experimentation in closed vessels, which includes particular attention to treatment of the (silica) vessel surface, the 'glow' may be manifest as a virtually isothermal oscillatory phenomenon, which is probably unique amongst gaseous reactions in a closed vessel [36]. Oscillatory glows can also be observed in well-stirred flow systems [185], the kinetic origins of which may be traced to two branching/termination interactions, comprising [186]

$$H + O_2 \rightarrow OH + O \tag{9.29}$$

$$H+O_2+M \rightarrow HO_2+M \quad (9.36)$$

and

$$O+H_2 \rightarrow OH+H \quad (9.31)$$

$$CO+O+M \rightarrow CO_2^*+M \quad (9.48)$$

In the glow reaction, the interaction between eqn (9.29) and eqn (9.36) remains in the non-branching mode. The oscillation is driven by the interaction between eqns (9.31) and (9.48), which switches from branching to non-branching and back again. The relevant kinetic equation is

$$\frac{v_{9.31}}{v_{9.48}} = \frac{k_{9.31}[H_2]}{k_{9.48}[CO][M]} \quad (9.50)$$

The ratio of $[H_2]/[CO]$ in the reactor at any given time determines whether or not chain branching via eqn (9.31) is the predominant process. The preferential consumption of hydrogen is controlled by the kinetics; its replenishment is controlled by the inflowing reactant mixture.

Oscillatory ignitions during carbon monoxide oxidation under well-stirred flow conditions are favoured by higher proportions of hydrogen, and the features are very similar to those of the hydrogen oscillations (section 9.5). The exothermicity of reaction is governed by the complete consumption of carbon monoxide, since the amount of water that is formed is very small indeed. In this case, the major part for the switch between branching and non-branching via eqns (9.29) and (9.36) is played by CO_2 formation, since it exhibits a higher efficiency than either the CO or the O_2 in the unreacted mixture (Table 9.1).

In closed vessels it must be presumed that hydrogen may be regenerated at the (active) surface in order to explain the oscillatory glow, which requires the water gas equilibrium at the surface to be involved [187],

$$CO+H_2O \Leftrightarrow CO_2+H_2 \quad (9.51)$$

An oscillatory glow in closed vessels can be simulated by adding a heterogeneous, water gas equilibrium to the main kinetic scheme, with a diffusion controlled term governing the forward reaction [4]. Whilst this may not yet be fully substantiated, there is rather better experimental evidence that the kinetic origins for the oscillatory glow reside in the H_2+CO interactions rather than in theories based on a pure $CO+O_2$ mechanism [177].

The 'steady glow' is easier to interpret kinetically. The transition into the glow region corresponds to net branching (($\phi > 0$) being established between the branching (eqn (9.31)) and termination (eqn (9.48)) involving the O atoms. The monotonic decay in intensity is associated with the consumption of the hydrogen (rather than of carbon monoxide), and reaction can effectively cease at the threshold ($\phi = 0$), which is determined by the depletion of H_2 (eqn (9.50)). The hydrogen is converted to water, which may be adsorbed on the vessel surface. Similar kinetic arguments apply to CH_4 that may present in the CO, since it can be a source of H and OH atoms and it can also enter into a chain branching reaction with O atoms which corresponds to eqn (9.31).

9.7 Spontaneous ignition and oscillatory cool flames of hydrocarbons

The $(p - T_a)$ ignition diagram for thermal ignition (Fig. 8.6) is characterised by the two regions of 'slow combustion' and 'ignition', which are separated by a critical boundary. Even in very early investigations of hydrocarbon oxidation in closed vessels (pre-1935) it was found that the ignition diagram was considerably more complicated (Fig. 9.8). This type of ignition diagram is constructed from the results of many experiments performed at different reactant pressures in a vessel at controlled temperatures. As in thermal ignition, there is a sharp boundary beyond which ignition occurs, but it comprises three branches, the middle one of which exhibits an *increase* of the critical pressure as the vessel temperature is raised. This results from the *negative temperature coefficient* (or NTC) in the heat release rate, the kinetic basis of which is described in section 7.4. The subcritical region of the ignition diagram is also subdivided into different types of behaviour. Part of this represents slow

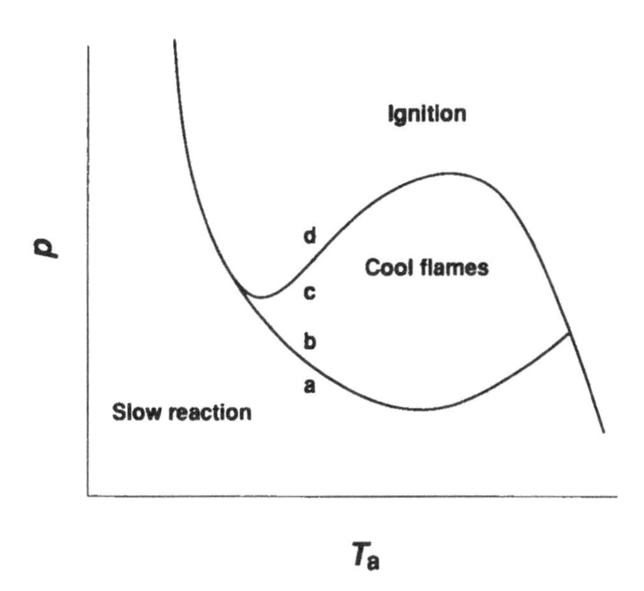

Figure 9.8 A representative $(p - T_a)$ ignition diagram for the gaseous oxidation of alkanes and other organic compounds in a closed vessel. Points a–d relate to Fig. 9.9.

combustion, but multiple (or oscillatory) cool flames occur in a closed region bounded by the NTC branch of the ignition boundary. In some combustion systems there is also further complexity, called multiple stage ignitions [142].

The qualitative structure in Fig. 9.8 is characteristic of alkanes and alkenes containing three or more carbon atoms, of acetaldehyde (ethanal) and higher molecular mass aldehydes, higher alcohols, ethers other than dimethyl ether, and a variety of other organic compounds which contain fairly large aliphatic groups. Although there are some resemblances of the behaviour, the features of Fig. 9.8 are not exhibited fully by methane, ethane, ethene or formaldehyde. Nor are they shown by the aromatics such as benzene, toluene or xylene.

The reactant pressures and vessel temperatures at which the different features occur are dependent not only on the particular fuel and its proportions with oxygen or air but also on the size and shape of the vessel [142]. Surface reactions may play some part also, especially in the initiation processes, so the quantitative details are usually specific to the particular system employed. The normal alkanes are more reactive than their isomeric, branched chain structures and the reactivity increases considerably in the range C_3–C_6. As an approximate guide, equimolar mixtures of alkanes with oxygen would exhibit ignition at temperatures above about 520 K, and within the ignition peninsula, at pressures above about 30 kPa. Substances such as aldehydes or ethers tend to be more reactive in oxygen, so cool flames and ignitions are observed at lower temperatures and pressures than those of alkane oxidation.

Most organic fuels mixed with air do not undergo spontaneous ignition in the low temperature range at pressures much below 100 kPa (1 atm). The lowest autoignition temperature at a given pressure is generally observed in mixtures that are in the range of molar proportions fuel: $O_2 = 1{:}2$ to 2:1 ($> 10\%$ by vol of the vapour in air, say). This is considerably more rich than the optimum conditions for the propagation of premixed flames. Methane is alone amongst the hydrocarbons in exhibiting its minimum ignition temperature in a mixture with air which is slightly leaner than the stoichiometric mixture (which is at $[CH_4]:[O_2] = 1{:}2$) [188].

The features of Fig. 9.8 are illustrated by referring to the temperature–time or pressure–time records that are obtained from the experimental measurements. Insofar that the phenomena are non-isothermal, knowledge of the temperature history is extremely important, but qualitative insight is gained from the pressure record.

The curves in Fig. 9.9 relate to a fixed vessel temperature with gradually increased initial pressures. Slow reaction takes place at the lowest pressures. The pressure time curves are sigmoidal (curve a in Fig. 9.9), as is characteristic of degenerate chain branching reactions (section 9.3). As the initial reactant pressure is increased (curves b and c

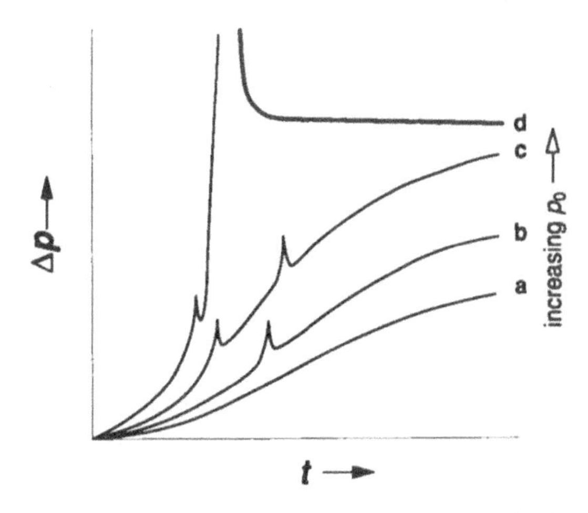

Figure 9.9 Typical pressure–time records obtained during the oxidation of alkanes in a closed vessel. a, slow reaction; b and c, single and multiple cool flames; d, two-stage ignition, as described in the text.

in Fig. 9.9) there are momentary pressure pulses, up to seven in number in propane oxidation, which interrupt the smooth growth in pressure. At slightly higher pressures, the transition to ignition is marked by a reaction in which there is a massive and very rapid pressure rise. Within the ignition peninsula the ignition is preceded by a cool flame. This process is known as two-stage ignition (Fig. 9.9, curve d). The multiple stage ignitions that can occur adjacent to the negative temperature dependent branch of the ignition boundary are detected as a succession of one or more cool flames preceding the two-stage ignition. A single stage event is observed elsewhere in the ignition region. The periodic pulses in the pressure records, caused by the cool flames, and the overshoot accompanying full ignition are a consequence of temperature increases.

The typical form of temperature changes measured by a very fine thermocouple during oscillatory cool flames is shown in Fig. 9.10 [189]. The temperature change in the cool flames may be up to 200 K, but if the vessel temperature is raised to the upper limit of the cool flame region in Fig. 9.8, their frequency increases and their amplitude decreases (Fig. 9.10). There is a damped oscillatory character in this representation which is observed only in well-stirred closed vessel conditions. Convection effects in unstirred reaction conditions cause considerable complications in the thermal records [189]. Normally a thermocouple is not sufficiently responsive to follow the ignition to the maximum temperature because the rate of heat release is extremely rapid during the final stages.

Visual observation, in a completely darkened laboratory, reveals a very faint, blue luminescence at the moment that each cool flame pulse occurs.

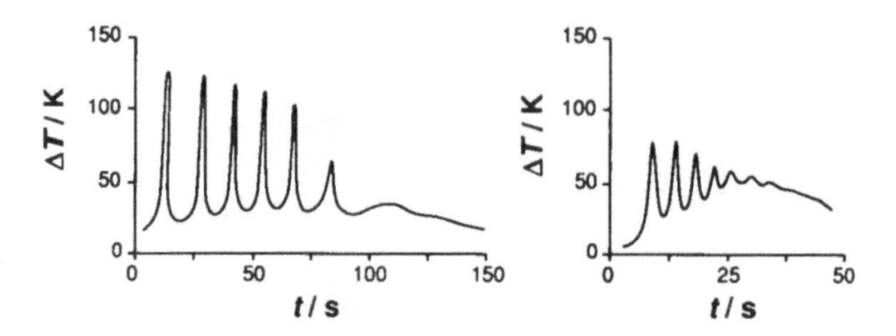

Figure 9.10 ΔT-time records obtained during the oxidation of propane in a well-stirred closed vessel (after Gray *et al.* [189]). The damped oscillatory character of the multiple cool flames develops as the vessel temperature is raised.

This luminescence originates from an electronically excited state of formaldehyde, which is formed in a chemiluminescent reaction mainly by the radical + radical reactions (section 6.6)

$$CH_3O + OH \rightarrow CH_2O^* + H_2O; \quad \Delta H^{\ominus}_{298} = -296.4\,kJ\,mol^{-1} \quad (9.53)$$

and

$$CH_3O + CH_3O \rightarrow CH_2O^* + CH_3OH; \quad \Delta H^{\ominus}_{298} = -278.5\,kJ\,mol^{-1} \quad (9.54)$$

On average only one photon is emitted per 10^3 molecules of fuel consumed. The thermokinetic origins of oscillatory cool flames are discussed in Chapter 10.

Most of the phenomena observed in closed systems have their counterparts also in flow systems. In fact, the first observations of cool flames were made in moving gases by Sir Humphry Davy (1815) and investigated further by Sir William Henry Perkin in 1888 [190]. The cool flame luminescence from diethyl ether combustion is the most intense, and also the most easily generated. Perkin made his observations by dropping liquid diethyl either on to a bath of hot sand. It was not possible to ignite paper from the 'will-o'-the-wisp', blue luminescence that was visible above the sand (in a darkened laboratory), nor was the hand burned when put close to it. The term 'cool flame' originates from these observations. There was a strong, acrid odour of partially oxidised products from the reaction. The extent of reaction is low in cool flames. For reasons of safety, repetition of Perkin's experiment is not recommended.

In more modern studies, cool flames have been stabilised in heated, laminar flow tubes or above heated burners. Their velocities are relatively low (0.05–0.1 m s^{-1} compared with a normal flame speed of 0.4 m s^{-1}). On increasing the proportion of oxygen in the gaseous mixture a second stage, hot flame is also stabilised downstream from the cool flame, giving a stabilised two-stage ignition. Thermocouple measurements and detailed chemical analysis can be made by continuous sampling from the regions below, between and above the flames by use of a moving probe system [130, 132, 191–193], thus building up a spatial profile which relates to the temporal evolution of the phenomena in closed systems.

In well-stirred flow reactors, oscillatory cool flames are sustained indefinitely at constant reactor temperature and gas flow. Two-stage or multiple-stage ignitions also become repetitive. The combustion of acetaldehyde [194, 195] and of a number of alkanes [196] have been studied in this way at atmospheric pressure or below, and experiments have also been performed at pressures up to 1 MPa using well-stirred flow reactors [197]. Investigations in this pressure range are related to conditions in the combustion chambers of spark ignition or diesel engines.

Further reading

Baldwin, R.R. and Walker, R.W. (1972). Branching-chain reactions: the hydrogen-oxygen reaction. In *Essays in Chemistry* (Vol. 3) (eds J.N. Bradley, R.D. Gillard and R.F. Hudson). Academic Press, London, UK, p. 1.

Barnard, J.A. (1977). The oxidation of compounds other than hydrocarbons. In *Comprehensive Chemical Kinetics* (Vol. 17) (eds C.H. Bamford and C.F.H. Tipper). Elsevier, Amsterdam, The Netherlands, p.441.

Dainton, F.S. (1966). *Chain Reactions, An Introduction* (2nd edn). Methuen, London, UK.

Dixon-Lewis, G. and Williams, D.J. (1977). The oxidation of hydrogen and carbon monoxide. In *Comprehensive Chemical Kinetics* (Vol. 17) (eds C.H. Bamford and C.F.H. Tipper). Elsevier, Amsterdam, The Netherlands, p. 1.

Gray, P. and Scott, S.K. (1985). Isothermal oscillations and relaxation ignitions in gas-phase reactions. In *Oscillations and Traveling Waves in Chemical Systems* (eds R.J. Field and M. Burger). John Wiley, New York, USA, p. 493.

Gray, P. and Scott, S.K. (1990). *Chemical Oscillations and Instabilities.* Oxford University Press, Oxford, UK.

Griffiths, J.F. (1995). *Prog. Energy Combust. Sci.*, **20**, 461.

Griffiths, J.F. and Scott, S.K. (1987). *Prog. Energy Combust. Sci.*, **13**, 161.

Lignola, P.-G. and Reverchon, E. (1987). *Prog. Energy Combust. Sci.*, **13**, 1.

Mulcahy, M.F.R. (1973). *Gas Kinetics.* Nelson, London, UK.

Pollard, R.T. (1977). Oxidation of hydrocarbons. In *Comprehensive Chemical Kinetics* (Vol. 17) (eds C.H. Bamford and C.F.H. Tipper). Elsevier, Amsterdam, The Netherlands, p. 249.

Problems

(1) Sketch the p–T_a ignition boundaries which would result from the following kinetic dependences for the branching and termination

steps in four isothermal reactions. Establish on which side of the limit the ignition region lies.

	Branching rate	Termination rate
1	$p \cdot A_b \exp(-E_b/RT)$	A_t
2	$p \cdot A_b \exp(-E_b/RT)$	$p^2 \cdot A_t$
3	$p \cdot A_b \exp(-E_b/RT)$	$A_t \exp(-E_t/RT)$, (where $E_t = E_b$)
4	$p \cdot A_b \exp(-E_b/RT)$	$A_t \exp(-E_t/RT)$, (where $E_t > E_b$)

p = total pressure; A = pre-exponential factor; E = activation energy.

(2) On the basis of the following elementary reaction scheme for hydrogen oxidation at low pressures, derive an expression for the minimum temperature at which spontaneous ignition is possible in the reactant mixture H_2+O_2. Assume that k_5 is independent of pressure. Hence calculate the minimum temperature and the critical reactant pressure at this temperature.

1. $H_2+O_2 \rightarrow 2OH$ initiation
2. $H_2+OH \rightarrow H_2O+H$ propagation
3. $H+O_2 \rightarrow OH+O$ branching
4. $H_2+O \rightarrow OH+H$ branching
5. $H \rightarrow$ wall termination 1
6. $H+O_2+M \rightarrow HO_2+M$ termination 2

$E_3 = 70\,\text{kJ mol}^{-1}$, $A_3 = 2\times 10^{14}\,\text{mol}^{-1}\,\text{cm}^3\,\text{s}^{-1}$, $k_5 = 140\ \text{s}^{-1}$, and

$k_6 = 4.59\times 10^{16}\,\text{mol}^{-2}\,\text{cm}^6\,\text{s}^{-1}$. Assume that $p(O_2) = p/2$.

(*Hint*: Derive the expressions which represent the p–T_a relationship at the first and second limit independently. Sketch these p–T_a limits to establish the next step.)

(3) In a cylindrical vessel whose surface destroys hydrogen atoms very efficiently, solution of the diffusion equation leads to an effective first-order 'rate constant' for removal of hydrogen atoms at the wall as [165]

$$\kappa_w = 23D/d^2$$

where d is the diameter of the vessel and D is the diffusion coefficient of hydrogen atoms through the reaction mixture. If, in the hydrogen + oxygen reaction, chain branching occurs by the reaction

$$H+O_2 \rightarrow OH+O$$

show that at the first explosion limit

$$kx(O_2)p = 11.5D/d^2$$

where p is the total pressure and $x(O_2)$ is the oxygen mole fraction. From this result show that for a fixed temperature at the first limit: (a) $d=$ constant in a particular reaction mixture, and (b) $x(O_2)p=$ constant in a given vessel. What effect will the addition of inert gas have on the explosion limit pressure?

(4) The diffusion coefficient of H atoms through an hydrogen + oxygen mixture is given by

$$D=0.0531T^{1.8}/(1+0.62x(O_2))/p$$

where $x(O_2)$ is the mole fraction of oxygen in the mixture and p is the total pressure in torr. The first explosion limit in a cylindrical vessel, $d=102$ mm, has been measured as follows [198]:

$2H_2+O_2$	T(K)	965	931	899	871	847
	p(torr)	2.43	2.72	2.94	3.29	3.64
$9H_2+O_2$	T(K)	997	959	896	859	833
	p(torr)	4.43	4.76	5.80	6.72	7.67

The vessel was coated with MgO, a surface on which hydrogen atoms are efficiently destroyed. Use these data, and those in the previous question, to calculate the Arrhenius parameters for the chain branching reaction.

(5) Measurements of the second explosion limit (p) in stoichiometric mixtures of hydrogen and oxygen gave the following results:

T(K)	761	782	800	820
p(torr)	40	56	75	100

If the activation energy of the termolecular terminating step is $-7\,\mathrm{kJ\,mol^{-1}}$, calculate the activation energy of the branching step on the assumption that all HO_2 radicals are destroyed.

(6) At a particular temperature, the second explosion limit (p) in mixtures of hydrogen and oxygen varied with composition as follows:

p(torr)	54	66	78	90
$p(O_2)$(torr)	20	40	60	80

Use these results to obtain the efficiency of oxygen relative to that of hydrogen in collisions with HO_2 on the assumption that all HO_2 radicals are removed at the wall.

10 Ignition, extinction and oscillatory phenomena

10.1 Introduction and background

Not only do the topics discussed in this chapter offer an interesting and varied intellectual challenge but also they underpin many facets of efficiency and safety in combustion applications. The topics relate to (i) the operation of industrial chemical reactors, through the principles of stirred flow reactors, (ii) the use of solid fuels or the application of catalytic converters in automobile exhaust systems, through the principles of heterogeneous combustion at surfaces, and (iii) the interpretation of cool flames and spontaneous ignition of hydrocarbons.

The background is that of Chemical Reactor Theory, which is much discussed by chemical engineers. This topic was developed as part of combustion theory in the 1940s, by the work of Zel'dovich [199,200]. Outside the former Soviet Union the subject took independent pathways of development from the mid 1950s in chemical engineering [201] and combustion applications [202], until about the 1980s. Thereafter a closer association was re-established, driven largely through the dramatic growth of interest in non-linear dynamics (manifest as 'chaos'), with considerable benefits to a mutual development of each subject. The associations were maintained throughout the period in Russia [31, 39]. Parts of non-isothermal theory are not amenable to a full, analytical treatment and so approximations have to be made or numerical methods invoked. However, there are analogous, isothermal, quadratic or cubic autocatalytic, reaction mechanisms, which can be analysed fully and thus yield lucid insights into the rich phenomonology of flow systems [36].

10.2 Combustion in a CSTR

A CSTR is a vessel into which there is a continuous supply of the reactant and continuous outflow of products (Fig. 10.1). The reaction takes place in the vessel, and mixing may be achieved either by mechanical means or by jet-injection (of gases). The vessel is set at a controlled temperature (T_a). The inflowing reactant is assumed to be preheated to T_a. The equations are more complicated with no preheating, but without qualitative differences in behaviour. A CSTR could be operated on an industrial scale or it may be restricted to a volume appropriate to

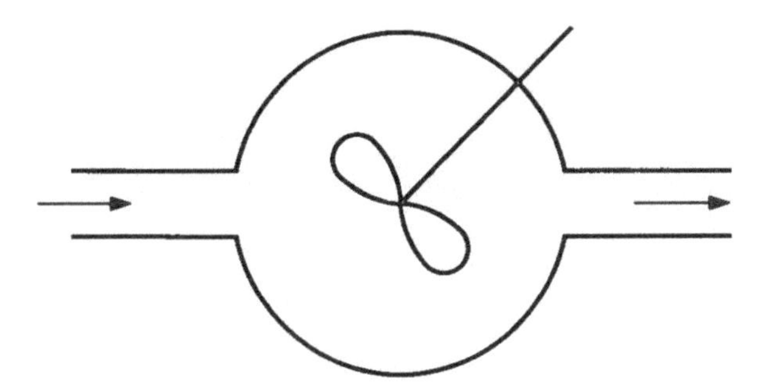

Figure 10.1 A schematic representation of a continuous, stirred tank reactor (CSTR).

laboratory operation ($0.5\,dm^3$, say). Adiabatic or non-adiabatic operation gives important qualitative distinctions in behaviour.

C_p replaces C_v in appropriate terms at constant pressure, energy changes are determined by ΔH rather than ΔU, and a continuous supply of reactant ensures that the chemical reaction can be maintained at a stationary state away from the final state of thermodynamic equilibrium. In idealised, spatially uniform conditions in the CSTR, the composition at the outflow is representative of the composition anywhere within the reaction volume.

Unifying links with thermal ignition theory are established through the non-dimensionalised terms and the natural timescales encountered in Chapter 8, such as the dimensionless temperature excess, $\theta(=E\Delta T/RT_a^2)$, and the dimensionless adiabatic temperature rise, $B(=Eqc_0/\sigma C_p RT_a^2$ in the present case). The latter is appropriate to adiabatic operation of a reactor, but a derivative of it, B^*, emerges with respect to non-adiabatic operation. The natural timescales were the characteristic chemical time, $t_{ch}(=1/A\exp(-E/RT_a))$, and the Newtonian cooling time, $t_N(=\sigma C_p V/hS)$. To these are added the mean residence time in a CSTR, t_{res}, which is given by V divided by the volumetric flow rate. This leads to the definition $B^*=B/(1+t_{res}/t_N)$. A dimensionless reactant concentration, $\alpha=a/a_0$ is also introduced, where a_0 is the inflow concentration and a the instantaneous concentration in the vessel.

When reaction is first order with respect to reactant A the mass conservation equation involving dimensionless terms takes the form

$$\frac{d\alpha}{dt}=-\frac{\alpha}{t_{ch}}\exp(\theta/(1+\varepsilon\theta)+\frac{1}{t_{res}}(1-\alpha) \tag{10.1}$$

It is possible to non-dimensionalise eqn (10.1) fully by relating the timescales to either t_{ch} or t_{res}. The first term on the right-hand side of eqn (10.1) is the reaction rate. The second term is the inflow minus the outflow rate. In a stirred flow reactor under stationary state conditions

($\mathrm{d}\alpha/\mathrm{d}t = 0$), the reaction rate (of any order) can be derived directly from the difference between the inflow and outflow reactant concentrations, and the mean residence time.

The non-dimensionalised energy conservation equation resembles eqn (8.24):

$$\frac{\mathrm{d}\theta}{\mathrm{d}t} = \frac{B\alpha}{t_{\mathrm{ch}}} \exp(\theta/(1+\varepsilon\theta) - \theta\left(\frac{1}{t_{\mathrm{res}}} + \frac{1}{t_{\mathrm{N}}}\right) \tag{10.2}$$

The second term on the right-hand side of eqn (10.2) represents heat transport from the vessel at the outflow and, in non-adiabatic operation, the losses through the vessel wall. The first term is the dimensionless chemical heat release rate. Stationary states of the system are determined when eqns (10.1) and (10.2) equal zero.

10.2.1 Adiabatic operation

When no heat is lost to the surroundings ($t_{\mathrm{N}} \rightarrow \infty$), in a stationary state the temperature excess is linked to the extent of reaction under all conditions through a (dimensionless) relationship obtained from eqns (10.1) and (10.2).

$$(1 - \alpha_{\mathrm{s}}) = \theta_{\mathrm{s}}/B \tag{10.3}$$

where subscript s denotes a stationary state. The dimensionless adiabatic temperature excess is reached in the vessel in the limit of complete reaction ($\alpha_{\mathrm{s}} \rightarrow 0$). Although both reactant concentration and temperature excess have been considered, there is only one independent variable and the system can be described by one equation, in the limit of $\varepsilon \rightarrow 0$, as

$$\frac{\mathrm{d}\theta}{\mathrm{d}t} = \frac{(B-\theta)}{t_{\mathrm{ch}}} \exp(\theta) - \frac{\theta}{t_{\mathrm{res}}} \tag{10.4}$$

Heat release rate $\boldsymbol{R}$ Heat loss rate $\boldsymbol{L}$

Ignition and extinction phenomena are possible because the condition for stationary states (eqn (10.4) = 0), even in the approximate form, can be satisfied by more than one value of θ_{s}. As in thermal ignition theory, the critical transitions occur at the 'tangency condition':

$$\frac{\mathrm{d}\boldsymbol{R}}{\mathrm{d}\theta} = \frac{\mathrm{d}\boldsymbol{L}}{\mathrm{d}\theta} \tag{10.5}$$

Solving (10.5), with the supplementary condition $\boldsymbol{R} = \boldsymbol{L}$, yields the criteria for ignition and extinction as solutions of the quadratic

$$\theta_{\mathrm{s}}^2 - B\theta_{\mathrm{s}} + 1 = 0 \tag{10.6}$$

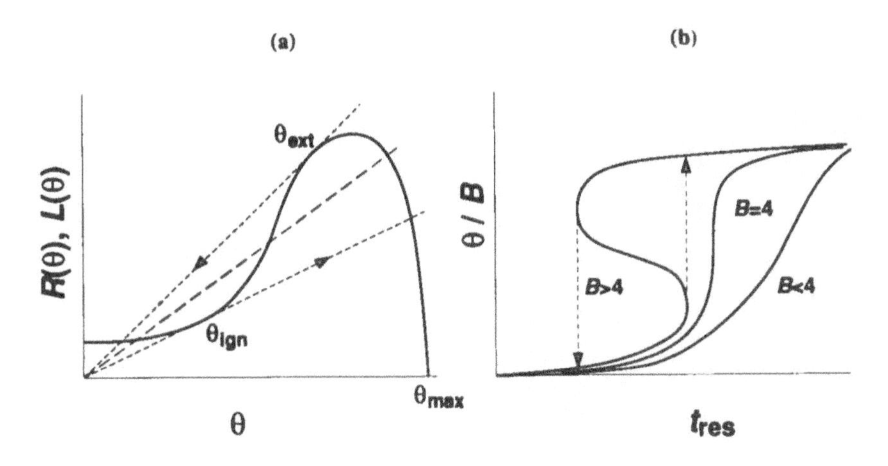

Figure 10.2 (a) Relationships between heat release, $R(\theta)$, and heat loss rates, $L(\theta)$, as a function of dimensionless temperature excess (θ) in an adiabatic CSTR. The ignition and extinction transitions are shown. (b) The corresponding dependence of θ_s/B as a function of t_{res} at $B>4$, $B=4$ and $B<4$.

from which it may also be shown that criticality is lost at $B<4$. Representations of multiple stationary states within a certain range of t_{res} for $B>4$ and the solutions for (θ_s/B) as $f(t_{res})$ are shown in Fig. 10.2. Since the heat release rate is independent of residence time, when the reaction is sufficiently exothermic (i.e. $B>4$), a change in the rate of reactant supply to the vessel (i.e. a change of gradient of L) can cause a sudden change in the reactant temperature. This could signal a damaging explosion in a vessel (ignition) or, for a reactor operating in a highly reactive state, a sudden loss of efficiency of the process (extinction). In combustion processes B is often in the range 50–150, so potential problems of ignition and extinction are not unusual, regardless of reaction order.

Following the analytical derivation of the Semenov criterion (eqns (8.9) and (8.10)) the lower root to eqn (10.6) yields the dimensionless critical temperature excess at ignition, to a good approximation, as

$$\theta_{ign} \approx 1 + (1/B) \ldots \quad (10.7)$$

With some qualification, the upper root represents the (physically realistic) case of extinction. There is no difficulty in obtaining θ_{ext} if a numerical method is used to solve eqn (10.4) without approximation, but in the approximate form the condition $\varepsilon\theta \ll 1$ may no longer be satisfied at high values of θ. The procedure is then to shift the reference temperature for ε and θ from T_a to a value in the region of interest. A natural reference temperature in Fig. 10.2a to determine θ_{ext} accurately would be the adiabatic temperature for the system, T_{ad}, yielding $\theta = E(T_{ad} - T)/RT_{ad}^2$, $\varepsilon = RT_{ad}/E$ and $B = Eqc_0/\sigma C_p RT_{ad}^2$. Solving the appropriate equation and then recalculating θ in the form $\theta = E(T - T_a)/RT_a^2$, gives

$$\theta_{ext} \approx B - 1 \ldots. \tag{10.7}$$

With reference to Fig. 10.2, as in thermal ignition theory, in the region of reaction multiplicity the upper and lower stationary states are readily established to be stable whereas the intermediate state is unstable. However, there are more rigorous approaches to determining these properties, as discussed next.

10.2.2 Non-adiabatic operation

When the term in eqn (10.2) which represents heat loss to the vessel walls is retained, the fixed relationship between extent of reaction and temperature excess (eqn (10.3)) no longer prevails. That is, the additional route for heat transport affects the reactant temperature without affecting the reactant concentration, so α and θ (or a and T) are not uniquely linked. Therefore, both parameters must be specified to characterise the system fully. It becomes a two variable problem, which means that although the stationary states may be located as intersections in a thermal diagram like Fig. 10.2, the local stability of the stationary states requires deeper investigation. The ground that is common to adiabatic conditions is that in a stationary state ($d\theta/dt = d\alpha/dt = 0$) the dimensionless temperature and concentration are linked by an expression similar in form to eqn (10.3), namely

$$(1 - \alpha_s) = \theta_s / B^* \tag{10.8}$$

in which B^* depends on the quotient (t_{res}/t_N):

$$B^* = B/(1 + t_{res}/t_N) \tag{10.9}$$

If normal operating conditions were such that $t_{res} \approx 0.1\, t_N$, the temperature excess reached during complete combustion of quite strongly exothermic reactions would be only several hundred K. In the fast flow limit ($t_{res} \rightarrow 0$) $B^* \rightarrow B$, whereas when $t_{res} \rightarrow \infty$, $B^* \rightarrow 0$.

The condition for stationary states also parallels that derived from eqn (10.4), which is (in approximate form)

$$\frac{(B^* - \theta_s)}{t_{ch}} \exp(\theta_s) - \frac{\theta_s}{t_{res}} = 0 \tag{10.10}$$

There can still be only one or three values of θ_s for given values of t_{res} and B^*, but there are implications for critical phenomena. Combining eqns (10.9) and (10.10) and rearranging the terms gives a quadratic equation in t_{res} of the form

$$t_{res}^2 + (t_N + t_{ch}\exp(-\theta_s) - Bt_N/\theta_s)\,t_{res} + t_{ch}\exp(-\theta_s)\,t_N = 0 \tag{10.11}$$

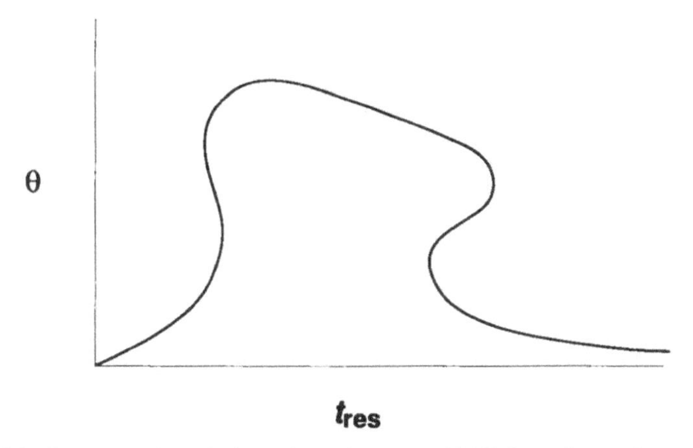

Figure 10.3 Representative solutions ($\theta_s(t_{res})$) to eqn (10.11) for a first order exothermic reaction in a non-adiabatic CSTR ($B^* > 4$).

There are seven qualitatively different patterns for the relationship between θ_s as $f(t_{res})$ which are governed by the magnitudes of t_N and t_{ch} [148, 203]. Their detailed discussion is beyond the scope of this introduction, but one of the more simple forms is exemplified in Fig. 10.3. There are two separate regions of t_{res} in which ignition and extinction phenomena may be observed and in which multiple stationary states can co-exist. In the limit $\varepsilon \rightarrow 0$, critical phenomena are possible as long as the condition $B^* > 4$ is satisfied, but this translates to the condition

$$\frac{t_{res}}{t_N} < \frac{B}{4} - 1 \tag{10.12}$$

Thus, under non-adiabatic conditions, criticality may cease to exist even at quite high values of B if t_{res} grows too large. The 'tangency conditions' for ignition and extinction in the non-adiabatic CSTR parallel those of the adiabatic system, taking the forms

$$\theta_{ign} \approx 1 + 1/B^* \ldots. \tag{10.13}$$

$$\theta_{ext} \approx B^* - 1 \ldots. \tag{10.14}$$

Stability of stationary states. The local stability of stationary states now has to be addressed. This serves as a useful introduction to methods of stability analysis [204, 205], which are also relevant to the interpretation of cool flame phenomena (section 10.5). The local stability is governed by the response of the system to small perturbations from the stationary state. The quantities that can be used to characterise this response in the two-variable problem are the partial derivatives of the conservation equations, pertaining to α and θ in the present case. The conservation equations, in limiting form ($\varepsilon \to 0$), are

$$\frac{d\alpha}{dt} = -\frac{\alpha}{t_{ch}} \exp(\theta) + \frac{(1-\alpha)}{t_{res}} \tag{10.15}$$

$$\frac{d\theta}{dt} = \frac{B\alpha}{t_{ch}} \exp(\theta) - \theta\left(\frac{1}{t_{res}} + \frac{1}{t_N}\right) \tag{10.16}$$

and their partial derivatives are given by

$$A' = \frac{d(d\alpha/dt)}{d\alpha} = -\frac{\exp(\theta)}{t_{ch}} - \frac{1}{t_{res}} \tag{10.17}$$

$$B' = \frac{d(d\alpha/dt)}{d\theta} = -\frac{\alpha\exp(\theta)}{t_{ch}}$$

$$C' = \frac{d(d\theta/dt)}{d\alpha} = -\frac{B\exp(\theta)}{t_{ch}}$$

$$D' = \frac{d(d\alpha/dt)}{d\theta} = -\frac{B\alpha\exp(\theta)}{t_{ch}} - \left(\frac{1}{t_{res}} + \frac{1}{t_N}\right)$$

subject to the stationary state equalities

$$\alpha_s = 1 - \frac{\theta_s}{B}\left(\frac{1}{t_{res}} + \frac{1}{t_N}\right) \tag{10.18}$$

$$\frac{(1-\alpha_s)}{t_{res}} = \frac{\alpha_s}{t_{ch}} \exp(\theta) \tag{10.19}$$

The partial derivatives are the coefficients in the equations

$$\frac{d\alpha'}{dt} = A'\alpha' + B'\theta' + \ldots. \tag{10.20}$$

$$\frac{d\theta'}{dt} = C'\alpha' + D'\theta' + \ldots. \tag{10.21}$$

where α' and θ' are small perturbations from the stationary state (α_s, θ_s) given by

$$\alpha' = \alpha - \alpha_s \text{ and } \theta' = \theta - \theta_s \tag{10.22}$$

When terms in α' and θ' of order greater than unity are neglected, the system of eqns (10.20) and (10.21) is linear and its solutions are linear combinations of the terms $\exp(\lambda_1 t)$ and $\exp(\lambda_2 t)$. The exponents λ_1 and λ_2 are roots to the characteristic equation

$$\begin{vmatrix} A'-\lambda & B' \\ C' & D'-\lambda \end{vmatrix} = 0 \tag{10.23}$$

Table 10.1 The nature of singularities determined by the solutions to the characteristic equation and their local stability

Sign and character of roots λ_1 and λ_2	Condition	Time-dependent response to perturbation	Nature of (α, θ) phase plane singularity
Both real and negative	$A' + D' < 0$ $(A' - D')^2 + 4B'C' > 0$	Monotonic decay	Stable node
Both real and positive	$A' + D' > 0$ $(A' - D')^2 + 4B'C' > 0$	Monotonic growth	Unstable node
Opposite sign	$A'D' - B'C' < 0$	Not accessible	Unstable saddle point
Complex conjugates with negative real parts	$(A' + D') < 0$ $(A' - D')^2 + 4B'C' < 0$	Oscillatory decay	Stable focus
Complex conjugates with positive real parts	$(A' + D') > 0$ $(A' - D')^2 + 4B'C' < 0$	Oscillatory growth	Unstable focus
Pure imaginary	$(A' + D') = 0$	Birth of oscillations	Stable limit cycle

Positive real values for either or both λ_1 and λ_2 signify a growth and departure from the singularity, which means that the stationary state is unstable. Negative real values guarantee decay of the perturbation, which means that the stationary state is stable. Complex and pure imaginary roots λ_1 and λ_2 signify oscillatory evolution in time associated with the singularity, which may be stable or unstable. Since there are only two variables the results may also be represented in a time independent phase plane which shows $\alpha(\theta)$. The conditions and some of their implications for the time dependent and the phase-plane properties are given in Table 10.1. These conditions are called the *Liapounov stability criteria* [204, 205] and they relate to the conditions at which *Hopf bifurcations* [36] may occur as a parameter of the system (such as T_a or t_{res}) is varied continuously. Computational 'path following' techniques, such as AUTO [206], have been developed to trace the stationary state solutions in the multivariable parameter space and to identify the bifurcation points.

These rules can also be used to prove the stability of the upper and lower stationary states for the adiabatic CSTR, and also to show that the intermediate state is a saddle point. Since adiabatic operation is a single-variable problem, oscillatory states cannot exist.

Computations of the time dependent evolution from an initial condition to attain selected stationary state conditions on the $\theta(t_{res})$ diagram for the two-variable, non-adiabatic system are shown in Fig. 10.4. The bifurcation into oscillatory reaction occurs where the stationary state becomes unstable, that is when the condition $A' + D' = 0$ is satisfied. This happens at two points on the upper branch of Fig. 10.4, giving an unstable region in which sustained oscillations occur. The stationary state is approached in an oscillatory manner at stable conditions which exist on either side of the unstable region. A full description of how the events unfold is beyond the scope of the present text.

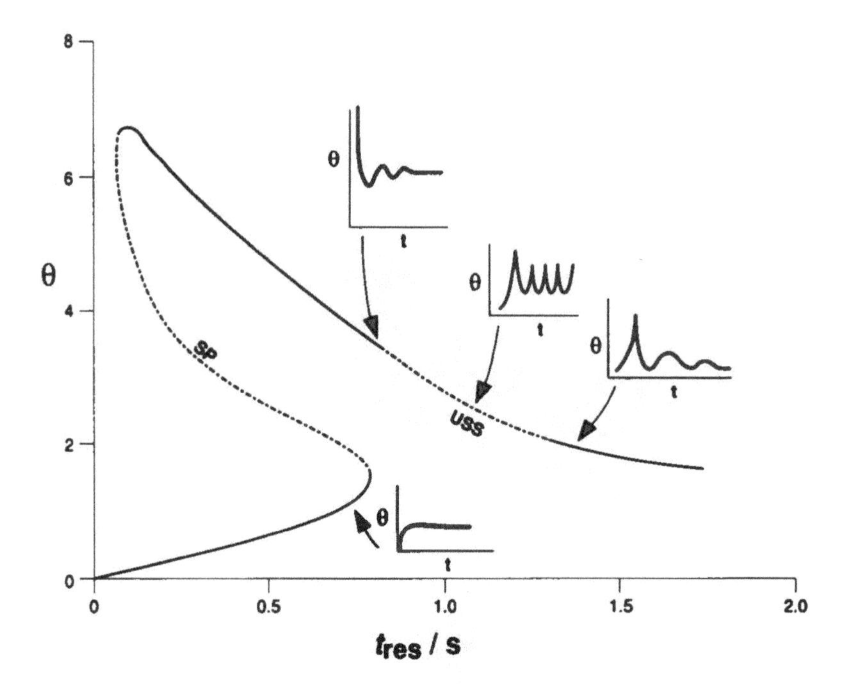

Figure 10.4 The locus and nature of stationary state solutions determined numerically from eqn (10.2) but without making the exponential approximation; $B = 8$, $\varepsilon = 0.027$, $t_{ch} = 7.018$ s and $t_N = 1$ s. The solid lines represent stable stationary states, the broken lines represent unstable stationary states (USS) and saddle points (SP). The time dependent approach to the stationary states are shown in the insets at various points. Integration of eqn (10.2) was made using the NAG library routine Dϕ2EBF. (Calculations courtesy of C. Hildyard).

10.3 Exothermic oxidation at a surface

In their simplest form, the principles of heterogeneous, exothermic oxidation have much in common with those of the adiabatic CSTR. An exothermic oxidation at a surface which is assumed to be first order with respect to the concentration of a gaseous component might be appropriate to the combustion of carbon

$$C(s) + O_2(g) \rightarrow CO_2(g); \quad \Delta H^{\ominus}_{298} = -393\,\text{kJ mol}^{-1} \tag{10.24}$$

It might also represent the oxidation of carbon monoxide by oxygen that is already dissociatively adsorbed on a catalyst surface. The reactive component is then the adsorbed O atom.

$$CO(g) + O(s) \rightarrow CO_2(g); \quad \Delta H^{\ominus}_{298} = -283\,\text{kJ mol}^{-1} \tag{10.25}$$

More detailed features of these types of reactions are discussed in Chapters 11 and 13.

To avoid having to consider diffusion of the gaseous reactant in a quantative way it is useful to represent the supply to the surface as a 'step change' in concentration which is controlled by a surface mass transfer coefficient $\beta(m^{-2}s^{-1})$, as shown in Fig. 10.5. Physically this is more like a convective mass transport process with a characteristic time $t_{tr} = 1/\beta S$, where S is the surface area. The gaseous component is at ambient temperature T_a. The heat loss rate from the surface will be determined by a thermal diffusivity which, for simplicity, we may also characterise as a convective heat transport time (t_{th}), the loss rate being proportional to the temperature difference between the surface and the gas. The heat and mass transport occur through the same physical processes, which means that t_{tr} and t_{th} are linked. Their ratio could be interpreted from the inverse of the Lewis number (section 2.5) which is used to relate thermal diffusivity to the diffusion coefficient. Heat loss by radiation may be important in practice, but this does not change the qualitative picture.

The maximum surface temperature is the adiabatic temperature, given in dimensionless terms as $B' = (t_{th}/t_{tr})Eqc_0/\sigma C_p RT_a^2$ when the ambient temperature is taken as the reference temperature. The ratio (t_{th}/t_{tr}) governs the magnitude of B'. If these times are equal then the surface can reach the thermodynamically determined adiabatic temperature, although this is unlikely in practice. The model is analogous to that for the adiabatic CSTR and is governed by similar conservation equations which take the dimensionless forms

$$\frac{d\alpha}{dt} = \frac{\alpha}{t_{ch}} \exp(\theta/(1+\varepsilon\theta)) - \frac{1}{t_{tr}}(1-\alpha) \tag{10.26}$$

$$\frac{d\theta}{dt} = \frac{B\alpha}{t_{ch}} \exp(\theta/(1+\varepsilon\theta)) - \frac{\theta}{t_{th}} \tag{10.27}$$

In this context the term $(1-\alpha)$ represents fractional extent of consumption of the gaseous reactant at the surface. The system has only one

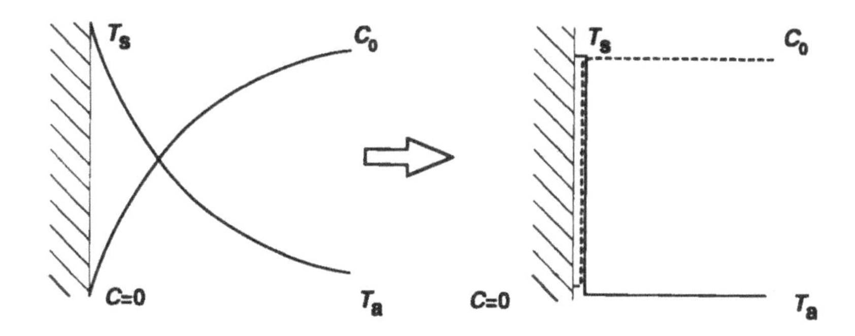

Figure 10.5 Schematic representation of heat and mass transfer at a surface in order to eliminate the spatial dependence of temperature and reactant concentration in the gas phase.

variable because the fractional extent of reaction and the dimensionless temperature excess are linked by the relationship

$$1-\alpha_s=\theta_s/B' \tag{10.28}$$

and the governing energy conservation equation takes the approximate form ($\varepsilon \rightarrow 0$)

$$\frac{d\theta}{dt}=\frac{(B'-\theta)}{t_{ch}}\exp(\theta)-\frac{\theta}{t_{th}} \tag{10.29}$$

Heat release rate ***R***   Heat loss rate ***L***

The behaviour of the system corresponds to that displayed in Fig. 10.2 for the adiabatic CSTR, and the criteria for ignition and extinction can be shown to be, in approximate form,

$$\theta_{ign}\approx 1+1/B'\ldots. \tag{10.30}$$

$$\theta_{ext}\approx B'-1\ldots. \tag{10.31}$$

With reference to the surface combustion of carbon, the reaction rate is given by

$$v=k[O_2] \tag{10.32}$$

where k is a first-order rate constant and $[O_2]$ is the gaseous oxygen concentration. In a stationary state, the rate of oxygen consumption is equal to its rate of arrival at the surface. Thus

$$k[O_2]=\beta S\{[O_2]_o-[O_2]\} \tag{10.33}$$

from which the gas-phase concentration of oxygen close to the surface is given by

$$[O_2]=\frac{[O_2]_o}{1+(k/\beta S)} \quad \text{or} \quad \frac{[O_2]_o}{1+[(t_{tr}/t_{ch}\exp(-(\theta/1+\varepsilon\theta))]} \tag{10.34}$$

Invoking the exponential approximation, the rate of heat release is given by

$$R=k[O_2](-\Delta H^{\ominus})=\frac{[O_2]_o(-\Delta H^{\ominus})}{(1/k+1/\beta S)}=\frac{[O_2]_o(-\Delta H^{\ominus})}{(t_{ch}\exp(-\theta)+t_{tr})} \tag{10.35}$$

In the limit $(t_{ch}\exp(-\theta))\gg t_{tr}(\text{low}\,T_a)$ the reaction rate is governed entirely by reactivity at the surface and the heat release rate is under kinetic control given by

$$R\approx\frac{[O_2]_o(-\Delta H^{\ominus})}{t_{ch}\exp(-\theta)} \tag{10.36}$$

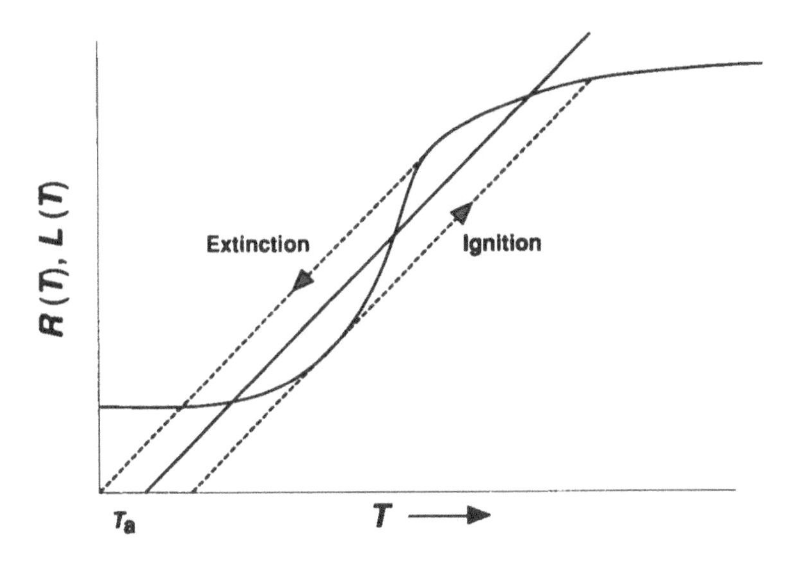

Figure 10.6 Relationships between heat release and heat loss rates ($R(T)$ and $L(T)$) during exothermic heterogeneous reaction as a function of ambient temperature. Ignition and extinction transitions are marked by arrows.

In the limit $t_{tr} \gg (t_{ch}\exp(-\theta))$ the reaction rate is governed entirely by diffusion of the oxygen to the surface and the heat release rate is relatively independent of temperature at its upper limit, given by

$$R = \frac{[O_2]_0(-\Delta H^{\ominus})}{t_{tr}} \tag{10.37}$$

These characteristics are displayed in Fig. 10.6, in which 'ignition' and 'extinction' are also presented. The gradient of the heat loss line relates to t_{th}.

10.4 Unifying links between thermal ignition, reaction in a CSTR and at surfaces

The associations that have emerged for first order exothermic reactions are shown in Table 10.2. The behaviour in thermal ignition, an adiabatic CSTR or at a surface is governed by two characteristic times which represent the heat or mass transport rates (given by t_N, t_{res} or t_{tr}) and the chemical heat release rate (t_{ch}). The non-adiabatic CSTR is distinguished by the decoupling of the heat and mass transport rates, giving three characteristic times which govern its behaviour (t_{ch}, t_N and t_{res}). This decoupling opens up the possibility of oscillatory states in a simple kinetic system with thermal feedback.

Table 10.2 Relationship between theories for thermal ignition and reaction in a CSTR or at surfaces[a]

System	Ignition (θ_{ign})	Extinction (θ_{ext})	Stationary states
Homogeneous (Chapter 7)	1	—	$\theta \exp^{-\theta} = \frac{B t_N}{t_{ch}}$
Heterogeneous	$1 + 1/B'$	$B' - 1$	$\frac{\theta \exp^{-\theta}}{1 - \theta/B'} = \frac{B' t_{th}}{t_{ch}}$
Adiabatic CSTR	$1 + 1/B$	$B - 1$	$\frac{\theta \exp^{-\theta}}{1 - \theta/B} = \frac{B t_{res}}{t_{ch}}$
Non-adiabatic CSTR	$1 + 1/B^{\bullet}$	$B^{\bullet} - 1$	$\frac{\theta \exp^{-\theta}}{1 - \theta/B'} = \frac{B^{\bullet} t_{res}}{t_{ch}}$

[a] $B' = \frac{t_{th} B}{t_{tr}}$; $B^{\bullet} = \frac{B}{(1 + t_{res}/t_N)}$

10.5 The interpretation of oscillatory cool flames and multiple-stage ignitions

The origins of oscillatory cool flames in a closed vessel are most easily understood by addressing the simplest formal structure which is able to account for these states in a unified thermokinetic theory [207, 208]. Spatially uniform concentrations and temperatures are assumed and reactant consumption (species A) is ignored for the purpose of analytical interpretation. The system of equations is set up in two-variables and the structure involves only linear interactions of the reactive intermediate species X, of the form

Reaction	Type	Exothermicity		
$A \rightarrow X$	initiation	q_i	(i)	(10.38)
$X \rightarrow$ inert	termination 1	q_1	(1)	(10.39)
$X \rightarrow 2X$	branching	q_b	(b)	(10.40)
$X \rightarrow$ inert	termination 2	q_2	(2)	(10.41)

This branching chain scheme does not explicitly consider any propagation processes, regarding them to be thermoneutral. The mechanism is subject to the condition for eqns (10.39)–(10.41) that $E_1 < E_b < E_2$, which is the only requirement to ensure that an overall negative temperature dependence of reaction rate is possible within a limited range of temperature. The temperature dependence of the rate coefficients is expressed in the Arrhenius form $k = A\exp(-E/RT)$. The equations representing conservation of mass of the intermediate species X and the reactant temperature are given by

$$\frac{dx}{dt} = v_i + [k_b - (k_1 + k_2)]x \qquad (10.42)$$

$$C\frac{dT}{dt}=v_i q_i + x(k_b q_b + k_1 q_1 + k_2 q_2 - l(T-T_a) \qquad (10.43)$$

The concentration of intermediate X is given by x, q_i and v_i are the exothermicity and rate of initiation respectively, C is a volumetric heat capacity, and l is a (Newtonian) heat transfer coefficient incorporating the vessel surface to volume ratio. $(T-T_a)$ is the difference between the reactant and vessel temperatures. The problem may be considered in terms of the two independent variables T and x if the concentration of the primary reactant A does not change significantly throughout reaction (the 'pool chemical approximation'), so that v_i is then regarded to be constant. The independent variables are linked non-linearly through the temperature dependence of reaction rates.

Yang and Gray [207, 208] enumerated the singularities of this system and investigated their properties. The stationary states satisfy the thermal equation

$$v_i q_i + v_i \frac{(k_b q_b + k_1 q_1 + k_2 q_2)}{[k_b - (k_1 + k_2)]} - l(T-T_a) = 0 \qquad (10.44)$$

and, when subject to the condition $E_1 < E_b < E_2$, they exist at the points of intersection between a heat release rate curve of the form shown in Fig. 10.7 [207, 208]. $\boldsymbol{R}(T)$ represents the combined heat release rate terms above, and $\boldsymbol{L}(T)$ represents the (linearly temperature dependent) heat loss rate. The local stability of the singularities is determined by application of the Liapounov stability criteria (Table 10.1). The singularities on the branch marked AB (at which a negative temperature dependence of heat release rate occurs) may be unstable states. This signifies that oscillatory cool flames or two stage ignition would take place when the system is driven into this regime. There is no simple representation to distinguish these types of behaviour in a thermal diagram like Fig. 10.7, since a third dimension is required to represent the behaviour of the singularity in the co-ordinate for x, and multiple dimensions when more complex representations of the chemistry involve the concentrations $x_1..x_n$. A two-dimensional analysis is made by examining the behaviour of the system in the time independent T–x (temperature–chain carrier concentration) phase plane. The thermokinetic stationary states correspond to singularities in this phase plane (there are up to five for the particular formal scheme under discussion) and the nature of each singularity furnishes much information about the way the system approaches that particular stationary state [209].

In addition to the analytical development, Yang and Gray [207] and Yang [209] simulated extremely successfully, by numerical computation, the main features of the typical p–T_a ignition diagram (Fig. 9.8) at conditions which closely resembled the experimental results for propane oxidation in a closed vessel. This numerical work, dated 1969, may be

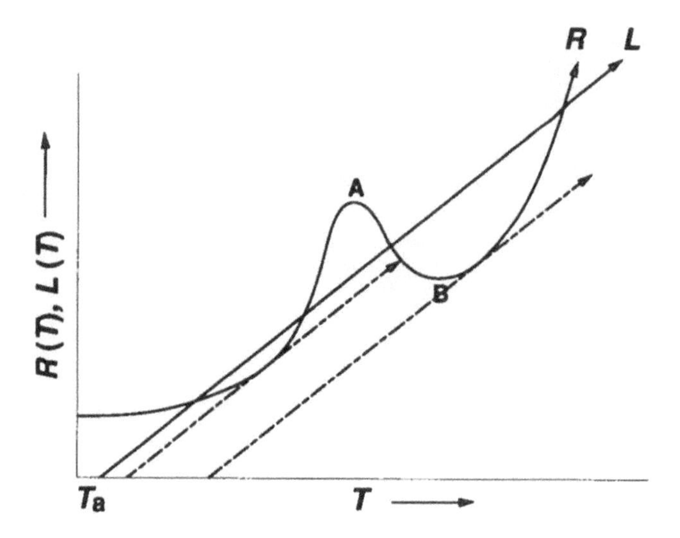

Figure 10.7 Dependence of heat release rate on temperature according to the Yang and Gray thermokinetic model [207, 208]. There can be up to five intersections between the heat release and heat loss rates (***R*** and ***L***) and two different types of critical transitions. AB marks the region in which the heat release rate exhibits a negative temperature dependence.

regarded as the forerunner to all numerical computation of thermo-kinetic phenomena involving structured kinetic schemes. With some modification the scheme can also be used to predict multistage ignition as a three-variable ([A], [X], T) numerical problem for simulating the behaviour in a CSTR [210].

Alkylperoxy radical isomerisation theories were only in their infancy in 1969 [119, 120], but the elementary reactions known today to be associated with hydrocarbon oxidation may not be readily identified in the Gray and Yang scheme, even in qualitative form. However, the interaction between chain branching and non-branching reaction modes which is controlled by the relative magnitudes of the activation energies and coupled to thermal feedback, so clearly represented in this scheme, is the cornerstone to the interpretation of cool flames and two-stage ignitions in current thermokinetic models regardless of their complexity. The simplicity of the structure, based on linear concentration dependences and reducible to two dimensions so that analytical insights could be sought, is the benchmark for 'reduced kinetic schemes' to represent hydrocarbon autoignition in the simulation of engine combustion [4].

The closest approach of 'real chemistry' is in the simplest thermo-kinetic model to represent acetaldehyde (ethanal) combustion, although even this is a multivariable problem, which is amenable only to numerical analysis. In acetaldehyde oxidation at about 500 K, acetyl radicals

(CH_3CO) give rise to a degenerate branching intermediate, peracetic acid, through the following reactions

$$CH_3CHO + O_2 \rightarrow CH_3CO + HO_2 \quad \text{initiation} \quad (10.45)$$

$$CH_3CO + O_2 \rightarrow CH_3CO_3 \quad \text{propagation} \quad (10.46)$$

$$CH_3CO_3 + CH_3CHO \rightarrow CH_3CO_3H + CH_3CO \quad \text{propagation} \quad (10.47)$$

Chain branching and OH radical propagation occur by

$$CH_3CO_3H \rightarrow CH_3 + CO_2 + OH \quad \text{branching} \quad (10.48)$$

$$CH_3CHO + OH \rightarrow CH_3CO + H_2O \quad \text{propagation} \quad (10.49)$$

The branching step has a high activation energy ($E > 150\,\text{kJ mol}^{-1}$) so the autocatalysis develops only slowly, but its rate may be enhanced considerably as a result of thermal feedback. The formation of the branching agent is governed by eqn (10.47) which has a relatively low activation energy ($E \approx 40\,\text{kJ mol}^{-1}$). That for eqn (10.46) is negligible. However, there is also a competitive, unimolecular decomposition of acetyl radicals,

$$CH_3CO + M \rightarrow CH_3 + CO + M \quad E > 80\,\text{kJ mol}^{-1} \quad (10.50)$$

which becomes increasingly influential as the reactant temperature rises. This may be regarded to be a non-branching reaction mode, although that depends on how the methyl radical reactions are constructed. A satisfactory termination takes the form

$$CH_3 + CH_3 \rightarrow C_2H_6 \quad (10.51)$$

and it may be necessary to consider a surface termination of the peracetyl radicals

$$CH_3CO_3 \rightarrow \text{inert} \quad (10.52)$$

This combination yields the condition $E_{10.52} \approx E_{10.46} < E_{10.50}$, in keeping with the Gray and Yang criteria, and has been used as a basis for the simulation of oscillatory cool flames during the non-isothermal oxidation of acetaldehyde [211].

Equations (10.45)–(10.52) are only a skeleton scheme. If the temperature is sufficently low, methyl radical oxidation can yield methylperoxy radicals and, with acetaldehyde able to provide a very labile H atom from the acyl group, methyl hydroperoxide can be formed (section 7.2). A satisfactory mechanism for acetaldehyde oxidation in the temperature range $T_a = 500$–750 K then becomes very much more complicated than appears to be necessary at first. The non-isothermal phenomena exhibited by acetaldehyde in a CSTR [194, 195] have been simulated from kinetic schemes which the reactions described here and in section 7.2 [137, 212]. Methane is unable to exhibit similar low temperature/low pressure spontaneous ignition and cool flame characteristics because the abstraction of an H atom from it is so difficult.

Further reading

Andranov, A.A., Vitt, A.A. and Khaikin, S.E. (1966). *Theory of Oscillators*. Pergamon Press, Oxford, UK.
Davis, H.T. (1962). *Introduction to Nonlinear Differential and Integral Equations*. Dover, New York, USA.
Denbigh, K.G. and Turner, J.C.R. (1984). *Chemical Reactor Theory, An Introduction* (3rd edn). Cambridge University Press, Cambridge, UK.
Frank-Kamenetskii, D.A. (1969). *Diffusion and Heat Transfer in Chemical Kinetics* (Translated by J.P. Appleton). Plenum Press, New York, USA.
Gray, B.F. (1974). Kinetics of oscillating reactions. In *Specialist Periodicals of the Chemical Society, Reaction Kinetics* (Vol. 1). The Chemical Society, London, UK, p. 309.
Gray, P. (1990). Chemistry and Combustion. In *Twenty-Third Symposium (International) on Combustion*. The Combustion Institute, Pittsburgh, PA, USA, p.1.
Gray, P. and Scott, S.K. (1990). *Chemical Oscillations and Instabilities*. Oxford Science Publications, Oxford, UK.
Griffiths, J.F. (1985). *Ann. Rev. Phys. Chem.*, **36**, 77.
Griffiths, J.F. (1986). *Adv. Chem. Phys.*, **LXIV**, 203.
Griffiths, J.F. (1995). *Prog. Energy Combust. Sci.*, **20**, 461.
Griffiths, J.F. and Scott, S.K. (1987). *Prog. Energy Combust. Sci.*, **13**, 161.
Lignola, P.-G. and Reverchon, E. (1987). *Prog. Energy Combust. Sci.*, **13**, 1.
Minorsky, N. (1962). *Nonlinear Oscillations*. Krieger, New York, USA.
Vulis, L.A. (1961). *Thermal Regimes in Combustion* (Translated by M.D. Friedman). McGraw Hill, New York, USA.
Zel'dovich, Ya. B., Barenblatt, G.I., Librovich, V.B. and Makhviladze, G.M. (1985). *Mathematical Theory of Combustion and Explosion*. Consultants Bureau, New York, USA.

Problems

(1) Use eqn (10.1) to derive the extent of conversion in a stationary state (α_s) as a function of t_{ch} and t_{res} during *isothermal* reaction in a CSTR. Hence plot α_s for the decomposition of di-*t*-butyl peroxide vapour as a function of vessel temperature over the temperature range 460–540 K in intervals of 10 K at a mean residence time of 3 s ($A = 2 \times 10^{15}\,s^{-1}$, $E/R = 18280$ K).

(2) Calculate the adiabatic temperature excess in an adiabatic CSTR (a) for di-*t*-butyl peroxide vapour at the inflow, (b) for 10% di-*t*-butyl peroxide vapour in nitrogen at the inflow. Hence, calculate for pure di-*t*-butyl peroxide vapour (a) B at 510 K in an adiabatic CSTR, (b) B* at 510 K in a non-adiabatic CSTR (0.5 dm^3 sphere) at a

mean residence time of 3 s. $E/R = 18280$ K, $\Delta H^{\ominus}_{510} = -180\,\mathrm{kJ\,mol^{-1}}$, $h = 15\,\mathrm{W\,m^{-2}\,K^{-1}}$, C_p(pure vapour)$=250\,\mathrm{J\,mol^{-1}\,K^{-1}}$, C_p(10% mixture$= 52\,\mathrm{J\,mol^{-1}\,K^{-1}}$, $\sigma = 0.63\,\mathrm{mol\,m^{-3}}$.

(3) Studies of a liquid-phase first-order exothermic reaction in a CSTR yield values for the rate constant as follows; at 333 K $k = 0.0101\,\mathrm{s^{-1}}$ and at 363 K $k = 0.0516\,\mathrm{s^{-1}}$. Estimate the activation energy assuming an Arrhenius temperature dependence. Given that the volume of the reactor is $10\,\mathrm{m^3}$ and the feed stream contains reactant at a concentration of $5\,\mathrm{kmol\,m^{-3}}$, at a feed rate of $0.032\,\mathrm{m^3\,s^{-1}}$ the stationary state temperature in the vessel is found to be 353.5 K. Calculate (a) the rate constant at 353.5 K, (b) the mean residence time of the fluid in the vessel, (c) the extent of conversion of the reactant, and (d) the concentration of reactant at the outlet.

11 Aspects of mixed and condensed phase combustion

11.1 Introduction

Aspects of heterogeneous combustion are considered in this chapter which relate to the combustion of droplets, pools of flammable liquids and kinetic aspects of carbon or char oxidation. Some practical applications which involve heterogeneous combustion are also discussed. Polymer combustion is discussed in the context of combustion hazards and fire retardancy (Chapter 12). The combustion of coal is used in an illustrative context in some of the following Sections, but it is not discussed as a major theme in this book. This is a substantial topic, which commands considerable attention worldwide because of the economic and environmental factors involved.

In liquid–gas and solid–gas systems, combustion may occur either at the surface or in the gas phase. The latter is likely to occur if the heat released by the reaction promotes vaporisation, melting, sublimation or degradation of the condensed phase material. Commonly the vaporisation of liquids precedes reaction so that combustion occurs in the gas phase, but solids may decompose quite readily also to form volatile components. In addition, the dispersion or the size of the liquid or solid particles is important. This factor determines the extent to which the gaseous fuel is able to mix with the oxidant. In heterogeneous systems, the products of combustion tend to separate the fuel from the oxidant but this protection is reduced if the gas is in motion relative to the condensed material. Such motion will tend to promote contact between the fuel and oxidant but, simultaneously, may reduce the rate of heat transfer to the fuel.

11.2 The burning of a liquid droplet

The different situations which may occur when droplets burn in air are illustrated in Fig. 11.1 [213]. If a liquid is dispersed in very fine droplets, say below 10 μm in diameter, it will vaporise completely in the preheat zone and the resultant flame will be a typical premixed flame, the fuel

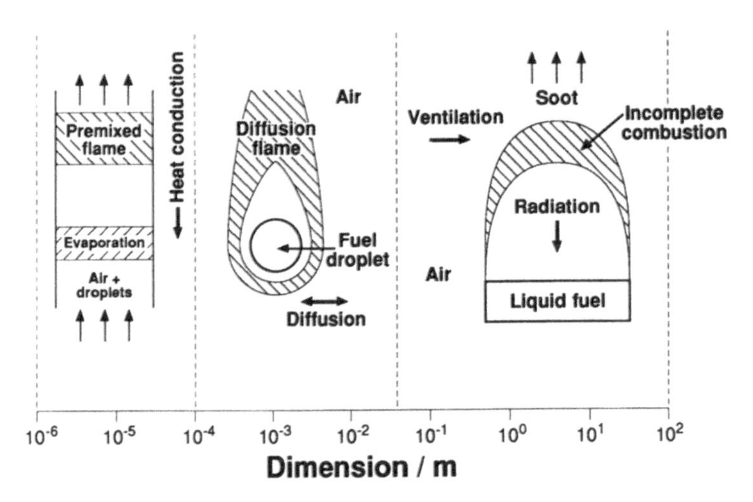

Figure 11.1 Different types of phenomena that are observed during the combustion of liquid droplets of different sizes. The bold arrows indicate the major physical processes that dominate the burning rate (after Fristrom and Westenberg [213]).

and oxidant being mixed intimately within the reaction zone. The critical dimension of a droplet below which a premixed flame may be regarded to occur depends on the rate at which the droplet is heated. This is governed by the final flame temperature and the time which the droplets spend in the preheat zone. Smaller droplets are required for complete vaporisation and mixing at higher burning velocities. The critical size depends also on the latent heat of vaporisation of the fuel, and on the thickness of the flame.

If the liquid droplet exceeds the critical size, but is less than about a millimetre in diameter, the combustion takes the form of a spherical diffusion flame around the droplet. The burning rate is then determined by the rate of evaporation from the surface of the droplet. The behaviour is closely related to the diffusion flame formed on the wick of a paraffin lamp or a candle. The degree of turbulence also may play a part since interdiffusion of reactants and convective heat transfer depend on such motion.

In simple theoretical analyses of droplet burning [214–218] it is assumed that the liquid is at uniform temperature close to its boiling point (which fixes the vapour pressure at atmospheric pressure and the gaseous mole fraction at unity). Fuel evaporates from the surface and diffuses into the surrounding atmosphere. A diffusion flame is established at the position where fuel and oxidant are in stoichiometric proportions and the heat released there is transported back to the surface to maintain the temperature of the droplet and hence the supply of fuel by evaporation. The chemical reaction rate is assumed to be sufficiently fast for it not to be rate controlling and radiant heat transfer is neglected.

Following Hayhurst and Nedderman [218], a droplet, radius a, is surrounded by a spherical flame, radius r_f. The flame is assumed to have negligible thickness and the oxygen concentration at the flame surface is zero. It is known that the radius of the droplet decreases with time according to a relationship of the form

$$\frac{-\mathrm{d}a^2}{\mathrm{d}t} = k_b \tag{11.1}$$

in which k_b is a burning rate constant. The objective is to derive an expression to predict the rate of combustion, k_b, and also to predict the flame temperature, T_f. If the surface temperature, T_s, of the droplet is assumed to be at the boiling point of the liquid, it is necessary to consider only (i) the mass transfer of oxygen into the flame, (ii) the heat transfer from the flame to the droplet, and (iii) the heat transfer away from the flame. Since both convective heat and mass transport are involved, a convenient simplification is to assume that the Lewis number, Le, is unity, although the following derivation can be concluded without it. In addition, the system is considered to have a temperature independent, effective diffusivity, D, which is a necessary simplification for this algebraic treatment.

Let the chemical reaction at the interface between fuel vapour and oxidant be represented by the formal stoichiometric equation

$$\mathrm{F} + s\mathrm{O}_2 = n\mathrm{P} \tag{11.2}$$

If the molar evaporation rate of the fuel (F) is $\dot{M}$(mol s^{-1}), then in a quasi-stationary state the molar outflow rate of product (P) is $n\dot{M}$(mol s^{-1}) and the molar supply rate of oxygen is $s\dot{M}$(mol s^{-1}).

11.2.1 Diffusion of oxygen outside the flame front

The conservation equation for the molar flux of oxygen through the surface of an imaginary sphere, radius r, is

$$\frac{-s\dot{M}}{4\pi r^2} = \frac{(n-s)\dot{M}x_{ox}}{4\pi r^2} - \frac{DN\mathrm{d}x_{ox}}{\mathrm{d}r} \tag{11.3}$$

Inward molar flux	Net convective molar flux of O_2	Diffusive transport

in which N is the number of moles per unit volume. It is assumed that the total flux of all species is $(n-s)\dot{M}/4\pi r$. Thus if ideal gas behaviour is assumed, at total pressure p and temperature T

$$\frac{-s\dot{M}}{4\pi r^2} = \frac{(n-s)\dot{M}x_{ox}}{4\pi r^2} - \frac{Dp\mathrm{d}x_{ox}}{RT\mathrm{d}r} \tag{11.4}$$

The mole fraction of oxygen is given by x_{ox}. The flux of nitrogen can be assumed to be negligible since the contraction of the droplet exerts negligible effect on its gaseous environment. Rearrangement of eqn (11.4) yields

$$\frac{\dot{M}}{4\pi}\int_{r_f}^{\infty}\frac{dr}{r^2}=\frac{Dp}{RT}\int_{0}^{x_\infty}\frac{dx_{ox}}{(n-s)x_{ox}+s} \qquad (11.5)$$

where x_∞ represents the limiting mole fraction of oxygen, which in air is 0.21.
Integration of eqn (11.5) gives

$$\frac{\dot{M}}{4\pi r_f}=\frac{Dp}{RT(n-s)}\ln\left\{1+\left(\frac{n-s}{s}\right)x_\infty\right\} \qquad (11.6)$$

11.2.2 Heat balance at $r < r_f$

Fuel vapour diffuses outward in the region $a<r<r_f$ and heat is conducted inward to the droplet surface. The total heat flux past a surface of radius r is given by

$$4\pi r^2\left(-\kappa_F\frac{dT}{dr}\right) \quad + \quad \dot{M}(H_F^{\ominus}+C_F T)$$

Conductive heat transport inward — Conductive heat transport outward (11.7)

In this equation κ_F is the (average) thermal conductivity of the fuel-rich gases in the temperature range T_s to T_f. The second term represents the molar enthalpy of the vapour leaving the droplet, which comprises its standard molar enthalpy, $H_F^{\ominus}$, and a component involving the (average) heat capacity C_F of the fuel. The heat balance at the droplet surface, taken to be at the liquid boiling point, is given by

$$\dot{M}(-\Delta H_{vap}^{\ominus})+M(H_F^{\ominus}+C_F T_s) \qquad (11.8)$$

in which $\Delta H_{vap}^{\ominus}$ represents the enthalpy of vaporisation.
Equating eqns (11.7) and (11.8), with some rearrangement, gives

$$4\pi r^2\left(-\kappa_F\frac{dT}{dr}\right)=\dot{M}[\Delta H_{vap}^{\ominus}+C_F(T-T_s)] \qquad (11.9)$$

or

$$\frac{\dot{M}}{4\pi}\int_{a}^{r_f}\frac{dr}{r^2}=\int_{T_s}^{T_f}\frac{\kappa_F dT}{\Delta H_{vap}^{\ominus}+C_F(T-T_s)} \qquad (11.10)$$

Integration of eqn (11.10) gives

$$\frac{\dot{M}}{4\pi}\left(\frac{1}{a}-\frac{1}{r_f}\right)=\frac{\kappa_F}{C_F}\ln\left\{1+\frac{C_F(T_f-T_s)}{\Delta H^{\ominus}_{vap}}\right\} \qquad (11.11)$$

11.2.3 Heat balance at r > r$_f$

If radiation is ignored, the rate of heat transport outside the flame (past an imaginary sphere of radius r) is also equal to the rate at which heat arrives at the droplet surface (eqn (11.8)). Thus, in the flame front, enthalpy is neither created nor destroyed, but there is heat released when a chemical reaction proceeds in this adiabatic fashion. In a stationary state the heat balance at $r>r_f$ is given by

$\dot{M}(-\Delta H^{\ominus}_{vap})$	+	$\dot{M}(H^{\ominus}_F+C_fT_s)=$		
Latent heat transported inwards		Enthalpy convected by vapour outwards		
$4\pi^2\left(-\kappa_{ox}\frac{dT}{dr}\right)$	−	$\dot{M}\{s(H^{\ominus}_{ox}+C_{ox}T)$	+	$n(H^{\ominus}_P+C_pT)\}$ (11.12)
Conductive heat transported inwards		Convective heat transported inwards		Convective heat transported outwards

The thermal conductivity, κ_{ox}, is an averaged value representing the fuel-lean environment. The enthalpy of combustion is given by

$$\Delta H^{\ominus}=nH^{\ominus}_P-sH^{\ominus}_{ox}-H^{\ominus}_F \qquad (11.13)$$

so eqn (11.12) simplifies to

$$-(\Delta H^{\ominus}+\Delta H^{\ominus}_{vap})=\dot{M}[T(nC_p-sC_{ox})-C_FT_s]-4\pi r^2\kappa_{ox}\frac{dT}{dr} \qquad (11.14)$$

It is helpful to simplify eqn (11.14) as follows. Suppose that

$$|\Delta H^{\ominus}|\gg\Delta H^{\ominus}_{vap}-C_FT_s \qquad (11.15)$$

as is borne out in practice, and that a mean heat capacity may be defined in the fuel-lean mixture at $r>r_f$, given by

$$nC_p-sC_{ox}=(n-s)C_o \qquad (11.16)$$

Then eqn (11.14) becomes

$$-\dot{M}\Delta H^{\ominus} = \dot{M}T(n-s)C_o - 4\pi r^2 \kappa_{ox}\frac{dT}{dr} \qquad (11.17)$$

Rearrangement of eqn (11.17) for integration gives

$$\frac{\dot{M}}{4\pi}\int_{r_f}^{\infty}\frac{dr}{r^2} = \int_{T_f}^{T_\infty}\frac{\kappa_o dT}{(n-s)C_o T + \Delta H^{\ominus}} \qquad (11.18)$$

from which

$$\frac{\dot{M}}{4\pi r_f} = \frac{\kappa_o}{(n-s)C_o}\ln\left\{\frac{(n-s)C_o T_\infty + \Delta H^{\ominus}}{(n-s)C_o T_f + \Delta H^{\ominus}}\right\} \qquad (11.19)$$

11.2.4 An expression for the flame temperature, T_f

Equating eqns (11.6) and (11.19) gives

$$\frac{Dp}{RT}\ln\left\{1 + \left(\frac{n-s}{s}\right)x_\infty\right\} = \frac{\kappa_o}{C_o}\ln\left\{\frac{(n-s)C_o T_\infty + \Delta H^{\ominus}}{(n-s)C_o T_f + \Delta H^{\ominus}}\right\} \qquad (11.20)$$

The terms $\kappa_o RT/C_o Dp$ represent the Lewis number which, if taken to be unity, gives a solution for the flame temperature as

$$T_f = \frac{sT_\infty}{s + (n-s)x_\infty} - \frac{\Delta H^{\ominus} x_\infty}{C_o[s + (n-s)x_\infty]} \qquad (11.21)$$

To illustrate a point of principle, supposing that $s \gg (n-s)x_\infty$, which is not a bad approximation for the combustion of hydrocarbons in air, then

$$\Delta T = T_f - T_\infty \approx \frac{-\Delta H^{\ominus} x_\infty}{C_o s} \qquad (11.22)$$

This expression shows that the temperature excess in the flame surrounding the droplet is governed by the mole fraction of oxygen in the external atmosphere and the stoichiometric coefficient for oxygen in the overall combustion process (eqn (11.2)). ΔT is appreciably lower than the adiabatic temperature excess.

11.2.5 An expression for the burning rate constant, k_b

Addition of eqns (11.11) and (11.19) gives

$$\frac{\dot{M}}{4\pi a}=\frac{\kappa_F}{C_F}\ln\left[1+C_F\frac{(T_f-T_s)}{\Delta H^\ominus_{vap}}\right]+\frac{\kappa_O}{(n-s)C_O}\ln\left\{\frac{(n-s)C_O T_\infty+\Delta H^\ominus}{(n-s)C_O T_f+\Delta H^\ominus}\right\} \quad (11.23)$$

which, from eqn (11.20) at $Le=1$ and eqn (11.21), may be written as

$$\frac{\dot{M}}{4\pi a}=\frac{\kappa_F}{C_F}\ln\left\{1+C_F\left[\frac{C_O T_\infty-\Delta H^\ominus x_\infty/s-C_O T_s(1+(n-s)x_\infty/s)}{C_O\Delta H^\ominus_{vap}(1+(n-s)x_\infty/s)}\right]\right\}$$

$$+\frac{\kappa_O}{(n-s)C_O}\ln\left\{1+\left(\frac{n-s}{s}\right)x_\infty\right\} \quad (11.24)$$

Since

$$\dot{M}=\frac{-\mathrm{d}((4/3)\pi a^3\sigma_{liq})}{\mathrm{d}t}=-4\pi a^2\sigma_{liq}\frac{\mathrm{d}a}{\mathrm{d}t} \quad (11.25)$$

or

$$\frac{-\mathrm{d}a^2}{\mathrm{d}t}=k_b=\frac{\dot{M}}{2\pi a\sigma_{liq}} \quad (11.26)$$

$$k_b=\frac{2\kappa_F}{C_F\sigma_{liq}}\ln\left\{1+C_F\left[\frac{C_O T_\infty-\Delta H^\ominus x_\infty/s-C_O T_s(1+(n-s)x_\infty/s)}{C_O\Delta H^\ominus_{vap}(1+(n-s)x_\infty/s)}\right]\right\}$$

$$+\frac{\kappa_O}{(n-s)C_O}\ln\left\{1+\left(\frac{n-s}{s}\right)x_\infty\right\} \quad (11.27)$$

In practice the second term in the right-hand side of eqn (11.27) makes only about 2% contribution to the predicted value of k_b.

The simplest expression of all for k_b takes the form

$$k_b=\frac{2\kappa_F}{C_F\sigma_{liq}}\ln\left\{1+\frac{C_F}{\Delta H^\ominus_{vap}}\left[(T_\infty-T_s)-\frac{\Delta H^\ominus x_\infty}{C_O s}\right]\right\} \quad (11.28)$$

when the same approximation as in eqn (11.22) is invoked and the second term of eqn (11.23) is neglected. Althought the quantitative accuracy of this relationship is limited, it serves to show which properties are likely to be most important in determining the burning rate of a droplet.

If the second term in eqn (11.23) is neglected but $Le = 1$ has not been assumed [218], a useful and more precise expression than eqn (11.28) for the burning rate constant is obtained:

$$k_b = \frac{2\kappa_F}{C_F \sigma_{liq}} \ln \left\{ 1 + \frac{C_F}{\Delta H^{\ominus}_{vap}} \left[(T_\infty - T_s) - \frac{\Delta H^{\ominus} x_\infty}{Le\, C_o s} \right] \right\} \quad (11.29)$$

In practical combustion systems, droplets are usually burned in sprays and unless the spray is very dilute the single droplet model may not be applicable. Furthermore, in high-speed diesel engines spray, combustion may take place under high pressures where the accuracy of the quasi stationary-state model is reduced. Theoretical treatments of droplet combustion under these more extreme conditions are available but further refinement is still needed before they can be regarded as completely satisfactory [217].

11.2.6 Droplet combustion in heavy fuel oil burners

In industrial fuel oil burners, the oil is 'atomised' or dispersed as fine droplets in a stream of air. The droplet size usually covers a wide range, but the majority of droplets are less than 50 μm in diameter in order to complete combustion rapidly in a sufficiently small chamber volume. In addition, with larger droplets of the heavier oils, there is a tendency to form a solid carbon residue which can take much longer than the liquid fuel to burn.

Atomisation of the fuel oil may be achieved in two ways. In the mechanical, or oil-pressure, atomising burner, the oil is forced at high pressure into a 'whirl' chamber through a series of tangential slots and then passes through an orifice into the combustion chamber. The static pressure is converted to rotational motion in the whirl chamber so that when the oil leaves the orifice the centrifugal force throws the oil outwards in a fine spray. In an air atomising burner, streams of both air and fuel are brought together, the necessary dispersion being produced on mixing. The two streams may either pass through separate orifices and converge in the combustion chamber or they may be mixed first and then flowed through a single orifice.

11.3 Pool fires

In quantitities exceeding about 1 cm^3, the fuel behaves as a burning 'pool', the rising volume of hot gas from which entrains air, restricts diffusion of reactants and reduces heat transfer by conduction (Fig. 11.2 [219]). The combustion zone becomes separated from the surface of the fuel and the temperature gradient is severely reduced so that eventually

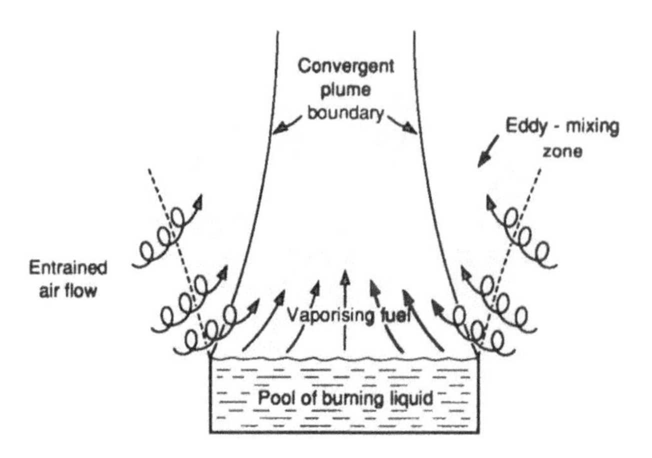

Figure 11.2 A schematic representation of a pool fire (after Herzberg [219]).

transport of heat occurs primarily by radiation. Because of the restricted access of the oxidant to the fuel, the flame becomes very rich and, in the case of organic materials, soot is produced. The soot provides the radiation source which transfers heat to the fuel. Fires on large pools of fuel are very much affected by gas motion (such as that set up by winds) and, when the fire becomes large enough to produce its own meteorological effects, it is termed a *mass fire*. When motion in the surrounding air causes rapid rotation of the hot gases, a *fire whirl* may be set up.

For pool diameters of less than 50 mm the burning is usually laminar and the burning rate per unit area is roughly inversely proportional to the diameter of the pool. With larger pools, the burning rate is independent of diameter. Methanol fires produce relatively little radiation and so their burning rate is constant even in pools of diameter greater than 1 m. For most liquid fuels, above diameters of about 0.3 m the radiative heat transfer and burning rate increase markedly with diameter.

In the intermediate region convective heat transfer is the controlling influence on the flame. The low density flame gases rising into the denser ambient air above cause mixing of air with fuel vapour supplied by the pool surface. The resulting turbulent motion encourages convective heat transfer to the vaporising surface and thus controls the rate of burning. While a complete theoretical treatment of pool fires has yet to be achieved, some useful semi-empirical equations have been developed which make it possible to predict burning rates for pools of various liquids [220, 221].

The burning rate also depends on heat transfer within the liquid. The surface of a burning pool is usually just below its boiling point and it is sometimes possible to extinguish a fire by stirring liquid from below so that cool liquid of low vapour pressure is brought to the surface [222,

223]. The *fire point* is the lowest temperature of a liquid surface at which sustained diffusional burning of the liquid can occur (see Chapter 12) and this is dependent on mass and heat transfer considerations which in turn influence the flame temperature [224].

11.4 Heterogeneous combustion of coal and carbonaceous chars

In a number of solid–gas systems, the combustion process occurs on the surface, the oxidant being adsorbed there prior to reaction. The heat released then normally causes the products to be liberated rapidly as gases. Coal combustion provides a good example of the way the controlling process is influenced by the conditions under which reaction is taking place. There is an initial stage during which very considerable yields of volatile combustible material are given off, and an involatile residue of porous carbon is left behind, which is called char, or coke [225]. The volatile material may constitute up to 90% of the original mass of the coal [226]. Three different regimes, or *Zones*, of the subsequent combustion of the char can be distinguished [227].

Although the chemical reaction takes place on the surface, it is not necessarily the rate-determining step. The complete process takes place by a sequence of events and any one of these may control the rate. The first step is the transfer of reactant (normally oxygen) from the bulk gas through the stagnant layer of gas adjacent to the surface of the solid particle. The reactant is then adsorbed and reacts with the solid before the gaseous products are released and diffuse away. If, as is frequently the case, the solid is highly porous then much of the available surface can be reached only by diffusion along the relatively narrow pores and this process may be rate controlling. The formation of a solid involatile combustion product (ash) may also complicate matters. Control may thus be exercised by either gas–film diffusion, adsorption and chemical reaction or pore diffusion. Diffusion of products away from the surface is seldom rate controlling.

As discussed in section 10.3, kinetic control will occur if the surface reaction is slow compared with the diffusion processes. Since the surface reaction depends on the Arrhenius term, $\exp(-E/RT)$, whereas diffusion is only weakly temperature dependent, kinetic control predominates at low temperatures while diffusion control becomes more important at higher surface temperatures.

In Zone I, the rate of diffusion to and away from the surface is very fast compared with the rate of the surface reaction. For a spherical particle the rate of reaction per unit external surface area is

$$\frac{dn_{O_2}}{dt}=\frac{1}{6}\,S\sigma_s a[O_2]^j A_s\exp(-E_s/RT) \qquad (11.30)$$

where n_{O_2} is the number of moles of oxygen per unit area of surface, S is the external surface area per unit mass of carbon, σ_s is the density of the particle, a is the particle diameter, j is the order of reaction, $[O_2]$ is the concentration of oxygen in the bulk gas, and A_s and E_s are the Arrhenius parameters of the surface reaction. Zone I kinetics are observed at low temperatures.

At very much higher temperatures, the rate at which oxygen molecules are transported from the bulk gas to the external surface is slow enough to be rate controlling and this is the region known as Zone III. Here the observed rate of combustion of the particle can be equated to the rate at which oxygen diffuses to the external surface of the particle. If the oxygen molecules react as soon as they reach the surface, the rate, corresponding to eqn (11.2), for example, is given by an adaptation of eqns (11.3)–(11.6), if diffusion of oxygen to a spherical particle surface is assumed. In this case $\dot{M}$ is defined as a molar rate of oxidation and $r_f = a$. Following eqns (11.3)–(11.6) the algebraic solution takes the form

$$\frac{\dot{M}}{4\pi a} = \frac{Dpx_\infty}{RT(n-s)} \ln\left\{1 + \left(\frac{n-s}{s}\right)x_\infty\right\} \tag{11.31}$$

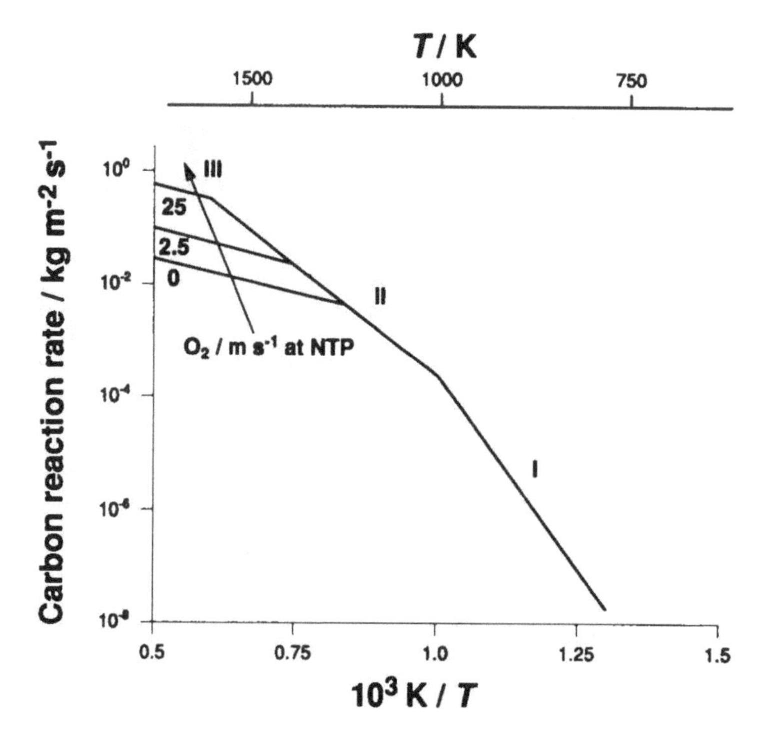

Figure 11.3 The dependence of reaction rate on temperature for a 10 mm sphere of porous graphite in air at 100 kPa, showing the existence of the reaction zones I, II and III (after Mulcahy [227]). The data are plotted in an Arrhenius temperature-dependent form.

If $n = s = 1$, as is the case for

$$C(s) + O_2 \rightarrow CO_2 \tag{11.32}$$

the reaction rate (mol s^{-1}) is given by

$$\dot{M} = \frac{4\pi Dapx_\infty}{RT} \tag{11.33}$$

In the intermediate Zone II, oxygen is transported quickly to the external surface, but diffuses relatively slowly down the pores before reaction occurs. By considering a simplified model of the pore structure it can be shown [226] that the reaction rate per unit area of external surface is independent of the particle size. The apparent activation energy is half the activation energy of the true surface reaction (E_s) and the apparent order of reaction is greater than the true order (j).

The three regimes in the reaction of a porous graphite are illustrated in Fig. 11.3 [227]. In practice there are enormous differences between the reactivities of different carbons. Some of these can be ascribed to variations in pore structure and some to the presence of impurities (e.g. alkali metal salts) which have a pronounced catalytic effect on the

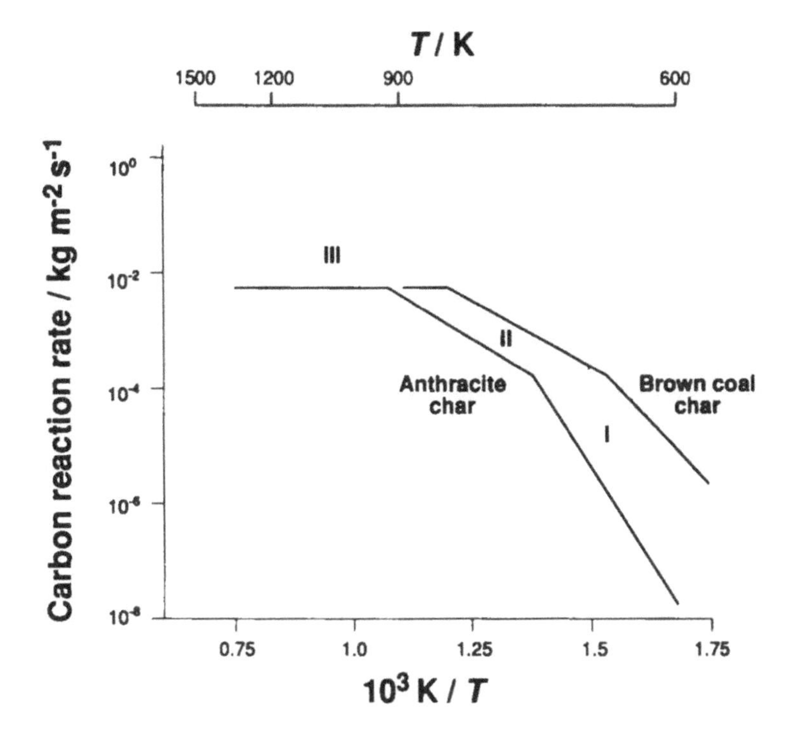

Figure 11.4 Experimental evidence for the similarities and differences between the reactivity of brown coal char and anthracite in Zones I, II and III (after Mulcahy [227]).

surface reaction. Consequently, the temperature ranges corresponding to the three Zones differ (Fig. 11.4 [227]). Whereas Zone I behaviour may be observed for a reactive brown coal char up to about 650 K, a less reactive anthracite char may follow Zone I kinetics up to nearly 800 K. At very high temperatures, both materials react at the same rate which is controlled by diffusion of oxygen to their external surface (Zone III kinetics). Temperature is not the only parameter characterising which combustion Zone is operative for a particle. The diameter can be important too: in general, very small particles never burn under diffusion control. However, one would expect a sphere greater than about 10 mm to burn under Zone III conditions.

11.5 Practical applications of particulate or solid combustion

Solids which volatilise prior to combustion show very much the same dependence on particle size as liquids, but with an added effect of increased radiative heat transfer because of their high absorptivities. Dispersions of solid particles in air are used in industrial burners in the same way as liquid fuels, as discussed next. Dust explosions are discussed in Chapter 13.

11.5.1 Pulverised coal burners and cyclone furnaces

A high proportion of electricity is generated by the combustion of pulverised fuel. The coal is ground to particles about 40 μm in diameter and blown by a stream of air into the hot combustion chamber, which is lined with steel tubes containing water. The tubes are heated principally by radiation from the burning particles and the water in the tubes is converted to steam.

In the first stage of combustion, the volatile material burns in a spherical diffusion flame surrounding the particle and, in the second, the solid residue burns at its surface. The combustion of the char is controlled by diffusion of oxygen in the porous structure (Zone II kinetics). Only with large particles at very high temperatures does external mass transfer have any influence [227]. Similar technology can be used for all types of pulverised coal since the times for burn-out (i.e. complete combustion) of typical particles do not differ by more than a factor of ten. If Zone I kinetics were obeyed, the burn-out times would vary by a factor of up to one thousand.

The flame speed in the dust cloud depends primarily on the combustion of the volatile portion of the material. The result is that flame speeds reach a maximum for fuel to air ratios three to six times more rich than the stoichiometric ratio (Fig. 11.5 [228]). In order to carry dust particles in a horizontal pipe, the air velocity must exceed the terminal velocity of the particles to a considerable extent, i.e. $>20\,m\,s^{-1}$, but as the cloud passes through the nozzle into the combustion chamber it will expand

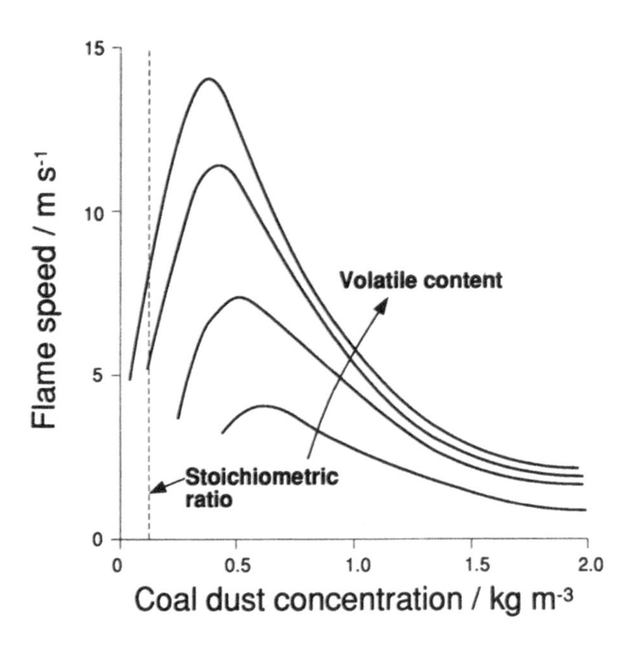

Figure 11.5 The dependence of flame speed on density in a coal dust cloud (after de Gray [228]).

and its speed will fall. Since flame velocities of dusts are always below $20\,m\,s^{-1}$ the flame will be stabilised somewhere ahead of the nozzle. In order to reduce the size of the combustion chamber and to operate under the most stable conditions, the fuel:air ratio is selected which gives the maximum flame velocity. This very high ratio compared with that for a stoichiometric composition requires that two separate air streams are used, the primary stream carrying the required amount of fuel and the secondary stream being used to complete the combustion process. The way in which the primary and secondary air streams are mixed is an important factor in burner design. An optimum location to introduce the secondary air is probably where the first stage, the combustion of volatiles, is complete.

Pulverised fuel burners suffer from certain disadvantages, the most important being that in order to obtain high heat release rates the mixtures of fuel and air have to be passed in rapidly so that, for complete burning, large combustion chambers are required if the above criteria are to be satisfied. In addition, the ash tends to remain in the gas stream and can cause choking of exhaust flues. In a cyclone furnace, the combustible mixture is injected tangentially at high velocity (100–$150\,m\,s^{-1}$). The burning particles are thrown outwards by centrifugal forces and adhere to the molten slag on the walls. Although the throughput of air is very high, the residence time of the solid is extended in this way so that combustion may go to completion. The slag can be tapped off the walls. The relative motion between gas and solid ensures efficient mixing.

11.5.2 Solid fuel beds

The best-known example of coal combustion is provided by the domestic fire and the majority of industrial coal burners operate on the same overfeed principle. Once combustion has begun, a number of separate reaction zones can be distinguished vertically through the bed. The fresh, or 'green', coal comprises the top layer and the heat transferred from the bed below causes evaporation of the volatile material which burns in the secondary air and leaves a residue of fixed carbon or coke.

Primary air enters the base of the grate and passes first through the lowest layers, the ash zone. The ash performs a useful function in providing an insulation between the high-temperature reaction zone and the grate. In the first oxidation zone, the oxygen reacts at the surface of the carbon to give carbon monoxide, represented as

$$2C(s)+O_2=2CO \tag{11.34}$$

Carbon monoxide is released from the solid and reacts on mixing with oxygen, in the next oxidation zone, to give carbon dioxide by the overall stoichiometry

$$2CO+O_2=2CO_2 \tag{11.35}$$

The more detailed kinetic mechanism of CO oxidation is discussed in Chapter 6. Each of these processes is strongly exothermic and some of the heat released is transferred back to promote the reaction. This is usually the hottest part of the fuel bed. At this stage the oxygen concentration is very much depleted and the carbon dioxide is reduced by the Boudouard reaction in the next layer,

$$CO_2+C(s)\rightarrow 2CO;\quad \Delta H^{\ominus}_{298}=283\,\mathrm{kJ\,mol^{-1}} \tag{11.36}$$

Since this reaction is endothermic, the temperature is reduced. The carbon monoxide mixes with secondary air and is oxidised to CO_2. The oxidation zone is quite small, corresponding to only a few particle diameters, whilst the reduction zone is thicker because the reaction rate there is much slower. The maximum temperature coincides roughly with the position at which the CO_2 concentration is highest [229].

11.5.3 Fluidised-bed combustion

Fluidised-bed technology has been in existence for about 50 years, its main application being in the chemical industry, such as for the catalytic cracking of hydrocarbon vapours into the fractions most suited for liquid fuels and other products. A finely divided, particulate solid acts as the catalyst, and contact between the catalyst and the gaseous reactants is brought about by passing the gas stream upwards through a vertical column (or bed) of material in a vessel at a sufficiently high rate that the particles are maintained in a 'boiling' motion, that is in a *fluidised* state [230]. In general, the fluid may be a liquid or a gas.

The present interest is in the application of fluidised-bed technology in combustion, the merits of which are the potential gains in efficient fuel utilization with low levels of pollutant emissions. Also, it is possible to produce a large amount of heat in a very small volume, so capital costs are lower than for pulverised fuel combustion. One possibility is the combustion of gaseous fuels, such as natural gas (methane), catalysed on a surface such as silica (in particulate form as sand). Exothermic oxidation is able to occur heterogeneously at least in part, according to the principles discussed in section 10.3, by adsorption of the reactants on the extensive surface. The combustion temperatures are comparatively low (1000–1300 K), largely because the high emissivity of the solid strongly favours radiative heat loss. In a practical sense, exceptionally efficient heat transfer to water or steam pipes is also possible from the agitated material within the fluidised bed. There can be commensurately low emissions of nitrogen oxides at the prevailing temperatures (see section 14.6). Application to gaseous fuel has not yet been fully developed, but the fluidised-bed combustion of crushed coal is a technology with a stable future in the steam generating market.

If a gas is passed slowly upwards through a bed of solid particles, it finds its way through the spaces between the particles and the pressure drop across the bed is directly proportional to the flow rate. If the flow rate is increased, a point (the minimum fluidisation velocity) is reached at which the frictional drag on the particles becomes equal to their apparent weight (i.e. weight minus any buoyancy force). The bed then expands a little as the particles adjust their positions to offer less resistance to the flow. The bed is then said to be *fluidised* and further increase in flow is not accompanied by an increase in pressure drop (Fig. 11.6).

The physical behaviour within the bed depends on the nature of the bed at flow rates above the minimum fluidisation velocity. Some materials tend to aggregate so that the distribution of the fluid becomes irregular and stagnation regions may develop. With materials that undergo better fluidisation, as would be achieved with approximately spherical particles, the excess gas (that is, additional to that required for fluidisation) passes through the bed in bubbles at speeds much greater than the mean velocity. These burst at the surface of the bed, giving it the appearance of a boiling liquid. This *bubbling* mode of operation (i.e. the creation of cavities that migrate up through the bed) is accompanied by a fairly fast, vertical mixing of the bed particles: thus particles move

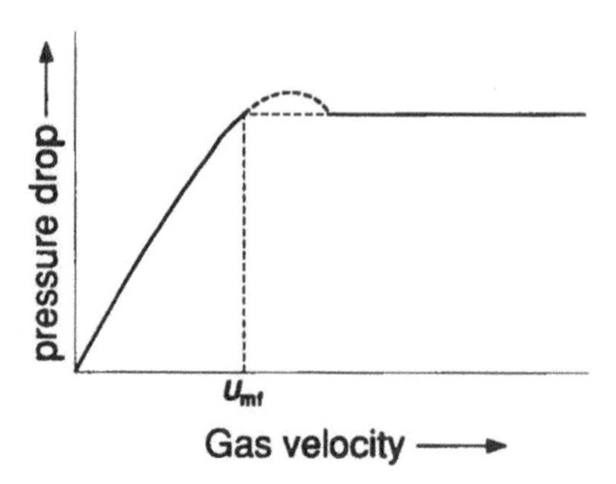

Figure 11.6 The pressure drop across a bed of particles as a function of gas velocity. The minimum fluidisation velocity corresponds to u_{mf}.

upwards in the wake of a bubble and create a 'gulf-streaming' motion of particles which ensures very high heat transfer. A distinctive feature of fluidised-bed combustion is the spatial uniformity of temperature that is achieved. Further increases of flow rate can become counter productive insofar that *slugging* may occur. That is, the bubbles merge into elongated 'slugs' as wide as the bed; they track through the bed having set up their own, albeit highly unstable, line of least resistance. Minimum fluidising air velocities at the bed temperature, in combustion applications, are below $1\,m\,s^{-1}$ and bubbling bed velocities may be up to $3\,m\,s^{-1}$ [231].

Despite very extensive experimental and theoretical investigation of fluidisation, the fluid mechanics of fluidised beds are, as yet, incompletely understood and so there remains considerable difficulty in the design and development of fluidised-bed combustors, especially with regard to scaling up from laboratory or pilot-plant operation. The problems are posed largely as a result of the numerous non-idealities of behaviour, and the very strong dependence on materials or geometric factors of the system. Since an idealised fluidised bed resembles a vigorously boiling liquid, the most widely used theory encapsulates two phases, a *lean* (or *bubble*) phase and a *dense* (or *emulsion* or *particulate*) phase which closely resembles the bed at the state of incipient fluidisation [232].

The potential for bubbling fluidised-bed combustion of crushed coal is not yet fully realised, and the present operations are mainly for industrial energy requirements of up to about 30 MW thermal output. This power generation is possible from a shallow bed (~ 0.3 m) in which coal graded below about 10 mm is burned. The heat release is normalised in terms of the surface area of the bed, typical outputs at atmospheric pressure being about $2\,MW\,m^{-2}$, which compares very favourably with the output from conventionally fired boilers. The output is linearly dependent on the fluidising velocity for a given excess of air [233]. Fluidised-bed combustors are usually operated at atmospheric pressure, but there are advantages of greater combustion intensities, i.e. heat outputs per unit volume, at higher pressures. Combustion systems have been studied at up to 2 MPa.

An alternative to the *bubbling* fluidised-bed for combustion processes is the *circulating* fluidised bed. In this case much higher flow velocities are used ($5–10\,m\,s^{-1}$) which causes significant carryover from the top of the reactor. Although all fluidised beds require some type of supplementary gas–solid separation unit in order to ensure dust-free flue gases, the equipment associated with the circulating fluidised-bed performs a much more important function in recycling the fuel. Circulating fluidised-bed combustors of up to 200 MW thermal output have been constructed. Intermediate between the bubbling and circulating fluidised-beds is the *spouted* bed. In this system air is injected through a restricted aperture at the base, which causes the fluidised medium to rise in a fountain at the surface and, thereby, create circulation within the combustion chamber.

The attraction of fluidisation technology for coal or other solid combustion is that the fuel, if suitably sized, becomes part of the fluidised bed. Air is brought into intimate contact with the solid fuel as a natural consequence of the fluidising process. Moreover, the residue (ash) from the solid combustion may remain as part of the bed until it is broken down sufficiently to be blown out and separated from the flue gas. Some bed cleaning or sieving is required if the fuel gives rise to more durable or coagulated ash lumps. Nevertheless, there is considerable scope for the combustion of low grade fuels in fluidised beds. Other materials may also be added to control pollutant emissions, as discussed below. However, an overall control of the particle size distribution by inclusion of inert materials is important for efficient fluidisation, since very marked changes accompany the devolatilisation and oxidation of the fuel particles.

The processes involved in fluidised-bed combustion of coal share the complexities of all other systems, involving the devolatilisation and char burn-out stages. The normal operating temperature is in the range 1000–1200 K, which is sufficiently high to minimise CO emissions owing to incomplete combustion and for rapid burn out of particles but sufficiently low to avoid melting or sintering of ash. The latter is exacerbated by the addition of limestone or dolomite, which is required to capture sulphur dioxide produced from the organic and inorganic sulphur content in coals. However, a very marked reduction in SO_2 emissions can be achieved by this method without recourse to flue-gas scrubbers.

The presence of fuel-bound nitrogen in solid fuels like coal means that nitrogen oxide emissions are inevitable following combustion, but at least the combustion temperatures are sufficiently low that oxidation of

Table 11.1 Nitrogen oxide emissions from circulating fluidised bed combustors [234]

Unit size (MW)	Fuel	NO (ppm at 7% O_2)
65	Bituminous coal	40–260
226	Bituminous coal	20–40
65	Brown coal	80–120
97	Petroleum coke	90–120
56	Wood waste	110–135

atmospheric nitrogen is not a significant issue (Chapter 14). Nitrogen oxide emissions are controlled as far as possible by staged combustion. That is, primary air is fed into the grid as required for fluidisation, but at approximately 60% of the rate required for the stoichiometric requirements in order to maintain an essentially reducing environment when devolatilisation of the coal predominates and reactions leading to N_2 are encouraged. Secondary air is then added further up the combustion chamber, which becomes an oxidising environment in which combustion is completed. It is believed that there may also be a catalytic reduction of nitrogen oxides on the ash or unburned carbon particles. Typical nitrogen oxide emissions from circulating fluidised-bed combustion are shown in Table 11.1 [234].

Further reading

Brame, J.S.S. and King, J.G. (1967). *Fuel, Solid, Liquid and Gaseous* (6th edn). Edward Arnold, London, UK.

Davidson, J.F. and Harrison, D. (1963). *Fluidised Particles.* Cambridge University Press, Cambridge, UK.

Denbigh, K.G. and Turner, J.C.R. (1984). *Chemical Reactor Theory, An Introduction* (3rd edn). Cambridge University Press, Cambridge, UK.

Howard, J.R. (ed.) (1983). *Fluidized Beds: Combustion and Applications.* Applied Science Publishers, Barking, UK.

Hoy, H.R. and Gill, D.W. (1987). The combustion of coal in fluidized beds. In *Principles of Combustion Engineering for Boilers* (ed. C.J. Lawn). Academic Press, London, UK, p. 521.

Jones, J.C. (1994). *Combustion Science, Principles and Practice.* Millennium Books, New South Wales, Australia.

Kanury, A.M. (1975). *Introduction to Combustion Phenomena.* Gordon and Breach, New York, USA.

La Nauze, R.D. (1985). *Chem. Engng Res. Design*, **63**, 3.

La Nauze, R.D. (1986). Combustion in fluidized beds. In *Advanced Combustion Methods* (ed. F.J. Weinberg). Academic Press, London, UK, p. 17.

Levenspeil, O. (1972). *Chemical Reaction Engineering* (2nd edn). John Wiley, New York, USA.

Mulcahy, M.F.R. (1978). The combustion of carbon in oxygen. In *The Metallurgical and Gaseous Fuel Industries*. The Chemical Society, London, UK, p. 175.

Problems

(1) By use of eqn (11.21) calculate the temperature in the flame front (T_f) around a droplet of toluene (C_7H_8) at values for the Lewis number of (a) 0.7, (b) 1.0 and (c) 1.3.

$$C_o = 100\,\mathrm{J\,mol^{-1}\,K^{-1}},\ \Delta H^{\ominus} = -3900\,\mathrm{kJ\,mol^{-1}},\ T_\infty = 300\,\mathrm{K}$$

(2) Compare the burning rate of a liquid droplet of n-C_8H_{18} in an atmosphere of air with that of a spherical particle of carbon of the same size. Assume that the diffusion controlled reaction $C(s) + 0.5O_2 \rightarrow CO$ occurs at the solid surface and that the gas round a burning sphere has a Lewis number of unity.

The relevant equations are eqns (11.31) and (11.24), for which only the first term on the right-hand side need be considered. It may be assumed in each case that $s \gg (n - s)x_\infty$. The most convenient simplification of eqn (11.31) is to expand the term in the form $\ln(1 + y)$ and truncate the solution.

$$T_\infty = 974\,\mathrm{K},\ T_s = 554\,\mathrm{K},\ C_F = 100\,\mathrm{J\,mol^{-1}\,K^{-1}},\ C_o = 30\,\mathrm{J\,mol^{-1}\,K^{-1}},$$

$$\Delta H^{\ominus}(\text{n-}C_8H_{18}) = -5000\,\mathrm{kJ\,mol^{-1}},\ \Delta H^{\ominus}_{vap}(\text{n-}C_8H_{18}) = 50\,\mathrm{kJ\,mol^{-1}}$$

(3) Show that the time required for complete combustion of a spherical graphite particle under diffusion controlled conditions may be expressed in the form

$$t = \frac{\sigma r_o^2}{Dc}$$

where $\sigma(\mathrm{mol\,m^{-3}})$ = molar density of particle, $r_o(\mathrm{m})$ = initial radius of particle, $D(\mathrm{m^2\,s^{-1}})$ is the diffusion coefficient and $c(\mathrm{mol\,m^{-3}})$ is the oxygen concentration in the gas phase. Hence calculate the time for complete combustion of graphite particles in air (21% O_2 in N_2, NTP), given that $r_o = 2\,\mathrm{mm}$, $\sigma = 183\,\mathrm{mol\,m^{-3}}$ and $D = 2.8 \times 10^{-4}\,\mathrm{m^2\,s^{-1}}$.

Hints

(a) The rate of consumption of graphite ($\mathrm{mol\,s^{-1}}$) can be expressed in terms of the rate of change of radius of the spherical particle (via the volume or surface area).

$$\frac{-\mathrm{dN(C)}}{\mathrm{d}t} = \frac{-\mathrm{d}(\sigma V)}{\mathrm{d}t} = \frac{-\sigma\mathrm{d}((4/3)\pi r^3)}{\mathrm{d}t}$$

The rate of change of particle radius is obtained via a conversion of the form in eqn (11.25).
(b) The rate of consumption of oxygen ($\dot{M}$(mol s^{-1})) is determined by eqn (11.33), following $C(s) + O_2(g) \rightarrow CO_2(g)$.
(c) For a stationary state $-dN(C)/dt = \dot{M}$.

12 Combustion hazards

12.1 Introduction

The initiation of a combustion hazard may constitute the exposure of a fuel + air mixture to particular conditions of temperature and pressure within a closed volume (Chapters 7–9). A related problem is that of a fuel + air mixture exposed to a hot surface in relatively free moving gas. Practical aspects of 'spontaneous criticality' are presented here. Ignition may be induced also by some sort of stimulus, such as a spark, and in such cases the minimum energy required for ignition must be established. A related problem is the possibility of initiation of ignition in potentially flammable atmospheres by laser light, as could happen if a fibre-optic conducting a powerful laser beam is fractured. There is also the potential for initiation of combustion at solid surfaces or of particles and dusts. Once ignition has been initiated, it is important to understand the characteristics of fire spread or explosion, especially on the large scale. These will be introduced in this chapter. Aspects of fire retardancy and extinguishing agents will also be discussed. Sources of information on the emerging topic of waste incineration, which includes the emission of polychlorinated biphenyls, are given in the suggested reading list.

In order to take preventative action, it is important to establish criteria which constitute 'conditions of safe operation' or which define the hazard in some way. The procedures may be part of safety legislation such as those set in the American Society for Test and Materials (ASTM), by British Standards (BS tests) or set by the EEC. Tests described in the 1984 Directives of the EEC (L 251) include the determination of 'flash point', 'flammability (solids)', 'flammability (gases)', 'explosive properties', 'autoinflammability of volatile liquids and gases (determination of the self-ignition temperature)', 'autoinflammability of solids (determination of the relative self-ignition temperature)' and 'oxidising properties'. Compilations of data obtained from these types of tests on many substances are already available. Certain compounds, such as pyrotechnics, may be sufficiently hazardous or of such specialised use that it is practicable to have them tested and certified only in designated laboratories, such as at those of the UK Health and Safety Executive, at Buxton. Within process industries, hazard and operability studies (HAZOP) represent a considerable fraction of the total design time for any new chemical plant. Procedures and codes of practice are discussed elsewhere [235, 236].

12.2 Minimum autoignition temperature of gases or vapours of volatile liquids

The purpose of the EEC test (formerly BS 4056 (1966)) is to measure the minimum temperature of a thermostatted vessel, a conical flask of volume 200 cm^3, at which spontaneous ignition (or autoignition) of a vapour can be brought about in air at 1 atm. Either a prescribed volume of the liquid (0.2 cm^3) or an excess of gaseous fuel is injected into the open flask. Evaporation and partial mixing in the air in the flask follows. If the vessel temperature is sufficiently high, ignition of the vapour occurs and is detected visually by the observation of a flame. The experiment is repeated at different vessel temperatures in order to establish the minimum temperature for spontaneous ignition. The composition of the vapour + air mixture at which ignition occurs is not determined.

The results for the minimum autoignition temperature of selected isomers of the alkanes, determined by this procedure, are shown in Fig. 12.1. The overall reactivity is considerably enhanced as the number of carbon atoms increases, but even at $n_C > 4$ certain isomeric structures are found to be far less reactive than their *n*-alkane counterparts under the test conditions. This type of test for autoignition may be regarded to give comparative data for the reactivity of different fuels, but there is only a limited direct application of the results. Furthermore, supplementary phenomena such as cool flames cannot be distinguished.

The spontaneous ignition temperature of mixtures cannot be derived from a linear relationship between the two temperatures representing the

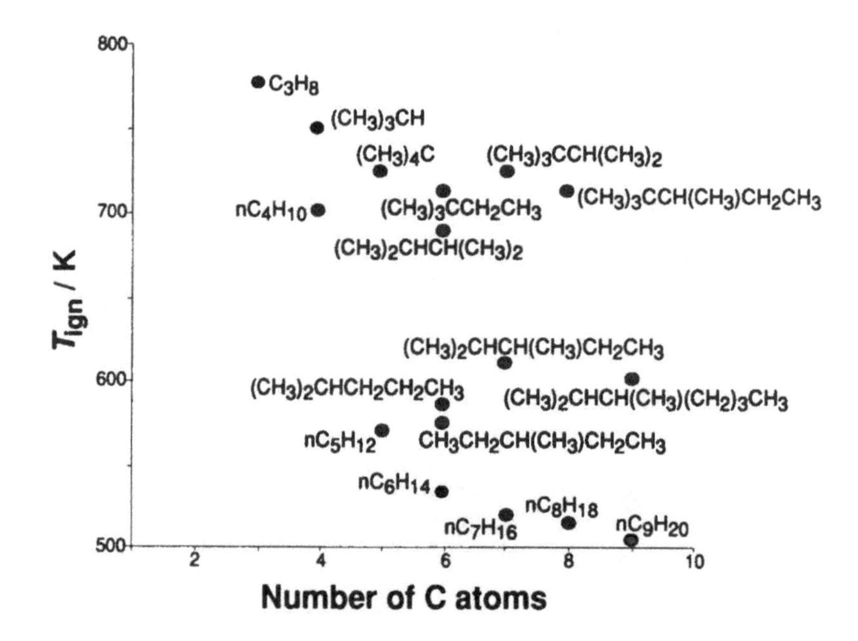

Figure 12.1 Minimum autoignition temperatures for various alkane isomers as measured by the ASTM and BS test methods.

pure components, as shown in Fig. 12.1, based on the mole or mass fractions, or from any other simple physicochemical basis. The behaviour is determined by the relative reactivities of the fuels in elementary reactions involving the main propagating free radicals. The effects on the autoignition temperature can be highly non-linear. This is especially relevant to the spontaneous ignition of natural gas, which may contain up to 8% of the lower alkanes, mainly ethane [237].

12.3 Flammability limit, flash point and fire point of vapours

The *flammability limit* is defined as the composition of fuel + air at which a flame just fails to propagate from a spark (or similar ignition source). As discussed in Chapter 3, flammable mixtures exhibit a lean (or lower) and rich (or upper) limit, which fall either side of the stoichiometric composition. It is important to measure the flammability limit in a sufficently large vessel that a self-sustained flame can be established. The EEC test requires a tube of minimum diameter 50 mm and minimum height 300 mm. The spark electrodes are located 60 mm from the bottom, with a spark gap of 3–5 mm. A standard induction spark, 0.5 s duration, is generated at 10–15 kV with maximum power input of 300 W. The test is performed at room temperature on mixtures that are successively enriched in fuel (in steps of 1% by volume) until ignition is obtained at the lean limit. The lean flammability limit (or LEL, lower explosion limit) is of greater practical importance than the rich limit, since it is leakage of a gas or vapour into the atmosphere that normally presents the ignition hazard. The LEL for the alkanes is at a molar volume in air which is approximately 60% of that in the stoichiometric mixture (Table 3.1 [23]).

The *flash point* of a liquid is the minimum temperature at which the vapour pressure is sufficiently high to cause a flame to 'flash' through the vapour + air mixture above its surface when a small flame source is applied to the mixture just above the surface. Although it is an experimentally measured property, obtained by a designated procedure, a relationship of the flash point to the lower flammability limit exists through the dependence of vapour pressure on temperature, given by the Clausius–Clapeyron equation

$$\frac{\mathrm{d}(\ln p)}{\mathrm{d}(1/T)} = \frac{-\Delta H_{\mathrm{vap}}}{R} \tag{12.1}$$

where ΔH_{vap} is the enthalpy of vaporisation of the liquid. The *fire point* of a liquid is the minimum temperature at which a flame is sustained above its surface when a small flame is applied to the vapour above the surface.

12.4 Minimum ignition energy

Spark ignition of a flammable mixture begins with the initiation of heat release in the reactants in the vicinity of the spark electrodes and is followed by growth of a flame kernel as a prerequisite to propagation of

Table 12.1 Experimentally measured minimum ignition energies (μJ) at ambient temperature and 100 kPa [23]

Fuel % by vol	Composition				
	$CH_4 + O_2$	$C_3H_8 + O_2$	H_2 + air	CH_4 + air	C_3H_8 + air
2.5	—	—	—	—	> 2200
5	25.1	—	—	> 750	247
7.5	—	5.8	—	420	> 400
10	7.9	—	—	420	—
12.5	—	2.0	—	> 1650	—
20	—	2.7	25.9	—	—
25	4.2	—	—	—	—
30	—	—	18.8	—	—
40	18.0	—	27.6	—	—

the flame away from the electrodes. These events occur on a submillisecond timescale.

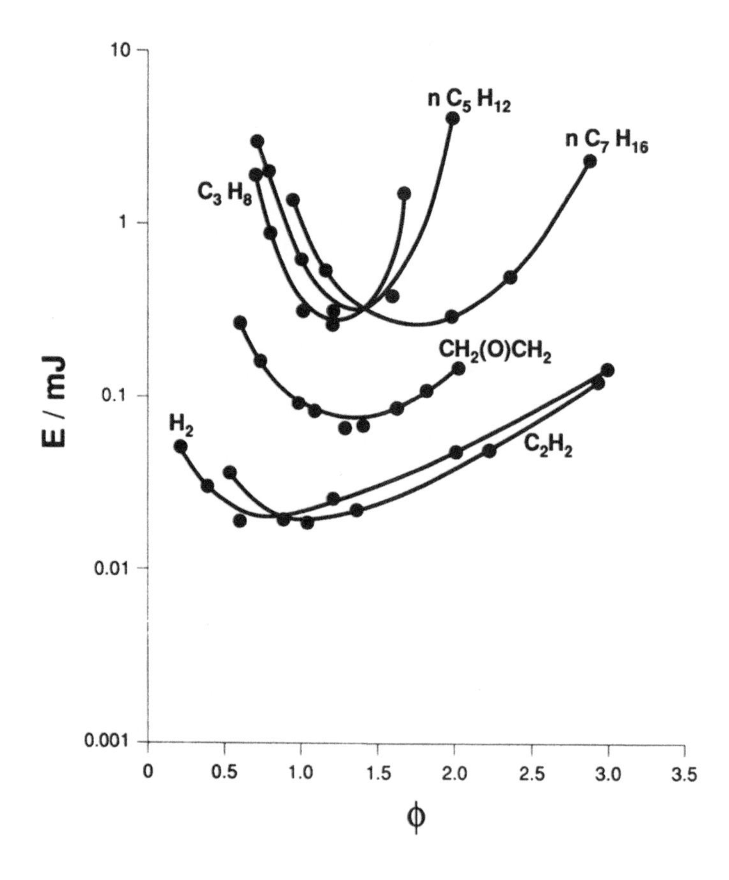

Figure 12.2 Dependence of the minimum spark ignition energy on composition for several fuels in air at 100 kPa (after Lewis and von Elbe [23]).

The spark energy that is required to initiate ignition in premixed fuel + air is dependent on the pressure, composition and temperature. These factors are governed mainly by chemical reactivity of the fuel. Mixtures in air are harder to ignite than mixtures in oxygen, and hydrogen is more easily ignited than the alkanes. Methane requires comparatively high energies for spark ignition (Table 12.1 [23]). The composition that ignites most easily is normally close to the stoichiometric mixture (Fig. 12.2 [23]). There is also a dependence on electrode size and spark gap, which arises from flame quench through heat loss or free radical destruction. The statutory test for minimum ignition energy is based on specified electrode size, separation and spark duration in a vessel that is sufficiently large for wall quenching to be negligible.

A unified interpretation of spark ignition is complicated by the need to identify how the energy is distributed in the spark. There is a thermal component which heats the gas locally, and could be regarded to be the cause of 'thermal ignition by a hot spot', but a considerable proportion of the energy creates free radicals and ionic species, which are also reaction initiators. Aerodynamic effects are also caused by the spark, which are governed by the current–voltage relationship, the spark duration and the

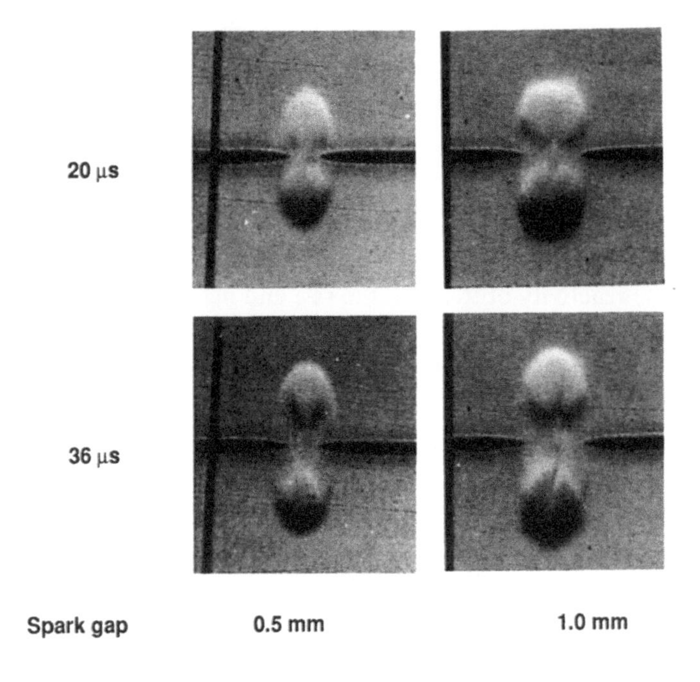

Figure 12.3 Schleiren photographs of the development of a toroidal, hot gas plume following spark generation (4.6 mJ, duration 0.3 μs) in air between electrodes (1.0 mm gap), at 20 μs and 36 μs, respectively, after spark initiation (after Kono *et al.* [238], by courtesy of The Combustion Institute).

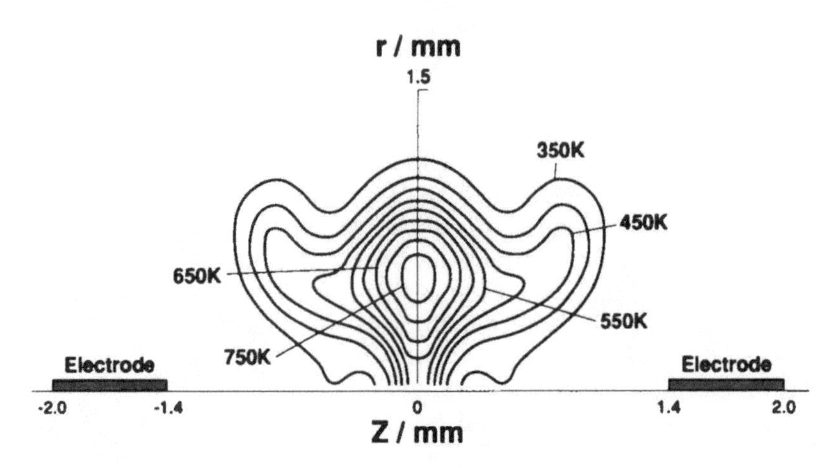

Figure 12.4 Simulated temperature profiles within one half of the toroidal cross-section following spark generation in inert gas at 100 μs after initiation. Lower maximum temperatures are predicted for smaller electrode gaps (after Kono *et al.* [238]).

size and shape of the electrodes. These effects can determine the ability of the flame kernel to develop. The toroidal growth of a hot gas plume away from the electrodes when a spark is struck in air is shown in Fig. 12.3 [238]. This Schleiren image shows the cross-section of the toroid [238]. Temperatures in excess of 750 K are predicted to occur within the toroid, even in the absence of chemical reaction (Fig. 12.4 [238]). Gas velocities within the active volume may exceed 5 m s^{-1}.

12.5 Large-scale explosion of gases and vapours

When significant quantities of a combustible gas liquid are released ignition may lead to a *sustained fire*, a *flash fire*, a *fire ball* or a *vapour cloud explosion*. Which of these occurs depends on factors such as the vapour pressure, the reactivity of the fuel, the size and manner of the release, the atmospheric conditions, the surrounding topography and the nature of the ignition source. The approximate timescales and rates of energy release associated with these phenomena are shown in Table 12.2 [239, 240].

Table 12.2 The approximate timescales and rates of energy release associated with various types of fire and explosion [239, 240]

Type of fire or explosion	Characteristic time (s)	Approximate rate of energy release (MW $tonne^{-1}$)
Sustained fire	>100	$< 10^3$
Flash fire	10–100	10^3–10^4
Fireball	1–10	10^4–10^5
Vapour cloud explosion	<1	$>10^5$

Although the effects of a sustained fire are felt over a smaller distance than those of other events, its longer duration may bring about more severe localised heating. The heat transfer rates are very high (up to 200 kW m^{-2}) and very serious damage to nearby equipment and storage vessels may result. A flash fire occurs when a release of flammable material dispersed over a wide area is ignited, and the flame then spreads along the premixed vapour-air layer. The flash fire persists for only a few seconds but, near the source, it may lead to a sustained fire.

If a container for pressurised liquids splits owing to localised weakening, then there may be a sustained discharge of liquid or vapour or a mixture of both. Ignition of a stream of flammable material under these circumstances would tend to give a diffusion flame rather like that on a burner and although the results of the fire might be quite serious, they would be confined to the immediate vicinity of the failure. By contrast, if the container is suddenly ruptured, the liquid will be violently ejected in the form of vapour and small droplets. Rapid mixing with air results, and if this mixture is ignited a rapidly expanding ball of flame rises from the ground entraining more air. The thermal radiation from such a fireball is extremely intense and its effects may be felt over several hundred metres. However, because of its relatively short duration, the heating effects on massive items of plant and equipment are negligible (Table 12.3 [241]).

The precise circumstances which lead to a vapour cloud explosion rather than a fireball are complicated, but when a premixed vapour air cloud is partially confined by major structures which promote turbulence and restrict the spread of the flame, or reflect pressure waves, there appears to be a greater likelihood of a vapour cloud explosion. The significance of the partial confinement is not clearly established and the phenomenon is often referred to as an *unconfined vapour cloud explosion*, or UVCE.

The damage from a UVCE can be well in excess of that to be expected from an ordinary deflagration. Although there is no clear evidence that detonation develops, flame speeds of 100 m s^{-1} or more are necessary to produce the overpressures required to account for the damage that can be caused, which may be consistent with an exceedingly high wind rather than a shock wave [242]. The rise-time and the duration of the positive pressure from a UVCE are long compared with those of a blast wave

Table 12.3 Main characteristics of a fireball as a function of the mass M (kg) of flammable material [241]

Maximum diameter (sphere m^{-1})	$5.8\,M^{1/3}$
Maximum diameter (hemisphere m^{-1})	$7.4\,M^{1/3}$
Duration as a ground level source of radiation[a] (s)	$0.45\,M^{1/3}$

[a] Radiation output 20–40% of heat of combustion of the fuel mass.

from a high explosive generating a similar overpressure. As with a normal blast wave, a UVCE is followed by a suction wave. The explosion at a large chemical plant at Flixborough, UK, in 1974 followed this pattern.

The operation at Flixborough involved the liquid-phase oxidation of cyclohexane to cyclohexanone at 430 K and a pressure of approximately 0.9 MPa in a cascade of six cylindrical stainless-steel reactors, each of about 45 m^3 capacity and containing around 35 tonnes of cyclohexane. A temporary bypass of one of these reactors ruptured during start-up after maintenance work, releasing cyclohexane to the atmosphere at a rate exceeding 1 tonne s^{-1}. This massive escape of cyclohexane produced a cloud of spray and vapour about 200 m in diameter and, in places, 100 m high. About 45 s after the rupture, the cloud ignited, destroying the plant and severely damaging the houses in two villages about 1 km away. Buildings up to 3 km away were affected.

The events at Flixborough have been the subject of a most detailed enquiry and also much subsequent discussion [243]. The very high flame speeds needed to explain the damage and other documented UVCEs imply burning velocities far in excess of those normally associated with laminar flow. Various mechanisms have been proposed to account for such high velocities including severe turbulence, absorption of radiation from hot gases, multipoint ignition and partial confinement [244, 245].

Turbulence is thought also to have been involved in an explosion of natural gas from a pipeline leak in the Ural mountains, in 1990. Some of the escaping gas accumulated in a railway cutting. Although an ignition source would have originated from only one train, that two trains arrived simultaneously and passed in the cutting, created very considerable turbulence between the carriages. This is believed to have dramatically increased the flame speeds and considerably worsened the damage and tragic loss.

The rapid formation of large quantities of vapour by flash evaporation can itself produce blast waves: this phenomenon is referred to as a *boiling liquid expanding vapour explosion*, or BLEVE, and can occur in any liquid under appropriate circumstances. The subsequent ignition of a rapidly expanding cloud of flammable vapour + air mixture would lead to very high combustion rates. Flame speeds as high as 50 m s^{-1} have been measured.

12.6 Mitigation of confined explosions

One of the most common ways to protect buildings and process installations from internal explosion is *explosion venting*. The unit is protected by blow-out panels to permit a sufficiently rapid escape of unburned gases and combustion products. This requires that vent openings are sufficiently large to keep the pressure below a structurally damaging threshold. Many correlations have been developed to determine appropriate vent opening areas. There can be marked differences in the recommendations, and appreciable discrepancies arise between model predictions and experimental test. Bradley and Mitcheson [246] have reviewed many of these empirical relationships and related experimental studies, and from these data have correlated the maximum pressure rise versus a venting parameter which represents the venting area divided by the laminar burning velocity.

The generation of turbulence appears to be a main contributory factor in the discrepancies, since the enhancement of the flame surface area or flame acceleration mechanisms may augment the burning rate, and hence the rate of pressure rise, very markedly. Moreover, the problems can be exacerbated by the venting procedure. The initial pressure increase as a result of an explosion blows open the vent; this reduces the pressure but strongly affects gas motion. For structural reasons the vent must burst at a low overpressure, at which stage only a fraction of combustible gases may have burned. Further pressure peaks may occur during the ensuing combustion, which can be considerably larger than the initial peak prior to bursting of the vent [247].

There are predictive models, such as the Design Institute for Emergency Relief Systems (DIERS) set up by AIChE amongst a consortium from the chemical process industries, for the safe venting of chemical reactors at the onset of 'thermal runaway'.

12.7 Dust explosions

The combustion of solid dispersions is important because it constitutes a serious explosion hazard in factories and warehouses. Almost all solid materials which are endothermic with respect to their oxides can undergo a self-sustaining reaction or explosion when dispersed in fine particles. This includes many organic materials [248] and virtually all metals. Dust explosions have been shown to be possible in several hundred different substances.

The rate of reaction per unit mass of fuel and therefore the rate of heat release depends on the surface to volume ratio. Furthermore, the importance of radiative energy transfer means that virtually all these solid materials can give violent explosions at sufficiently small particle sizes. Consequently, great care is required in handling powders on an industrial scale, even including such innocuous substances as wheatflour. The situation is further complicated by the fact that in motion such materials rapidly build up a charge of static electricity in very dry atmospheres and the discharge which may result can initiate the dust explosion.

The oxides of metal frequently have a higher sublimation temperature than the metal so that the products form a smoke which may be intensely luminous. This is exploited in photographic flash-lamps and in pyrotechnics. If the oxide products adhere to the surface, they can have a mitigating effect of protection against continued combustion.

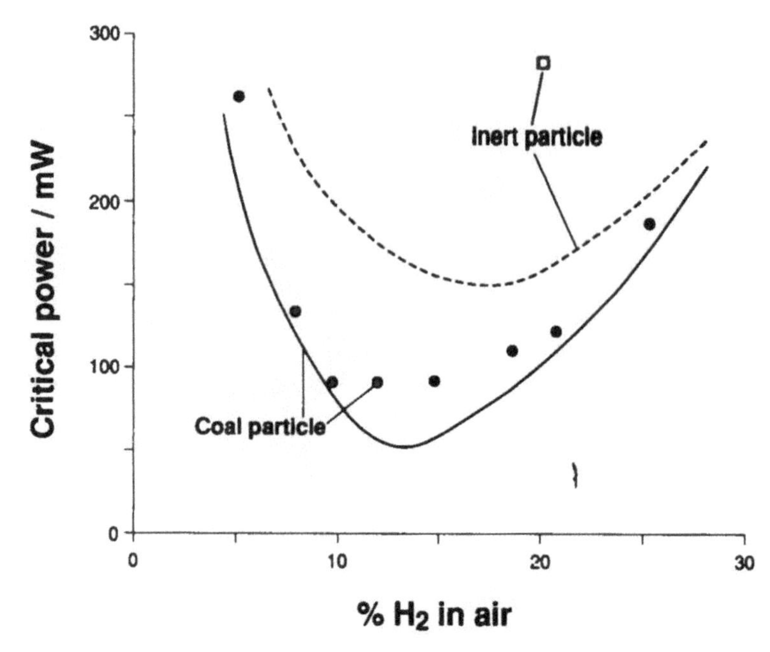

Figure 12.5 Experimental laser powers required to ignite H_2 + air mixtures by energy from an optical fibre incident on coal dust particles (40 μm dia). The lines represent numerical simulations (after Zhang *et al.* [249]).

12.8 Laser initiation of ignition

Although also applicable to flammable gases, the main hazard is likely to be that dust or other particulates in a gaseous environment are heated by incident radiation such as from a fractured fibre optic used for the transmission of laser light in sensor and communications. Recent experimental and theoretical studies [249] have shown that there are similar qualitative dependences for the ignition energy as a function of gaseous reactant composition as are observed in spark ignition (Fig. 12.5 [249]). Continuous radiation from an optical fibre (50 μm) incident on an individual coal particle (40 μm) was used to initiate ignition of a mixture containing hydrogen, as might happen in a coalmine. Rather higher powers were required to initiate a gaseous explosion by heating of inert particles. The interpretation of this type of behaviour relates to 'hot spot' thermal ignition theory, but there are complications connected with absorption efficiencies of the incident energy and the surface temperature reached. Experimental and theoretical developments of this topic are still in their infancy.

12.9 Smouldering combustion

The propagation of a combustion wave within porous or fibrous material combustion, or sometimes at the surface may be manifest as a smouldering process. This is a term used to describe the heterogeneous combustion of a solid without the appearance of flame. It is widely accepted that only materials which are capable of forming a char in the early stages of combustion can exhibit smouldering [250]. The char is expected to be highly porous. The difference between smouldering and flaming combustion then lies in the fate of the volatiles. The volatiles which are liberated by thermal degradation either ignite in the gas phase, the radiation from which promotes further degradation and flame propagation, or the volatiles escape without further oxidation. In the latter, the heat release is confined to the oxidation of the remaining char. Heat transfer is largely by conduction into the solid material. The smouldering propagation rate is an order of magnitude lower than that of firespread by flaming combustion because heat conduction is so much slower than heat transfer by radiation.

The destructive effects of smouldering fires are neither as rapid nor as complete as those of flaming fires, but there are severe problems nevertheless. Smouldering fires are extremely difficult to detect and they may proceed unchecked for very long periods. This applies particularly to underground fires, which may occur in refuse tips and landfill sites, coalmine workings or in the substrata following forest fires, for example. Moreover, smouldering can propagate even in a very poor oxygen supply, so it becomes extremely difficult to extinguish. There are cases of underground smouldering fires that have propagated for decades, having resisted all attempts at extinguishment. Smouldering fires are often initiated by spontaneous ignition of stockpiled or buried material, as described in section 8.6.

Smouldering need not occur on a large scale. Fires in property can develop from smouldering that has been initiated in material surrounding overheating electrical cables, switchgear or other electrical equipment. Polymeric materials that are used in the construction of furniture are often capable of smouldering. The volatiles that are released from smouldering propagation may be toxic, especially when man-made polymers are involved.

12.10 Firespread

Images of large-scale fires in property or forestry convey the destructive power of fire and the rate at which it can spread. The fundamental interpretation is extremely complex. Every physical and chemical principle involved in combustion is brought into play at some stage, there are enormous numbers of variables, and all conceivable ramifications of scale and orientation have to be considered. Even meteorological conditions may be relevant, and the fire itself may have local effects; winds created in forest fires can exceed 90 mph. Firespread is a subject that commands the long-term attention of many specialist and specialist organisations worldwide. The present Section can serve only to introduce a very limited part.

The phase following the initiation of many major fires in confined spaces (e.g. rooms) is that of *flashover*, and often constitutes the watershed between controllability of the fire and uncontrolled devastation. During flashover, events escalate dramatically and extraordinarily rapidly from a small fire (such as a chair) to all combustible surfaces in the enclosure. It is an almost instantaneous transition from a local fire to a complete conflagration throughout the entire space. This is contrary to the spread of fire to unburned material from adjacent burning sites, as usually occurs in the open air.

The major contributory factors to flashover are radiant and convective heat transfer from hot combustion products to other combustible bodies, whereby their temperature is raised sufficiently for ignition to be initiated. The high emissivity from dense smoke which accumulates in a thick layer below a ceiling, is particularly important in flashover. The hot ceiling itself can also contribute to radiative transfer.

The suddenness of the change in scale of the fire that occurs at flashover may arise from the incubation period required for the evolution of hot products from the primary source and the subsequent heat transfer from them to other bodies. Direct radiation from the initial burning object is comparatively limited in its effect. Flashover is a complex dynamic problem which involves the rate of combustion of the primary source and the extent to which the hot products are confined above other objects (such as, the height and area of the ceiling). Whilst ventilation is important for the fire to be sustained in the fully developed stage and determines the rate of combustion, overventilation, especially if it is close to the ceiling, may limit the extent to which smoke and other products can accumulate and so can have a mitigating effect on flashover. Sometimes the primary fire source is not able to develop to a sufficient extent for enough products to be generated for flashover to occur.

The simplest conceptual way of interpreting flashover is by analogy to the Semenov treatment of thermal ignition (Fig. 12.6 [251]). Although similar to Fig. 8.5, the heat release rate is regarded, ultimately, to be limited by the rate of air supply (upper branch of $\boldsymbol{R}$, in Fig. 12.6), as would happen in a large scale fire in a room involving many objects when ventilation is restricted.

In Fig. 12.6, $\boldsymbol{R}$ represents the fire plume and the hot layer gases. The heat loss rate ($\boldsymbol{L}$) is defined as a combination of heat transfer to the enclosure walls and convective heat transport by loss of hot combustion gases through ventilators. Quasi-stationary states are then considered in which either or both $\boldsymbol{R}$ and $\boldsymbol{L}$ are considered to be explicitly time dependent [251]. One vindication of $\boldsymbol{R}(t)$ is that the rate of the heat release in the source fire, located initially at point A (Fig. 12.6), may grow in time because of radiative feedback which raises the temperature of the source and thereby increases the rate of production of combustible, degradation products (or fuel vapour in the case of a liquid fire). There is a critical transition at the tangency condition brought about by

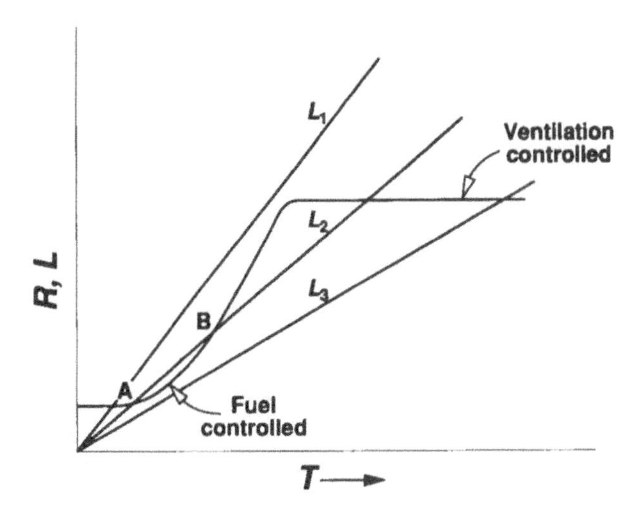

Figure 12.6 The relationship of heat release (*R*) and heat loss rates (*L*) in a thermal model for flashover, as described in the text (after Thomas *et al.* [251]).

the coalescence of A and B. In a typical room fire the effective heat transfer coefficient (proportional to the gradient of *L*) may be expected to decrease asymptotically with time as less heat is conducted into the enclosure surfaces [251]. The critically criteria depend on the nature of *R*(*T*, *t*) and L(*T*, *t*).

The quantitative, dynamic interpretation of flashover requires many contributory factors to be taken into account, including all modes of heat flux and also mass transport. Geometric factors may also be included. A fluid dynamic model including radiative heat transfer, developed by Lockwood and Malasekara [252], is representative. The quantitative interpretation of flashover is reviewed by Drysdale [253]. Radiation plays an extremely important part in many aspects of fire dynamics and in many combustion applications, such as in furnaces.

12.11 The combustion of polymers

There is considerable usage of polymers in building materials or in furniture and, in the absence of any fire retardant treatment, most of these materials can burn easily. The products of combustion are often toxic and there are many instances of fires in buildings where the burning polymers have added greatly to the damage caused and to the loss of life. The burning process is very complex and chemical reactions may take place in three interdependent regions, namely, the condensed phase, the gas phase and the interface between them. Some polymers, such as polystyrene, pyrolyse with the formation of large volumes of volatile degradation products which subsequently burn in the gas phase above the polymer surface. In the case of polystyrene, the presence of oxygen does not appear to have any effect on the breakdown of the polymer itself, whereas the decomposition of other polymers, polypropylene for example, is catalysed by oxygen in the atmosphere. Polymer combustion is a vast topic [254].

One criterion to assess the flammability of polymers is a measure of the limiting oxygen index (LOI). This is defined as the percentage of oxygen in the atmosphere surrounding the polymer that is just capable of supporting a flame, and is determined in a specified test apparatus [255]. Self-extinguishing polymers have an LOI of greater than 21. Sustained burning of such polymers can occur only in air that is enriched in oxygen. Whilst giving some rank of reactivity, this particular test is now regarded to be of limited relevance to the application and use of polymers.

The cone calorimeter, developed primarily for making heat release rate measurements and to act as a flammability test for radiant ignitability, has become accepted as a more appropriate tool for testing the performance of polymeric materials for domestic and commercial use [256]. The radiant source in the test apparatus is located in a cone, located above the sample. It is designed to give a uniform incident flux on the surface of the test material ($10\,cm^2$). A supplementary spark igniter is also incorporated so that the ignitability of decomposition products emitted from the heated surface may also be investigated. The combustion products are extracted via the cone to various monitoring systems. These include air flow measurements, sample mass loss and the extent of oxygen consumption. These data are then used to determine the heat release rate. Supplementary information may also be obtained, such as the yields of toxic substances or the density of smoke generation, both of which are often the major contributors to fatalities in fires.

The flammability is determined not only by the ease of thermal degradation of the polymer, but also by the nature and properties of the products. There can be extremely complicated mixtures close to the polymer surface (Table 12.4 [257]). Although a wide range of lower hydrocarbons are formed, the oxygenated products indicate that there has been a considerable extent of oxidation very close to the hot surface. Thus air has not been excluded from the zone between the surface and the flame, and may interact with the hot, liquid phase. The surface temperature may exceed 675 K [258].

Some polymers have inherent flame inhibition or retardancy characteristics which may be induced chemically or physically. Sustained burning can be discouraged by the melting and retraction of combustible material from the hot zone. Such is the distinction between a polyester cotton fabric, which retains its integrity, and a polyalkene which can melt.

The most effective constraint on initiation is that of thermal stability of the polymer structure. The typical high performance materials, such

Table 12.4 Volatile products (mol%) formed at the surface of burning polyalkenes [257]

Product	Polyethylene	Polypropylene
Methane	0.65	1.1
Ethene	5.8	2.7
Ethane	—	0.5
Propene	1.3	3.4
Allene	1.2	0.1
Butenes	1.7	2.0
Pentenes	0.7	7.4
Hexenes	1.3	—
Heptenes	0.4	1.1
Aromatics	0.2	1.0
Octenes	—	1.9
Carbon monoxide	3.4	3.2
Carbon dioxide	11.0	8.8
Water	14.6	12.1
Nitrogen	55.5	52.9
Oxygen	1.6	1.2

as Nomex or Kevlar, are aromatic polyamides. These can resist degradation at temperatures beyond 800 K, and hence are used in the manufacture of fire-protection clothing and other specialised equipment. Chemical retardancy tends to be conferred on the halogen-containing polymers, of which polyvinyl chloride (PVC) is the most familiar. Although its degradation takes place from about 500 K, the release of HCl into the gas-phase confers a retardancy on the ignition processes (section 12.12). A supplementary problem is that the degradation products are extremely toxic. PVC is noted also for its excessively dense smoke, which is generated because the complete oxidation to final products is inhibited by the halogen content.

Chemical methods of increasing flame retardancy of polymers include surface (or topical) treatment, but their use on clothing is restricted, especially children's nightwear, because of possible dangers of ingestion. Industrial or other utility polymers can be treated by incorporation of retardant materials during manufacture. There is always a constraint on how much to include before the tensile strength or physical durability of the material is affected. Additional considerations are the aesthetics of the 'finish', and unit cost also matters since polymers are extremely cheap materials made in enormous quantities in a highly competitive industry. Nevertheless the chemical industry is extremely responsive to the need for fire resistant materials, and considerable resources are put into the design and development of new retardant systems. 'The industry', rather than legislative bodies, was the prime mover in the development of the cone calorimeter.

Chlorine and bromine, or phosphorus-based materials are known to confer flame retardant characteristics. A synergism of retardancy characteristics occurs between chlorine and antimony, probably through an antimony oxychloride derivative, so complex compounds are formulated to exploit this. The general belief is that halogen-based additives act in the gas phase. For reasons connected with toxicity, their use is becoming much more restricted. The role of phosphorus-based additives is usually to encourage char formation on degradation of the polymer. The enhancement of char yield means not only that much lower proportions of flammable, organic vapours are generated but also that a temporary resistance to decomposition is provided by the thermal insulating effect of the carbonaceous layer. Another surface treatment, an intumescent coating, also exploits this last property. Coating materials can be designed to swell under the action of heat so that they give an enormously increased, protective depth in a hostile thermal environment.

12.12 Fire extinguishing agents

Water is able to act in a physical sense as a heat sink through its heat capacity and latent heat requirement for vaporisation. The vapour may then act as an inert gas which excludes oxygen. This can be interpreted as dilution of a fuel + air mixture to take the overall composition outside the flammability limits. This also is the predominant role of carbon dioxide as an extinguishing agent (Fig. 3.3).

Aqueous foam extinguishing agents, usually incorporating carbon dioxide, are often used on flammable liquid fires. They are designed to provide a long-lasting blanket which prevents evaporation of the liquid and absorbs radiation during the cooling period after the fire has been extinguished. Radiation absorption by foams is also important for the protection of structures close to a large-scale fire. The structure and integrity of the foam is extremely important. Foams have been developed to retain a skeletal form even after much of the water content has drained out.

Dry powders extinguishers usually comprise finely divided sodium bicarbonate. They are effective on flammable liquid fires, probably acting both as a radiation barrier and a free radical trap on the extensive particle surface area. Solid powders, such as sodium or potassium bicarbonate or ammonium phosphate, are used as explosion inhibiting agents in coal mines. They are dispersed into the atmosphere as a countermeasure against the initiation of coal dust explosions or ignition of flammable gases throughout the galleries if a fire or explosion occurs elsewhere in the mine.

The most widely used chemical agent to date has been the halogenated hydrocarbon, trifluorobromomethane (CF_3Br, Halon 1301). However, its use as an extinguishing agent is being phased out in the developed countries because of its adverse environmental effects: at normal temperatures this compound is sufficiently stable to be able eventually to diffuse into the stratosphere and there become involved in ozone layer depletion (see Chapter 14). Other materials of similar retardant effectiveness are being investigated. CF_3I is a possible substitute, and has an appreciably shorter lifetime in the troposphere, so is unlikely to present the same atmospheric problem. A similar compound is CF_3H (FE-13), but fire tests have shown that rather larger amounts are required for equally effective extinguishment as Halon 1301 [259]. In consequence, dangerously high yields of HF, which presents a health hazard, are produced in the vicinity of the fire.

The role of chain branching reactions in hydrocarbon flames via

$$H + O_2 \rightarrow OH + O \qquad (12.2) = (6.7)$$

and

$$O + RH \rightarrow OH + R \qquad (12.3) = (6.15)$$

has been discussed in Chapter 6. The retardancy by halogenated materials stems from their interference with these branching sequences by reactions of the kind

$$H + Br_2 \rightarrow HBr + Br \qquad (12.4)$$

$$O + HBr \rightarrow OH + Br \qquad (12.5)$$

$$OH + HBr \rightarrow H_2O + Br \qquad (12.6)$$

$$O + CF_3 \rightarrow CF_2O + F \qquad (12.7)$$

$$OH + CF_3 \rightarrow CF_2O + HF \qquad (12.8)$$

Competition between the main chain branching modes and these propagation or termination steps is possible because HBr is able to offer an H atom which is much more labile than those associated with the primary fuel. HF does not react with H, O or OH as readily as HBr (eqns (12.4)–(12.6)) because the activation energies of its reactions are much higher. However, following from eqns (12.7) and (12.8) there are effective propagation and termination reactions which involve CF_2O and its derivatives. Representative reactions that generate CF_3, Br_2 and HBr are

$$CF_3Br \rightarrow CF_3 + Br \qquad (12.9)$$

$$H + CF_3Br \rightarrow CF_3 + HBr \qquad (12.10)$$

$$Br + RH \rightarrow HBr + R \qquad (12.11)$$

$$H + HBr \rightarrow H_2 + Br \qquad (12.12)$$

$$Br + Br + M \rightarrow Br_2 + M \qquad (12.13)$$

Equations (12.4)–(12.13) are included to give an indication of the nature of the free radical processes involved. A comprehensive scheme would be considerably more complex. 'Flame inhibition' involving these types of kinetic interactions has been tested numerically [87].

Further reading

Bartknecht, W. (1981). *Explosions*. Springer-Verlag, Berlin, Germany.
Benuzzi, A. and Zaldovar, J.M. (eds) (1991). *Safety of Chemical Batch Reactors and Storage Tanks*. Kluwer Academic Press, Dordrecht, The Netherlands.
Bond, J. (1991). *Sources of Ignition*. Butterworth Heinemann, Oxford, UK.
Cullis, C.F. and Hirschler, M.M. (1981). *The Combustion of Organic Polymers*. Clarendon Press, Oxford, UK.
Drysdale, D.D. (1985). *An Introduction to Fire Dynamics*. John Wiley, Chichester, UK.
Fire, F.L. (1991). *Combustibility of Plastics*. Van Nostrand, New York, USA.
Harris, R.J. (1983). *Investigation and Control of Gas Explosions in Buildings and Heating Plant*. E. & F.N. Spon, London, UK.
Hester, R.E. and Harrison, R.M. (eds) (1994). *Waste Incineration and the Environment*. The Royal Society of Chemistry, Cambridge, UK.
Kashiwagi, T. (1994). Polymer combustion and flammability: Role of the condensed phase. In *Twenty-Fifth Symposium (International) on Combustion*. The Combustion Institute, Pittsburgh, PA, USA, in press.
Lees, F.P. (1980). *Loss Prevention in the Process Industries*. Butterworths, London, UK.
Lewis, B. and von Elbe, G. (1987). *Combustion, Flames and Explosions in Gases* (3rd edn). Academic Press, New York, USA.
Mullins, B.P. (1954). *Spontaneous Ignition of Liquid Fuels*. Butterworths, London, UK.
Penner, S.S. and Mullins, B.P. (1959). *Explosions, Detonations, Flammability and Ignition*. Pergamon Press, London, UK.
Sarofim, A.F. (1986). Radiative heat transfer in combustion: Friend or foe? In *Twenty-First Symposium (International) on Combustion*. The Combustion Institute, Pittsburgh, PA, USA, p. 1.
Seeker, W.R. (1990). Waste combustion. In *Twenty-Third Symposium (International) on Combustion*. The Combustion Institute, Pittsburgh, PA, USA, p. 867.
Strehlow, R.A. (1976). *Prog. Energy Combust. Sci.*, **2**, 27.
Symposium proceedings (1989). *International Symposium on Runaway Reactions*. AIChE, New York, USA.
Thomas, P.H. (1981). Fire modeling and fire behavior in rooms. In *Nineteenth Symposium (International) on Combustion*. The Combustion Institute, Pittsburgh, PA, USA, p. 503.
Wendt, J.O.L. (1994). Combustion science for incineration technology. In *Twenty-Fifth Symposium (International) on Combustion*. The Combustion Institute, Pittsburgh, PA, USA, in press.

Yoshida, T. (1987). *Safety of Reactive Chemicals*. Elsevier, Amsterdam, The Netherlands.

Problem

(1) Calculate the 'flash point' of toluene vapour. Assume that the proportion of toluene in air at the lower flammability limit is 50% of that in a stoichiometric mixture. The vapour pressure of toluene as a function of temperature is given by the expression

$$\ln(p/\mathrm{Pa}) = -\Delta H_{vap}/RT + 22.723$$

where $\Delta H_{vap} = 36.1\ \mathrm{kJ\,mol^{-1}}$.

13 Internal combustion engines and fuels

13.1 Introduction

Internal combustion engines may be divided into two types depending on whether they deliver their energy via a reciprocating piston or a rotating turbine. The majority of piston engines operate either on the Otto cycle, in which the combustion is spark-initiated, or on the diesel cycle, in which autoignition occurs in the hot, compressed charge. The combustion processes involved in the gas turbine, the turbojet and the ramjet engines, are similar to each other and will be discussed collectively. The rocket engine, which produces direct thrust from the combustion process, will not be discussed here.

The principles of operation of internal combustion engines are addressed as required in order to appreciate the principles of the combustion processes involved and the attainment of optimum performance, including the environmental impact. The main discussion of pollutant emissions is deferred to Chapter 14. Common transportation fuels and how they affect the performance of engines are also considered.

13.2 Spark ignition (s.i.) engines

13.2.1 Principles of operation, design and performance

In the s.i. engine, a fuel–air mixture is introduced into a cylinder, closed at one end, which contains a movable piston. The mixture is compressed by the motion of the piston and is then ignited by a spark, the timing of which is optimised for the most efficient power generation. The hot, burned gases at high pressure provide the energy which drives the piston back down the cylinder, the linear motion being converted to a rotary action by a crankshaft and flywheel. A valve system is arranged so that subsequent movements of the piston can be used to remove the burned gas and to introduce and compress fresh reactants prior to the next firing. Most s.i. engines operate on a *four-stroke cycle*, that is four strokes of the piston, down and up. Combustion occurs on each alternate compression stroke within the cylinder, the cycle comprising induction of charge, compression, combustion and power generation, exhaust of combustion products, as summarised in Fig. 13.1.

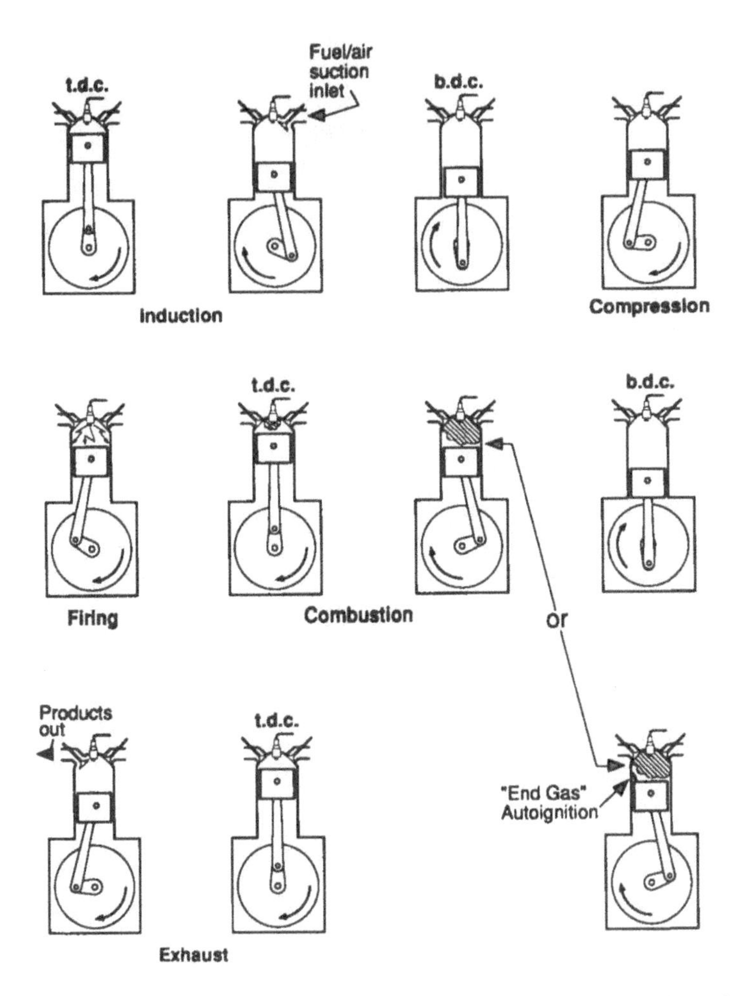

Figure 13.1 Schematic representation of the the four-stroke cycle in an s.i. engine.

In most production automobiles the engine comprises four to six, eight or twelve cylinders connected to the crankshaft, the smaller units being either 'in line', in a symmetrical 'V' or 'flat' (horizontally opposed) formation. The eight and twelve cylinder units are always set in a 'V' formation. The engine capacity is the sum of the swept volume of each cylinder, given by the product of the cross-sectional area and the piston stroke. *Two-stroke cycle* operation, which means that induction, compression, combustion and exhaust occur in each up and down stroke, is exploited in motorcycle, marine and horticultural applications. For reasons of (i) the compactness and dynamic balance of the machine, (ii) a reduction of the number of moving components, and (iii) the higher specific power output of two-stroke engines, interest in their potential for use in cars has been rekindled recently amongst some manufacturers.

Spark ignition engines operate normally between *c.* 500 and 5000 revolutions per minute, so each compression and expansion firing cycle (Fig. 13.1) occurs within an approximate timescale diminishing from 100 to 10 ms. Since the time allowed from the spark initiation to the completion of combustion is only a fraction of one cycle, combustion must occur extremely quickly, but it must also be achieved by smooth flame propagation. Laminar flame propagation of a stoichiometric mixture of a typical hydrocarbon fuel in air is far too slow (see Table 3.1), so the combustion chamber and the ports and valves have to be designed so that gas flow into the combustion chamber is directed in such a way and at such a rate that the induction and compression of the charge following closure of the inlet valve are able to create sufficient gas motion to promote turbulent flame propagation. This also aids mixing of the vaporising fuel and air during induction. Very stringent combustion requirements must be met in motorcycle engines and those used in car and motorcycle racing, which commonly run beyond 10^4 revolutions per minute. Multivalve and variable valve-timing systems are used in current engines to optimise the gas motion for the most efficient combustion under given operating conditions.

The greater the expansion of the post-combustion gases, the greater is the work extracted. Thus, fast combustion at the peak of compression stroke (t.d.c.) and a high compression ratio (CR) are desirable features.

$$CR = \frac{\text{Cylinder volume at b.d.c}}{\text{Cylinder (or clearance) volume at t.d.c}} \tag{13.1}$$

The overall efficiency can be interpreted, first on an idealised thermodynamic basis, then taking into account the non-idealities that contribute to reduction in the efficiency. The thermodynamic principles and description of both ideal and real *p–V indicator diagrams* of the Carnot cycle are discussed elsewhere [260–262]. The thermal efficiency of the engine (η) is defined as the fraction of the chemical energy released which is converted to (p–V) work. On the basis of an isentropic compression in an ideal cycle it can be shown that

$$\eta = 1 - \frac{1}{(CR)^{\gamma-1}} \tag{13.2}$$

A high value of $\gamma(= C_p/C_v)$, which enhances η, is favoured in lean fuel + air mixtures. Lean mixtures correspond to an air:fuel ratio which is higher than the stoichiometric ratio of about 15:1 by mass. Lean burn can also lead to reductions in the emission of pollutants (Chapter 14). However, lean mixtures burn more slowly and are prone to misfiring, but these problems can be counteracted by increasing the turbulence of the fuel + air mixture in the cylinder to enhance the burn rate (section 13.2.5).

Some of the chemical energy released by combustion may be lost in dissociation at high temperatures so the net useful energy released reaches a maximum with mixtures which are richer than stoichiometric, although the maximum temperature is slightly lower. Burning velocity is also at a maximum in mixtures which are slightly richer than stoichiometric. Thus, rich mixtures give maximum power, whereas maximum economy is obtained with lean fuel + air mixtures. There are additional environmentally related factors and their means of control that have a considerable bearing on the conditions at which engines are tuned to operate, especially with regard to the mixture control.

The power generation of an s.i. engine is governed by control of the mass of fuel and air admitted at each cycle. Thus, there is a *throttling* device in the induction system to restrict air flow, which reduces mechanical efficiency as a result of the so-called *pumping losses* [262], or *pumping work* [261], incurred by the partial vacuum created in the induction system on part throttle. This is an unavoidable design feature of conventional s.i. engines. Regardless of whether a carburettor or fuel injection is employed, the fuel + air mixture is generally prepared in the inlet manifold. In the idealised operation of an s.i. engine, it is assumed that the fuel is fully evaporated prior to admission to each cylinder and that the fuel vapour is mixed homogeneously with air. These conditions are not readily achieved in practice, and the failure to do so may contribute to exhaust pollutant emissions, especially of unburned hydrocarbons. These and other practical problems of engine design are discussed elsewhere [261, 262].

13.2.2 Lean burn and stratified charge

Lean burn signifies combustion of mixtures beyond 20:1 air to fuel, by mass, rather than the typical 15:1 for stoichiometric gasoline combustion. The potential advantages lie in a higher thermal efficiency and lower nitrogen oxide emissions than for engines operating at normal fuel + air mixtures. The main problems that have to be overcome are the comparative difficulty of ignition of fuel lean mixtures and their slow rate of combustion (section 13.2.5).

One route to overcoming these difficulties has been via stratified charge combustion. An additional advantage is a reduction of pumping losses by minimising the throttling at part load. In order to improve the ignition of lean mixtures, in early stratified charge engines (Honda CVCC, *c.* 1975), combustion chambers were designed with a small side chamber (or prechamber) within which the spark plug was located and into which some fuel was injected directly in order to produce an enriched mixture. This part of the charge then produced a 'flame torch' which issued from the side chamber to ignite a much weaker fuel to air mixture that had been drawn into the main combustion chamber. The torch ignition also tended to overcome the slow burn rate by virtue of its penetration into the chamber. This mode of combustion has fallen foul of the increasingly stringent requirements in air pollution control (Chapter 14), largely in connection with unburned hydrocarbon emissions, and there was found to be a reduction of efficiency incurred by enhanced heat losses within the prechamber.

Improved control over fluid motion within the combustion chamber itself has enabled a revision of combustion chamber designs in later development of engines. The separate chamber has been eliminated, usually by charge stratification in the main combustion chamber within a bowl in the piston crown. This fluid dynamic approach is much more efficient. The enriched fuel supply in the vicinity of the spark plug is generated by direct injection of liquid fuel into the combustion chamber, hence termed direct-injection stratified-charge (DISC), rather than into the inlet manifold. Control of the mixing processes is obtained within the main combustion chamber by swirl generated during the air induction and compression stroke. Satisfactory atomisation of the charge is vital but not easily achieved. There are potential advantages with regard to mixture control and to the minimisation of losses of unburned fuel to the exhaust, with attendant gains in fuel economy. However, the pitfalls include premature dispersion of the fuel cloud giving mixtures too lean for satisfactory ignition, or slow evaporation leading to over-rich pockets which then cause particulate emissions. Unburned fuel may be emitted in the exhaust if any part of the charge is outside the flammable range.

13.2.3 'Engine knock' and fuel octane rating

Even with the most efficient engine designs there is a limit to how far the compression ratio may be raised before *engine knock* is encountered. Engine knock arises from autoignition of the unburned part of the fuel + air charge while the spark ignited flame is propagated through the combustion chamber. The rapid compression of the fuel + air mixture in the firing cycle raises both the pressure and the temperature of the charge. The expansion of the spark ignited flame contributes to these effects in the *end gas*, which is the term given to the mixture that is most remote from the ignition point and, therefore, survives for the longest time in the combustion chamber (see inset Fig. 13.1). Hydrocarbon + air mixtures at high temperature and pressure can undergo spontaneous ignition, but whether or not this occurs in the end gas and how long it takes to develop is governed by the temperature, pressure, composition and nature of the fuel.

Engine knock may be audible as a light 'rattling' sound. Not only is smooth and efficient operation of the engine upset but when severe knock occurs, (i) very high amplitude pressure waves are propagated back and forth in the chamber at acoustic frequencies as a result of the end gas autoignition, (ii) local hot spots can be created as a result of enhanced heat transfer owing to increased turbulence, which are able to melt the piston or other aluminium alloy components, and (iii) very high pressure zones and developing detonation waves are set up, especially in the crevices around the top of the piston, which are capable of eroding the metal surfaces. These effects can lead to mechanical failure.

Much can be achieved in optimising performance with modern engine management systems. Nevertheless, the upper limit on the compression ratio of engines designed for normal commercial operation is set by the fuels that are currently available at about 10:1 if engine knock is to be avoided. A relatively high resistance to spontaneous oxidation and autoignition of the fuel is central to the avoidance of engine knock, and is one of the most important factors which governs the blend of the fuel. Another recent consideration has become the part that may be played by the fuel blend in controlling pollutant emissions (reformulated gasolines), which is discussed later.

A particular grade of petrol (or gasoline) does not signify a specific chemical composition. The composition is complex (Fig. 1.1) and is affected by the oil source and its refining process. There are also supplementary considerations such as the need for low-boiling constituents for easier 'cold start', especially in cold climates, or a predominance of higher boiling components in hotter climates to avoid vapour locks in the fuel system. Factors which relate to an efficient and long engine life, such as freedom from corrosive components, or from substances which produces deposits, can be dealt with by additives in the fuel. Whatever the supplemetary needs, the fuel must be capable of showing a particular knock resistance, specified by its *octane number*.

The *octane number* of a particular fuel blend is determined under specified operating conditions in a CFR single cylinder engine which has a variable compression ratio. (CFR stands for Co-operative Fuels Research, and refers to a Committee set up to standardise the test method.) Two common standards are used for the operating conditions, giving either the *research octane number* (RON), which is normally used to characterise fuels in Europe, or the *motor octane number* (MON), which is often used to characterise fuels in the USA, as the combination (RON+MON)/2. The test is performed by increasing the compression ratio while the engine is running, until engine knock is detected by sensors. The octane number is assigned as the percentage by volume of 2,2,4-trimethyl pentane ('iso-octane') in a binary mixture with *n*-heptane that gives rise to engine knock at the same compression ratio as the test fuel under the same operating conditions. The *octane number* assigned to pure *n*-heptane is 0 and to pure iso-octane is 100. At present a 'standard unleaded petrol' in the UK corresponds to 95 RON, whereas a 'super

Table 13.1 Compositions of typical gasolines (vol%)

Unleaded gasoline		Leaded four-star gasoline	
Component	vol%	Component	vol%
n-C_6H_{14} to n-C_8H_{18}	12.13	n-C_6H_{14} to n-C_9H_{20}	12.19
Isomeric alkanes and C_4H_{10} isomers	18.90	Isomeric alkanes and n-C_4H_{10}	11.13
c-C_6H_{12} and derivatives	6.22	c-C_6H_{12} and derivatives	4.98
C_4H_8–C_6H_{12} alkenes	20.31	C_4H_8–C_6H_{12} alkenes	24.61
		1-Nonene	11.73
Toluene	24.58	Toluene	0.94
Xylenes	17.86	Xylenes	22.00
		Higher aromatics	10.97

unleaded petrol' corresponds to 98 RON, as also does 'leaded four-star petrol'. Simplified compositions of representative unleaded and leaded fuels are shown in Table 13.1.

The octane numbers of various pure hydrocarbons are listed in Table 13.2 [263, 264]. Amongst the alkanes and alkenes, the knock resistance is greater for shorter carbon chains or a higher degree of branching in the fuel structure. In general, the alkenes have a higher knock resistance than the corresponding alkanes. The aromatic hydrocarbons have a very high knock resistance. Although not shown, the behaviour of alcohols is found to parallel that of alkenes. There is a relationship between knock resistance and the increasing difficulty of low temperature oxidation to take place, as discussed in Chapter 7. Another reference might be the ignition delay measured in closed vessels, or in other experiments [265].

An overview of the chemical reactivity is much more difficult to assess for fuel mixtures because their propensity to undergo spontaneous oxidation is controlled by complex kinetic interactions. Nevertheless, a high aromatic content and a predominance of alkenes and highly branched structure alkanes must be strongly favoured for good knock resistance of gasoline. Current 'unleaded petrol' (Fig. 1.1 and Table 13.1) contains a high proportion of aromatics (~40% by volume). Other, non-hydrocarbon components are also being included to an increasing extent because not only to they boost the octane rating but they may confer additional environmental benefits. These are the oxygenates, such as ethanol, methyl-*t*-butyl ether (MTBE) or ethyl-*t*-butyl ether (ETBE), and may be found at up to 15% by volume in some gasolines [266, 267]. They are 'components' of the fuel since they enter into the combustion process, rather than 'additives' to perform some very specific function. MTBE is knock-resistant because it does not readily form the highly reactive OH radicals during its low temperature oxidation, or intermediates that are favourable to degenerate chain branching [268].

A distinction of the composition of a leaded 4-star fuel from that of an unleaded fuel is the absence of toluene and the presence of 1-nonene, which has a comparatively low knock resistance (*cf.* 1–octene, Table

Table 13.2 Critical compression ratios (CCR) and octane numbers for some typical hydrocarbons [263, 264]

Fuel	Formula	CCR (engine at 600 rpm and 550 K)	RON	MON
Propane	C_3H_8	8.8	101.6	97.1
n-Butane	$CH_3(CH_2)_2CH_3$	5.3	94.0	89.1
i-Butane	$CH(CH_3)_3$	6.5	—	—
n-Pentane	$CH_3(CH_2)_3CH_3$	3.2	61.8	63.2
2-Methylbutane	$CH_3CH(CH_3)CH_2CH_3$	5.1	92.3	90.3
n-Hexane	$CH_3(CH_2)_4CH_3$	3.0	24.8	26.0
c-Hexane	C_6H_{12}	4.7	83	77.2
2-Methylpentane	$CH_3CH(CH_3)(CH_2)_2CH_3$	3.6	73.4	73.5
2,2-Dimethylbutane	$CH_3C(CH_3)_2CH_2CH_3$	5.1	91.8	93.4
n-Heptane[a]	$CH_3(CH_2)_5CH_3$	2.2	0	0
2,2-Dimethylpentane	$CH_3C(CH_3)_2(CH_2)_2CH_3$	5.5	92.8	95.6
2,4-Dimethylpentane	$CH_3CH(CH_3)CH_2CH(CH_3)CH_3$	4.3	83.1	83.8
2,2,3-Trimethylbutane	$CH_3C(CH_3)_2CH(CH_3)CH_3$	8.9	101.8	100.1
2,2,4-Trimethylpentane[a]	$CH_3C(CH_3)_2CH_2CH(CH_3)CH_3$	6.6	100.0	100.0
1-Butene	$CH_2CHCH_2CH_3$	5.3	97.5	79.9
2-Butene	$CH_3CHCHCH_3$	6.0	100	83.5
1-Pentene	$CH_2CH(CH_2)_2CH_3$	4.6	90.9	77.1
2-Methyl-2-butene	$CH_3C(CH_3)CHCH_3$	6.5	97.3	84.7
1-Hexene	$CH_2CH(CH_2)_3CH_3$	3.5	—	—
1-Octene	$CH_2CH(CH_2)_5CH_3$	2.7	—	—
Benzene	C_6H_6	—	—	102.7
Toluene	$C_6H_5CH_3$	11.4	105.8	100.3
o-Xylene	$C_6H_4(CH_3)_2$	11.5	—	—
Ethylbenzene	$C_6H_5C_2H_5$	8.2	100.8	97.9

[a] Primary reference fuel components

13.2). There is also a lower proportion of isomeric alkane structures in the leaded fuel. The potentially low knock resistance of this hydrocarbon mixture is counteracted by the presence of lead alkyl additives, that is, lead tetraethyl, $Pb(C_2H_5)_4$, or lead tetramethyl, $Pb(CH_3)_4$, which confer a pronounced increase in knock resistance, even when present in only trace amounts (~150 ppm).

These substances act by destroying free radicals and hence they prevent branching-chain reactions from developing in the end gas. The most likely mechanism is via the oxidation of the lead alkyl to lead oxide, which is formed as a very fine, solid suspension in the hot gases within the combustion chamber, as shown by electron microscopy [269]. This suspension presents an extremely large and chemically destructive surface which is able to inhibit spontaneous oxidation of the fuel + air mixture [270–272]. The use of lead alkyls in gasoline for use in many countries has declined considerably during the last decade. The solids derived from the lead additives are also able to deposit round the exhaust valve seat, which offers protection of the surface from erosion in that particularly hostile, high temperature environment. Thus, engines designed in the last decade are fitted with hardened valves and seat surfaces to combat this problem when unleaded fuels are to be used.

13.2.4 Reformulated gasolines

More careful control over the constituents of gasoline can have tangible benefits in the reduction of potential health hazards or atmospheric pollution both in direct and indirect ways. The physical and chemical properties of gasoline that are to come under control in 1996 in California are shown in Table 13.3 [273] with reference also to the present position and what is already achievable at the refinery.

Table 13.3 Reformulated gasolines for California [273]

	1992 'on the street'	1996 requirement	Reformulation tests 1992
Reid vapour pressure (kPa)	65	48	46
T-90 (K)	453	422	418
T-50 (K)	377	372	—
Sulphur (ppm)	300	40	41
Benzene (vol%)	1.6	1.0	0.7
Aromatics overall (vol%)	32	25	21.6
Oxygen content (wt%)	—	1.8–2.2	
Methyl-*t*-butyl ether (vol%)			14.9[a]

[a] Equivalent to 2.1% O by weight in $i\text{-}C_8H_{18}$.

A problem, especially in hotter climates, is the escape of fuel vapour, which may have direct effects on inhalation or may contribute to damage in the upper atmosphere. The volatility is monitored by the Reid vapour pressure measurement, a reduction of which is most easily achieved by decreasing the highest volatility fraction. By contrast, a volatility that is too low can affect the 'unburned hydrocarbons' that are emitted in the exhaust. The volatility is determined as the temperature by which a certain proportion of the fuel has distilled, and is measured at 50% (T_{50}) and 90% (T_{90}) fractions. The incentive to reduce the proportions of benzene and toluene in the fuel arises from particular concern about the potential health hazard of the aromatic components as well as their contributions to atmospheric pollution (Table 13.4 [273]).

Any sulphur content is undesirable because it is emitted as sulphur dioxide. A reduction of its content in reformulated gasolines also causes reduction of other emissions (Table 13.4 [273]). Although the inclusion of oxygenated compounds, measured as the oxygen present by mass %, is related to the knock resistance of the fuel, carbon monoxide and unburned hydrocarbon emissions are also reduced because the engine is effectively operating at slightly leaner conditions than at the equivalent fuel mass to air ratio of a 100% hydrocarbon blend (Table 13.4 [273]). Aldehyde emissions may increase when oxygenated compounds are present.

Table 13.4 Effect of gasoline reformulation on exhaust emissions, measured as % change for 1996 specification fuel relative to 1992 specification fuel, as given in Table 13.3 [273]

Specification	% change		
	NMHC[a]	CO	NO_x
T-90	−20	+2	+5
Sulphur	−15	−10	−8
Aromatics	−12	−14	+2
Oxygenates[b]	−5	−11	+4

[a] Non-methane hydrocarbons.
[b] Average from 15% MTBE, 17% ETBE or 10% ethanol.

13.2.5 The optimisation of combustion

Of primary interest is the potential of *lean burn engines* for reduced pollutant emissions (section 14.4) and improved fuel economy commensurate with a higher thermodynamic efficiency. To compensate for the lower laminar burning velocity in fuel lean mixtures it is necessary to increase the charge turbulence in order to attain a sufficiently rapid burn rate. However, as is shown in the relationship between S_t/S_l and u'_k/S_l (Fig. 4.4), although an increase in the effective r.m.s. turbulent velocity, u'_k, initially enhances the turbulent burning velocity through flame wrinkling, there is a diminishing return and, ultimately, the flame is quenched. Flame quenching occurs at high flame stretch rates and more readily at higher values of the Lewis number (*Le*), associated with lean mixtures (section 4.7).

Since flame wrinkling can enhance the burning velocity quite considerably, the cycle to cycle variation in turbulence may contribute to differences in burn rate and, therefore, be a cause of cyclic dispersion in engine combustion. The flame initiation from a point source is crucial to engine performance since it takes about one third of the combustion time to burn about 1% of the charge. The wrinkling experienced by the reaction surface of the developing flame is governed first by the higher frequencies of the turbulence spectrum, but progressively by the lower frequencies. The frequency band affecting flame propagation may be those frequencies above a threshold given by the reciprocal of the elapsed time from spark initiation, represented by u'_k (section 4.7).

Furthermore, a predominantly laminar flame is initiated by spark ignition and, in an isotropically turbulent regime, the flame kernel from it is convected randomly in space with a velocity close to u'. Flame development and propagation may be impeded if the kernel is swept towards a solid surface, as could easily happen in an engine and may be a cause of misfire if the kernel is quenched.

Imposed on Fig. 4.4 is a trajectory which is a simulation of typical behaviour in an s.i. engine [274]. Initially, the burning velocity is close to the laminar value. As the kernel grows it experiences an increasingly wide range of the turbulence spectrum (Fig. 4.5), so u'_k increases faster than the increase of S_l that occurs as a result of the higher unburned gas pressure and temperature owing to compression and combustion. The locus of the trajectory moves rapidly to the right in consequence, on a line of approximately constant *Ka. Le.* S_l continues to rise in response to cylinder gas pressure and temperature during the combustion process but, towards the end of the developing turbulence phase, the increase in u'_k is small (Fig. 4.5). However, the locus is driven sharply to the left, in the wake of a diminishing value of *Ka*, and away from the quench region.

Although enhanced turbulence may be beneficial at low engine speeds, at high engine speeds excessive flame stretch may force the apex of the trajectory 'upwards and outwards' into the quench region. The effect of lean burn may be interpreted as a displacement of the trajectory to the right, which originates in a reduction of S_l in a leaner mixture at any given u'_k, again with the possibility of causing flame quench. The superimposition of simulated s.i. engine cycles on Fig. 4.4, or other similar analyses, is important to the prediction of how best to match turbulence, mixture strength or other control factors to obtain optimum efficiency without the penalty of misfire or flame quench.

13.3 Diesel engine combustion and fuels

13.3.1 Principles of operation, design and performance

Most diesel (or compression–ignition) engines depend on the same mechanical, four-stroke cycle as the s.i. engine, namely induction, compression, combustion and exhaust. However, air alone is taken into the cylinder and compressed adiabatically, rather than a premixed fuel + air mixture. A liquid fuel spray is injected into the cylinder when the piston is close to the top of the compression stroke, and it ignites spontaneously as evaporation begins and mixing takes place with the hot air (Fig. 13.2). To sustain smooth burning, the injection of fuel continues into the combustion phase, although the injection is usually completed by about t.d.c.

In an idealised diesel engine cycle, combustion occurs smoothly at constant pressure during the first part of the power stroke and is followed by expansion only in the later stages. This contrasts with the s.i. engine in which the idealised combustion occurs very rapidly, and therefore essentially at constant volume. The ideal diesel engine cycle is not fulfilled in practice, and it is often difficult to distinguish the observed p–V indicator diagram from that of the s.i. engine.

Multicylinder units are similar to those adopted for s.i. engines. Turbocharging is often used to increase the power output of diesel engines,

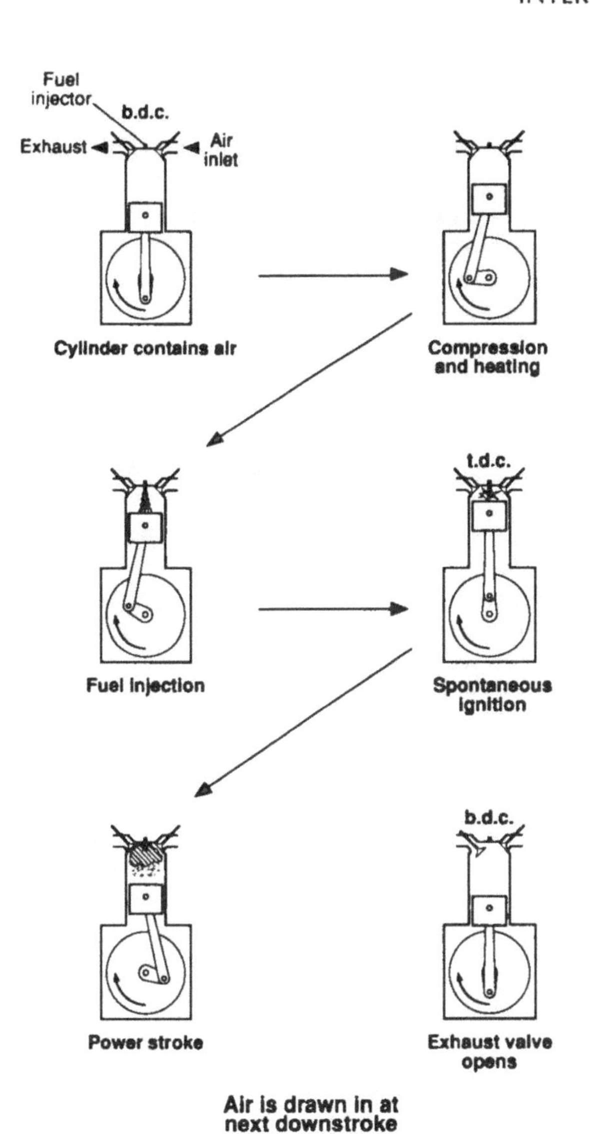

Figure 13.2 Schematic representation of the the four-stroke cycle in a diesel engine.

which raises the air density at induction by a boost of the inlet manifold pressure. This permits more fuel to be injected in each cycle while maintaining the appropriate overall air to fuel ratio.

For spontaneous ignition to occur, the air must be raised to a much higher temperature in the compression stroke than occurs in the s.i. engine, so diesel engines have a higher compression ratio, typically 14:1 to 20:1. The higher loadings imposed and the less smooth combustion that may occur require a very rugged construction. However, the diesel engine is able to operate on more crude fuels than the highly refined gasoline of the s.i. engine, which, helped also by better fuel consumption, can mean considerably lower operating costs. Diesel engines have considerable advantages over s.i. engines for industrial and commercial applications, as in continuous, heavy-duty operation such as is required in locomotives, heavy road transport and marine engines, or where robustness is at a premium as in tractors and earth-moving vehicles. Stationary applications for power generation are also important. Large capacity diesel engines operate at engine speeds below about 3000 revolutions per minute. Engine speeds of up to 5000 revolutions per minute are achieved in smaller diesel engines.

The thermal efficiency of the diesel engine, η, is given by an expression similar to, but slightly more complicated than that shown for the s.i. engine (eqn (13.2)). Both the higher compression ratio and the weaker fuel to air mixtures associated with diesel engines (pure air in the early part of compression) contribute to a greater thermal efficiency of the diesel cycle over that of the Otto cycle. From the corrected form of eqn (13.2) for the diesel cycle it is shown that, for identical values of CR and γ, there is a slightly greater ideal thermal efficiency of the Otto cycle [260]. However, the power generation in a diesel engine is controlled solely by the amount of fuel injected directly into the combustion chamber. There is no preliminary preparation of the fuel + air mixture, and there is no throttling of the air flow. Consequently, the mechanical efficiency is much improved over that of s.i. engines, especially on part load, because the pumping losses are comparatively low.

13.3.2 Fuel injection and combustion processes

Both premixed vapour phase combustion and the diffusion controlled burning of fuel droplets occur in a diesel engine. These have to be optimised in order to achieve efficient and clean combustion. Instantaneous evaporation of the fine spray of liquid droplets in the hot, compressed air to a uniform fuel + air mixture might seem desirable but, since the mixture is lean of stoichiometric overall, richer pockets are essential to promote initiation of autoignition as quickly as possible. There is an ignition delay following the onset of injection, which is a combination of the physical dispersion and evaporation process and the chemical factors discussed in Chapter 7.

Initiation of ignition is followed by a steep pressure rise as burning develops. Smooth combustion, controlled to some extent by the fuel injection rate profile, is rather more desirable than a violent ignition. However, the initial rapid rate of pressure rise can lead to a form of engine knock which may interfere with the smooth motion of the piston.

Owing to the accumulation of unburned fuel, this knocking becomes more severe the greater the ignition delay. The greatest hindrance to satisfactory operation occurs once steady burning has begun. In this stage of reaction, the incoming fuel has insufficient time to vaporise and mix completely so that local regions of very rich mixture occur, soot tends to form in the gas and pyrolytic residues are deposited directly on the walls. Both will eventually prevent satisfactory engine operation. In most applications a compromise compression ratio has to be adopted which, in order to alleviate difficulties associated with cold starting conditions, is higher than that required for optimum performance.

Since physical properties of the liquid state, such as viscosity, density and volatility, affect the performance of the fuel in the fuel spray and dispersion stage of the combustion, there is a substantial commitment to experimental investigations in which fuel sprays are measured and characterised. Non-reactive sprays coupled to turbulent flows in chambers simulating those in engines, or reactive sprays in engine combustion chambers that have been adapted for diagnostic access have been studied extensively. Non-invasive laser and other optical techniques are often used [275]. Numerical modelling (computational fluid dynamics) is used to develop a better understanding of the fluid mechanical processes involved in spray development and combustion. Different types of injector and injection flow patterns are adopted for different applications.

For relatively small, high-speed diesel engines, as used in passenger cars and light goods vehicles, a satisfactory way of achieving the short ignition delay and high degree of mixing is by indirect injection (IDI diesel engines). Controlled bulk air motion to enhance the mixing of the initially separated fuel and air is generated by forcing part of the air under compression by the piston into a separate prechamber in the cylinder head. Fuel is injected within the prechamber and spontaneous combustion starts in this rich mixture. A single orifice injector is often used, break up of the fuel spray being helped by impingement on the wall of the prechamber. The pressure rise associated with the initial combustion forces fluid into the main combustion chamber. The fuel jet is entrained in this flow, mixes with the main chamber air and the combustion process spreads into the leaner mixture in the cylinder as the piston descends. There are similarities between this system and the stratified charge principle in s.i. engines. Low emissions and noise are obtained with peak pressures not much above those encountered in s.i. engines. Cold starting may be aided by the strategic location of a glow plug in the prechamber wall, which is heated prior to cranking the engine at start up in order to promote fuel evaporation. The fuel economy of IDI diesel engines is relatively poor compared with other designs because of increased pumping losses and high heat transfer associated with the prechamber.

Direct injection (DI diesel engines) is used almost exclusively for larger, slower speed engines, for which longer combustion times are available. In these cases the fuel is forced into the main chamber at high pressure and distributed by a series of fine sprays, the air being simultaneously 'swirled' to assist in turbulent mixing. Liquid pressures of 20–40 MPa, rather higher than those required for the indirect injection, are used to promote good atomisation. The combustion chamber usually comprises a bowl in the piston crown. Direct injection is beginning to find favour also in high-speed diesel engines, as a result of improvements in spray profiles and air motion in the combustion chamber.

13.3.3 Composition of diesel fuels and their characterisation

The end-gas reactions which cause unwanted 'knock' in the s.i. engine are precisely those which are required for the initiation of diesel engine combustion. Thus it is extremely desirable to have a predominance of the *n*-alkanes in diesel fuel, although the components are of much lower volatility than gasoline, being mainly in the carbon number range C_{12}–C_{22} (Fig. 1.2). Branching-chain structure alkanes or alkenes, and small proportions of aromatics are also present. Diesel engines can be operated using more volatile fuels.

Diesel fuels are characterised by their *cetane number*, which is measured in a test similar in concept to the assignment of octane number to gasoline blends. The binary mixture scale used to characterise the cetane number is based on *n*-hexadecane (*n*-cetane, $C_{16}H_{34}$), assigned the cetane number 100, and α-methyl-naphthalene, assigned the cetane number 0. However, for practical reasons the binary mixture that is used in the comparative tests of commercial fuels is related to heptamethylnonane (an isomeric structure of cetane) at the lower end, with a cetane number of 15. Cetane numbers of some hydrocarbons are given in Table 13.5 [276]. The main but not necessarily the only parameter that is used to assess a diesel fuel performance in a test engine is the ignition delay, given by the time interval from initial injection to the maximum rate of pressure rise during combustion. The standardising binary mixture reflects the comparative ease of autoignition of the straight chain structure of *n*-hexadecane compared with its branched chain isomer. A high cetane number is compatible with very short ignition delay. A cetane number of 50 is typical for current diesel fuels.

In order to make more extensive use of the crude oil, or to exploit lower grade oil resources there is a desire to produce fuels of a naturally lower cetane number, and then to boost the performance with additives. The function of a cetane number improver is primarily to promote the initiation of autoignition, which may be achieved by thermally unstable organic compounds that are able to undergo exothermic oxidation and

Table 13.5 Cetane numbers of some typical hydrocarbons [276][a]

Fuel	Cetane number		Fuel	Cetane number	
n-Heptane	56.3	(0)			
n-Octane	63.8		1-Octene	40.5	(28.7)
n-Decane	76.9		1-Decene	60.2	
n-Dodecane	87.6		1-Dodecene	71.3	
n-Tetradecane	96.1		1-Tetradecene	82.7	
n-Hexadecane (cetane)	100.0		1-Hexadecene	84.2	
Heptamethylnonane	15.0				
Octadecane	102.6		1-Octadecene	90.0	
Methylcyclohexane	20.0	(74.8)			
Dicyclohexyl	47.4				
Decalin	42.1				

[a] Octane numbers (RON) of selected compounds are shown in parentheses.

generate free radicals. The favoured compounds are either organic peroxides or nitrates, which are added in small proportions by volume to the fuel [277, 278].

13.4 Alternative fuels for spark ignition and diesel engines

The alternative fuels which have attracted most attention are methanol, which can be synthesised from CO and H_2, and ethanol, which can be derived from natural sources by fermentation. Other possible substitute fuels include liquefied petroleum gas (LPG) which comprises mainly propane and butanes, compressed natural gas (CNG), vegetable oils and synthetic fuels derived either from coal or natural gas (methane). There is a substantial penalty in each case, since the mass or volume of the substitute fuel and its container is greater than that of the comparable amount of gasoline or diesel fuel and its tank for a given journey. LPG and CNG suffer particularly badly in this respect and also require specialised equipment for distribution and refuelling.

It is possible to run existing s.i. engines on some of these alternative fuels, with only slight modification. Indeed, methanol and ethanol have good knock-ratings and engines using them run cool and emit relatively small amounts of pollutants, but engine corrosion is a quite serious problem. LNG and CNG have high octane numbers and low emission characteristics, but also low burning velocites. Their advantages are best exploited in specially designed high-compression engines. The present generation of diesel engines cannot run on alcohol fuels and for these, fuel oils derived from vegetable sources such as soya beans, sunflower seeds and coconuts seem promising in those parts of the world where petroleum products are costly or in short supply.

The widespread adoption of new fuels is influenced by many non-technical factors. Taxation rates or other government policies can easily be used to distort the market economy and marked fluctuations in oil prices can force an unhealthy stop–go approach to the development of new technologies. The widespread use of any alternative fuel requires the setting up of a new and very costly distribution system.

13.5 Gas turbine, turbojet and ramjet engines

In these engines, air is first compressed and then fuel is injected and burned continuously in a combustion chamber. The burned gas at high pressure and temperature then provides the energy source. The propulsion of a turbojet is gained from the thrust at the exhaust, which is generated by increasing the momentum of the fluid passing through the engine. The exhaust nozzle has an extremely important function in accelerating the fluid to high velocity. The sole function of the turbine is to drive the air compressor which, in the simplest design, is a *single spool* system; the turbine downstream from the combustion chamber drives the upstream, air compressor (Fig. 13.3). Propulsion by a turboprop aircraft engine relies on an independent *power turbine* to drive a propeller. The power turbine is driven by the high velocity, combustion exhaust gases and is capable of rotating at different speeds from the compressor/turbine shaft. In industrial and marine gas turbine applications the power turbine follows the turbojet type of compressor/burner/turbine combination. Some industrial and marine gas engines are derived from turbojets.

In the ramjet engine, the compression is provided by shock waves formed in the intake owing to forward motion of the engine, supplemented by subsonic diffusion. The system cannot be brought into action before an aircraft is already in high speed flight. The ramjet has no rotating parts, but combined turbo/ramjet systems also exist. An alternative is the pulse jet, which has a self-sustained periodic combustion cycle created by mechanical valves at the air inlet duct. The valves are closed automatically by the rise in pressure at combustion, but open again to admit more air as a result of the pressure drop in the combustion chamber created by the gases expanding through the exhaust nozzle.

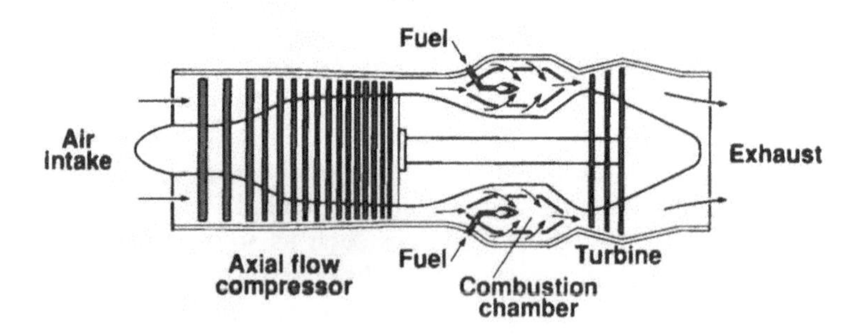

Figure 13.3 Schematic representation of a single spool turbojet with an axial flow compressor and annular combustion chamber.

13.5.1 Design and performance considerations of gas turbines

In many current turbojet designs the combustion takes place in an annulus downstream from the compressor (Fig. 13.3). Compactness and aerodynamic efficiency of the entire unit are not primary considerations of industrial applications, so the combustion chamber may be located on the side of the engine, with appropriate ducting for the compressed air from and combustion products back to the turbine. The combustion efficiency of turbojets is nearly 100% at take-off and more than 98% in high altitude flight. The combustion intensity is enormous, being well in excess of 100 MW in the most powerful turbojets at take-off. This sort of performance can be achieved only by ensuring that the primary combustion takes place close to the stoichiometric composition which is at air to fuel ~15:1 by mass. However, the gas turbine operates at extremely lean mixture overall because the total throughput of air required to generate thrust far exceeds that for combustion of a stoichiometric mixture. Thus the combustion takes place in a staged process by apportioning the intake air in order to maintain a stable flame in a high-velocity stream of gas, typically 150 m s^{-1} (Fig. 13.4). The possible composition range corresponds at its widest air to fuel ratio of ~40:1 to nearly 200:1. This range diminishes considerably as the air velocity through the combustion chamber is increased, and eventually extinction is inevitable.

The fuel is atomised on entry from one of many continuous injectors within the annulus to give a fine spray. It is common to inject the fuel in the same direction as the overall flow but to entrain the air in such a way that the air flow is reversed within the chamber (Fig. 13.4). A reversal of flow is created so that burned gas is recirculated through the flame to give continuous combustion, which occurs in a region of vortex flow. The primary air (20–30%) is fed into the combustion chamber. The secondary air is diverted around the combustion chamber to offer some cooling protection and to dilute the exhaust gas to temperatures that the turbine

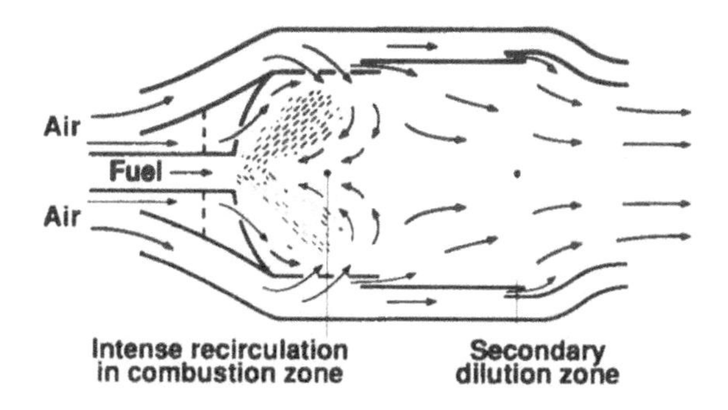

Figure 13.4 Schematic representation of fuel injection and air flow reversal to stabilise combustion in a gas turbine. The air flow is indicated by arrows.

and other components can withstand. Both the primary and secondary air flows are controlled and directed to ensure flame stability within the combustion zone and to minimise exhaust emissions.

In axial flow the compression of air is brought about by a multistage unit comprising alternate rows of rotating (rotor) blades and stationary (stator) vanes. Staged compression may also occur, in which there are two or three of these units in succession. These are called *double spool* or *triple spool* compressors. Each is driven by its own, concentrically mounted, shaft from its own turbine. The first stage, or low pressure, compressor is driven by the turbine most remote from the combustion chamber, and is the largest diameter and slowest rotor. The compression ratio of air admitted to the combustion chamber from staged compressors in the current generation of aircraft engines is as high as 40:1. The air temperature after compression can exceed 800 K, and is considerably higher than this in the most advanced designs. The increases of performance that are obtained in increasingly hostile combustion environments have become possible only through the advances made in materials science and tribology.

13.5.2 Gas turbine fuels

Industrial gas turbines that are derivatives of aircraft engines, are restricted to fuels that have similar properties to aviation fuels. Many other gas turbines are designed and built specifically for industrial or other stationary applications, and they can then accommodate a variety of different fuels.

The major commercial fuel used in gas turbines is kerosene, which comprises petroleum fractions in the boiling range 430–550 K, such as the aviation fuels 'Avtur' or 'Jet A'. Other blends, either lower or higher boiling fractions, are selected for special duties. The main constituents of kerosene are C_{10}–C_{16} alkanes, with aromatics being up to 25% by volume. Alkenes do not usually exceed 5% by volume. The sulphur content is limited to 0.3% by mass. Many different additives are included to serve a variety of purposes, such as corrosion inhibitors, antifreeze for water contaminant, and emulsifiers or thickening agents in the event of a crash. Trace metals, such as sodium or vanadium, are present as naturally occurring residuals from oil. Serious problems of inorganic deposits on turbine blades and corrosion at high temperatures can be caused by these components.

Amongst important physical properties of liquid fuels are the relative density and viscosity, and their temperature dependences. Fuel temperatures may vary considerably as a result of speed and altitude changes of an aircraft. Low external temperatures can cool the fuel appreciably, but air friction at the external surfaces in high speed flight can compensate for this. Nevertheless, the fuel temperature may fall nearly to the freezing point. By contrast, the liquid fuel temperature entering the combustion chamber may exceed 450 K under certain operating conditions. Vaporisation, thereby causing air-locks, or thermal degradation to form deposits must be controlled. With fire safety in mind, the flash point is also important, which is related to the vapour pressure (section 12.2).

Natural gas is an extremely common fuel for stationary gas turbines, including those derived from aircraft engines, given suitable adaptation. Natural gas is normally used directly as supplied but, for environmental reasons, purification is required if the sulphur content is unacceptably high. Other liquid hydrocarbon fuels can easily be burned in stationary gas turbines, including heavy residual oils. The combustion of solid fuels in turbines, either in a preliminary gasification process prior to the main chamber or as pulverised coal within the chamber, is under development and is destined to be a technology available to the electrical power generation industry.

Methanol and ethanol are potentially attractive fuels. They are 'environmentally clean', and have advantages for agriculturally based rather than oil-based economies. The low specific energy of alcohols, ~50% of that for kerosene, presents no problem for stationary applications, but is unsuitable when payload is important, as in aircraft. Corrosion of metals has to be controlled.

Engine design is extremely important with regard to control over the pollutant emissions from gas turbines, but the fuel content itself must come under scrutiny, especially when there is no potential for exhaust clean-up, such as in an aircraft. Emissions from aircraft have a direct effect both on urban environments and on the upper atmosphere.

Further reading

AGARD (1988). *Combustion and Fuels in Gas Turbine Engines*. Conference 422 Proceedings, Paris, France.

Agnew, W.G. (1984). Room at the piston top. In *Twentieth Symposium (International) on Combustion*. The Combustion Institute, Pittsburgh, PA, USA, p. 1.

Goodger, E.M. (1980). *Alternative Fuels*. Macmillan, London, UK.

Heywood, J.B. (1988). *Internal Combustion Engine Fundamentals*. McGraw Hill, New York, USA.

Lefebvre, A.W. (1983). *Gas Turbine Combustion*. McGraw Hill Series in Energy and Combustion, New York, USA.

Longwell, J.P. (1976). Synthetic fuels and combustion. In *Sixteenth Symposium (International) on Combustion*. The Combustion Institute, Pittsburgh, PA, USA, p. 1.

Lovell, W.G. (1948). *Ind. Engng Chem.*, **40**, 2388.

Mellor, A.M. (ed.) (1990). *Design of Modern Turbine Combustors*. Academic Press, New York, USA.

Rolls-Royce (1986). *The Jet Engine* (4th edn). Technical Publications Department, Rolls-Royce, Derby, UK.
Sawyer, R.F. (1992). Reformulated gasoline for automotive emission reduction. *Twenty-Fourth Symposium (International) on Combustion*. The Combustion Institute, Pittsburgh, PA, USA, p. 1483.
Stone, R. (1992). *Introduction to Internal Combustion Engines* (2nd edn). Macmillan, London, UK.
Weaving J.H. (ed.) (1990). *Internal Combustion Engines: Science and Technology*. Elsevier, London, UK.

14 Combustion and the environment

14.1 Introduction

The most familiar pollutant species are carbon oxides, nitrogen oxides and sulphur dioxide, but the low-molecular-mass hydrocarbons and chlorofluorocarbons (CFCs) are also important. Polycyclic aromatic hydrocarbons (PAH) occur in connection with soot formation (section 6.10). How these compounds affect the environment, their combustion sources, and the measures that are being taken to control emissions are addressed in this chapter. The sources of principal pollutants in UK cities are given in Table 14.1 [279].

14.2 Reactions of pollutants in the atmosphere

The topics discussed in this section are (i) the formation of atmospheric smog in the troposphere above urban environments, (ii) the atmospheric reactions that involve airborne sulphur emissions, (iii) the destructive processes that occur in the stratospheric ozone layer, and (iv) the 'greenhouse' effect in the stratosphere [280, 281]. The troposphere is that part of the atmosphere which extends from Earth's surface upwards for about 15 km in tropical regions and about 8 km in polar regions. The stratosphere lies above the troposphere, extending to a height of about 50 km.

The wavelength range of visible radiation is approximately $\lambda = 700$ nm (red) down to 420 nm (violet), corresponding to the wavenumbers $(1.43–2.38) \times 10^{-4}\ cm^{-1}$ and frequency $\nu = (4.3–7.1) \times 10^{14}$ Hz. Ultraviolet radia-

Table 14.1 Sources of principal pollutants in UK cities 1992 [279]

	% total emissions				
Source	SO_2	NO_x	CO	NMHC[a]	Soot
Road transport	2	51	90	37	47
Electricity supply industry	70	25	1		5
Other industry	25	21	5	62	20
Domestic	3	3	4	1	28

[a] Non-methane hydrocarbons including volatile organic compounds from solvent evaporation.

tion is associated with shorter wavelengths, and it is the range down to about 300 nm that is transmitted to Earth's surface. Electromagnetic radiation of short wavelengths is capable of photochemical initiation of reaction if the energy is absorbed by gaseous species. Chemical bonds for which the strength is approximately 400 kJ mol^{-1} can be broken by light of wavelength ~300 nm or less. There is an inverse relationship between the two.

14.2.1 Photochemical smog

In the presence of sunlight, carbon monoxide, nitrogen oxides (NO_x) and unburned hydrocarbons lead to photochemical smog. This problem was identified during the 1960s in California. The main contributory factors were considerable yields of emissions from car exhausts and very intense sunlight, but particular geographical features also exacerbated the effect. Photochemical smogs are now encountered to varying extents within very many urban environments worldwide as a result of increased combustion emissions. The effects of a photochemical smog are eye and respiratory irritation, reduced visibility and damage to plant life, peroxyacyl nitrates (PAN) being amongst the offensive components.

The primary pollutants emitted by combustion appliances undergo reaction in the atmosphere to produce secondary pollutants which give rise to smog. The secondary pollutants are nitrogen dioxide (NO_2), ozone (O_3), aldehydes, ketones, PAN and alkyl nitrates (Table 14.2 [282]). In 1994 the Petrochemical Oxidants Review Group reported that average daily concentrations of ozone in the UK were 0.1–0.3 parts per million (ppm), rising to 0.6–0.8 ppm on sunny days. There are a number of complex interacting chains involved in the processes leading to photochemical smogs which are propagated by the OH and HO_2 radicals and their organic analogues RO and RO_2. R can be alkyl, acyl or any other carbon-containing fragment. Some of the major propagating processes are discussed below.

Complex mathematical models have been set up to simulate the reactions taking place when photochemical smog is formed and these

Table 14.2 Typical concentrations of the major pollutants in photochemical smog [282]

Pollutant	10^2 concentration (ppm)	Pollutant	10^2 concentation (ppm)
Carbon monoxide	200–2000	Alkenes	2–6
Nitric oxide	1–15	Aromatics	10–30
Nitrogen dioxide	5–20	Ozone	2–20
Total hydrocarbons (excluding methane)	20–50	Aldehydes	5–25
		Peroxyacyl nitrates	1–4

have successfully reproduced the observed behaviour in smog-chambers where artificial urban atmospheres are irradiated under controlled conditions. Experimental measurements, kinetic foundations and simple model structures are discussed elsewhere [281].

The primary pollutant, nitric oxide, is oxidised to nitrogen dioxide in the atmosphere by

$$2NO + O_2 \rightarrow 2NO_2 \qquad (14.1)$$

which initiates a chain reaction involving ozone (at much lower altitudes than its normal, stratospheric existence) as follows. Nitrogen dioxide is photodissociated by the absorption of solar radiation ($\lambda < 398\,nm$) to produce ground-state oxygen atoms:

$$NO_2 + h\nu \rightarrow NO + O(^3P) \qquad (14.2)$$

This is followed by the reactions

$$O(^3P) + O_2 + M \rightarrow O_3 + M \qquad (14.3)$$

and

$$O_3 + NO \rightarrow O_2 + NO_2 \qquad (14.4)$$

to establish the overall equilibrium

$$NO_2 + O_2 \Leftrightarrow NO + O_3 \qquad (14.5)$$

Oxygen atoms and ozone reacting with hydrocarbons, especially alkenes and alkylbenzenes present in a polluted atmosphere, as a source of undesirable secondary pollutants such as aldehydes, acetaldehyde being the precursor to peroxyacetyl nitrate.

OH radicals are always present in the troposphere as a result of a chain involving electronically excited, $O(^1D)$ atoms

$$O_3 + h\nu \xrightarrow{\lambda < 310\,nm} O_2 + O(^1D) \qquad (14.6)$$

$$O(^1D) + H_2O \rightarrow 2OH \qquad (14.7)$$

but supplementary reactions, through an interconversion of HO_2, increase their concentrations quite considerably. For example,

$$NO + HO_2 \rightarrow NO_2 + OH \qquad (14.8)$$

$$O_3 + HO_2 \rightarrow 2O_2 + OH \qquad (14.9)$$

The major sources of HO_2 involve other trace components, such as the reaction

$$CO + OH \rightarrow CO_2 + H \qquad (14.10)$$

and also

$$OH + H_2 \rightarrow H_2O + H \qquad (14.11)$$

followed by

$$H + O_2 + M \rightarrow HO_2 + M \qquad (14.12)$$

The combination of H and O_2 to form HO_2 is favoured rather than

$$H + O_2 \rightarrow OH + O \qquad (14.13)$$

at the temperatures and moderate pressures of the troposphere. HO_2 radicals can also be formed by photolysis of formaldehyde, traces of which are derived from naturally formed methane.

$$CH_2O + h\nu \xrightarrow{\lambda < 333\,\text{nm}} H + CHO \qquad (14.14)$$

followed by

$$CHO + O_2 \rightarrow HO_2 + CO \qquad (14.15)$$

Methane is normally present in the atmosphere as a result of natural gas leakage, natural fermentation and other biological processes, as well as being produced by incomplete combustion, which promotes the reaction

$$OH + CH_4 \rightarrow CH_3 + H_2O \qquad (14.16)$$

followed by

$$CH_3 + O_2 + M \rightarrow CH_3O_2 + M \qquad (14.17)$$

$$CH_3O_2 + NO \rightarrow CH_3O + NO_2 \qquad (14.18)$$

and

$$CH_3O + O_2 \rightarrow CH_2O + HO_2 \qquad (14.19)$$

or the formation of methyl nitrate

$$CH_3O + NO_2 \rightarrow CH_3ONO_2 \qquad (14.20)$$

A route to PAN from acetaldehyde is

$$CH_3CHO + OH \rightarrow CH_3CO + H_2O \qquad (14.21)$$

$$CH_3CO + O_2 \rightarrow CH_3CO_3 \qquad (14.22)$$

$$CH_3CO_3 + NO_2 \rightarrow CH_3CO_3NO_2 \ \text{(PAN)} \qquad (14.23)$$

14.2.2 Airborne sulphur emissions

The main sources of sulphur doixide emissions are the burning of coal, the combustion of oil, refinery operations and smelting of non-ferrous ores. Prior to the Clean Air Acts in the UK (1956 and 1968) inefficient coal burning, mainly in domestic fires, contributed directly to smogs which formed in urban environments from the combination of solid particulate and sulphur dioxide emissions when the ambient temperature was close to 0°C and the relative humidity was high. The most noticeable personal effect was severe bronchial irritation, but there was also long-term damage to buildings and other property. The swift improvement in air quality that was brought about once the primary cause had been identified must be regarded as one of the triumphs of government legislation with respect to pollution control, of coal-burning fires in this instance. There is considerable evidence of a remaining problem associated with damage to plants by sulphur dioxide emissions from power stations and some industrial operations, often borne tens or hundreds of kilometres from the source. This is generally classified as the effect of 'acid rain'.

Much of the sulphur present in the atmosphere is converted to sulphates which can be detected as a sulphuric acid aerosol and particles of salts such as ammonium sulphate. These materials are returned to Earth's surface in rain or by dry deposition involving adsorption or chemical reaction. While they remain in the atmosphere, sulphates have a significant effect on the range of visibility, and erosion of buildings, especially those constructed from carbonate stone, results largely from the corrosive effects of sulphuric acid. The mechanism by which sulphur dioxide is converted to sulphate is complex but it appears that two main routes, governed by ambient conditions, are by homogeneous gas-phase oxidation, initiated photochemically, and heterogeneous oxidation, either in aqueous droplets or on aerosol particles.

During the winter, photochemical reactions are very slow and removal of sulphur dioxide is mainly by dry deposition and precipitation. In the summer, especially in polluted urban atmospheres, sulphur dioxide is oxidised by a homogeneous process sustained by the steady-state concentration of atoms and radicals, such as

$$OH + SO_2 \rightarrow HOSO_2 \tag{14.24}$$

$$HO_2 + SO_2 \rightarrow OH + SO_3 \tag{14.25}$$

$$CH_3O_2 + SO_2 \rightarrow CH_3O + SO_3 \tag{14.26}$$

The sulphur trioxide produced in eqns (14.25) and (14.26) can hydrolyse extremely rapidly to form a sulphuric acid aerosol, which also includes $HOSO_2$ and nitrylsulphuric acid, $HOSO_2ONO_2$. In the absence of alkenes there is no detectable dark reaction between sulphur dioxide and ozone, but when these are present there is a fairly rapid reaction, the precise mechanism of which is obscure but the products from which include aldehydes and sulphur trioxide [283].

14.2.3 Destruction of the ozone layer

The small concentration of ozone in the stratosphere absorbs harmful ultraviolet solar radiation in the wavelength range 280–320 nm, which would otherwise reach Earth's surface and would have an adverse effect on plant and animal life. Most of the sources of pollution are located close to Earth's surface, but aircraft may generate their combustion products in the stratosphere. Their emissions can catalyse the conversion of the ozone in the 'shield' to molecular oxygen. Other man-made materials which are stable in the troposphere and reach the stratosphere by diffusion include halogenated hydrocarbons (e.g. CFCs) and there they undergo photochemical decomposition, and may deplete the ozone layer.

Ozone originates in the stratosphere mainly from the dissociation of molecular oxygen to ground state oxygen atoms

$$O_2 + h\nu \xrightarrow{\lambda < 242\,\text{nm}} O(^3P) + O(^3P) \qquad (14.27)$$

followed by eqn (14.3). The destruction of ozone may be exemplified with respect to trichlorofluoromethane $CFCl_3$ as a pollutant in the stratosphere

$$CFCl_3 + h\nu \rightarrow CFCl_2 + Cl \qquad (14.28)$$

$$Cl + O_3 \rightarrow ClO + O_2 \qquad (14.29)$$

$$ClO + O(^3P) \rightarrow Cl + O_2 \qquad (14.30)$$

NO, H, OH may react in a way similar to Cl in eqns (14.29) and (14.30) and there may also be a synergism between some of these pollutant derivatives, such as

$$ClO + NO \rightarrow Cl + NO_2 \qquad (14.31)$$

The overall chain cycle, comprising either eqns (14.29) or (14.30), or when the regeneration of NO by

$$O + NO_2 \rightarrow NO + O_2 \qquad (14.32)$$

is taken into account (eqns (14.29), (14.31) and (14.32)), is

$$O(^3P) + O_3 \rightarrow 2O_2 \qquad (14.33)$$

There have been very sophisticated attempts at modelling the ozone layer in the stratosphere but the task is difficult because of lack of knowledge of the detailed chemistry and many of the rate constants, and also because both chemistry and transport have to be taken into account.

14.2.4 The 'greenhouse' effect

Even carbon dioxide has an important effect on the environment. Infra-red radiation is emitted from Earth's surface with a peak of energy distribution in the region of 1318 μm, which is absorbed by CO_2 as vibrational bond energy and re-emitted by fluorescence in the (short) lifetime of the excited species. The temperature of Earth's surface is raised by the radiation that is directed back to earth, and is called the *greenhouse gas effect*. As a result of the combustion of fossil fuels, it is thought that the concentration of carbon dioxide in the atmosphere may approach 400 ppm by the end of the century, which is about 50 ppm in 30 years. This is predicted to cause an average temperature change at the surface of nearly 1 K, which would bring about an appreciable reduction in the polar ice-caps. If by the end of the 22nd century most of the world's known fossil fuel supplies are consumed, it is estimated that the amount of carbon dioxide in the atmosphere will have risen by a factor of between four and eight from its present level. A change of this magnitude must bring about substantial alterations in Earth's climate [284, 285].

Of more recent concern is that other gases that are able to absorb infra-red radiation, such as methane and nitrous oxide (N_2O), are being generated at increasing rates and would also contribute to the greenhouse effect. Although methane is generated mainly from naturally occurring sources, there are increasing rates of production as a result of intensive agricultural activity. Naturally occurring sources also contribute mainly to the bulk of atmospheric nitrous oxide, but the evidence of a consistent increase of its concentration in the atmosphere over the last century would suggest that there are also contributory man-made sources. Nitrous oxide is known to constitute about 2% of the nitrogen oxides emitted in flue gases from stationary power sources.

14.3 Combustion-generated pollutants from engines

The products of complete combustion of hydrocarbon fuels are carbon dioxide and water. In addition, carbon monoxide, nitrogen oxides (NO_x) [286], unburned hydrocarbons (HC), PAH and solid particulates (soot) may be present in the exhaust gases from any practical device. Other materials may also be found, such as lead compounds from antiknocks added to motor fuel, or sulphur and nitrogen containing organic compounds from diesel fuel.

14.3.1 Carbon monoxide

Although some spark ignition (s.i.) engines have been developed to operate under 'lean burn' conditions, most current engines operate on

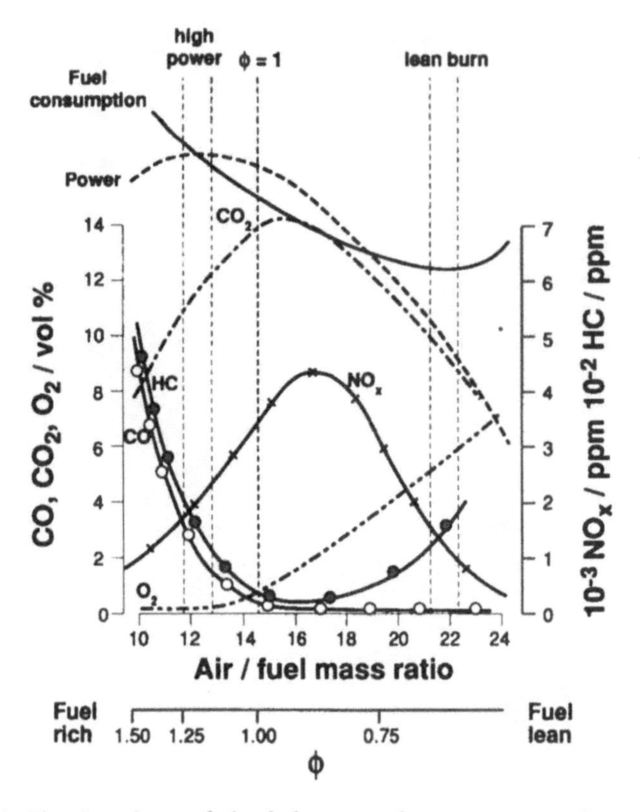

Figure 14.1 The dependence of the fuel consumption, power generation and exhaust composition on the fuel + air mixture in an s.i. engine (after Newhall [286], Starkman [287] and Suzuki [288]). The equivalence ratio (ϕ) is defined in Chapter 3.

stoichiometric fuel + air mixtures, with slight enrichment for power generation in acceleration. The exhaust gases from such engines contain about 1% by volume of CO but this level may rise under idling or decelerating conditions. These CO concentrations can be considerably reduced if the mixture is made more lean (Fig. 14.1 [286–288]). At an air to fuel ratio of 17:1 by weight, the CO concentration falls below 0.5 vol%. Considerable attention has been directed towards the development of s.i. engines which can utilise these and still more fuel-lean mixtures.

Under the conditions obtaining at the peak of the engine cycle (*c.* 2800K and ~40 atm) the equilibrium

$$CO + OH \Leftrightarrow CO_2 + H \tag{14.34}$$

is established, and yields of CO match this prediction. As the temperature falls in the expansion stroke the position of the equilibrium moves to the right, but the measured concentrations of CO do not fall so rapidly. Part of the reason for this 'frozen' equilibrium is that high radical concentrations are maintained by bimolecular equilibria (section 6.5) and third-order recombinations such as

$$H+OH+M \rightarrow H_2O+M \qquad (14.35)$$

become rate controlling. These processes are greatly slowed by the fall in pressure in the exhaust gas expansion. Other contributory factors to CO formation also include flame quench near the cylinder walls and partial oxidation of unburned hydrocarbons during the exhaust part of the cycle [289].

Since virtually all of the CO_2 formed during the combustion of hydrocarbons arises from the oxidation of CO (Chapter 6), the excess CO formed in rich conditions arises from the lack of oxygen to complete the final stage of combustion of the fuel.

14.3.2 Nitrogen oxides

The nitrogen oxide in the exhaust gases is almost entirely nitric oxide, NO. The nitric oxide reaches a maximum concentration of about 5000 ppm (0.5 mol%) in mixture compositions that are just on the lean side of stoichiometric, which conflicts with the conditions for minimum carbon monoxide formation (Fig. 14.1). Nitric oxide is formed mainly in the post-flame gases where the O atom concentration and the temperature are high, especially in the vicinity of the sparking plug where the temperature remains high for the longest period.

Thermal origins of NO formation from atmospheric nitrogen may be interpreted by the Zel'dovich mechanism. The equilibrium

$$O_2 \Leftrightarrow 2O \qquad (14.36)$$

is established in the combustion chamber, and is followed by the reactions

$$O+N_2 \Leftrightarrow NO+N \qquad (14.37)$$

$$N+O_2 \Leftrightarrow NO+O \qquad (14.38)$$

There is also a contribution from

$$N+OH \Leftrightarrow NO+H \qquad (14.39)$$

The predicted NO concentration as a function of air:fuel ratio, derived from the equilibria of eqns (14.37) and (14.38),

$$[NO]=\{K_{14.37}K_{14.38}[N_2][O_2]\}^{1/2} \qquad (14.40)$$

follows the form shown in Fig. 14.1. An excess of oxygen in the system favours the augmentation of the equilibrium concentration of NO, but the overall temperature dependence of the product $K_{14.37}K_{14.38}$ (which is dominated by the activation energy, ~320 kJ mol^{-1}, for the forward reaction $O + N_2$ (eqn 14.37)) forces a reduction in the calculated equilibrium concentration as the flame temperature associated with increasingly lean mixtures is reduced. Lower NO concentrations are predicted in increasingly rich mixtures as a result of the reduction of the molecular oxygen concentration and of the flame temperature.

However, higher concentrations of NO are measured than those predicted from eqn (14.40) as the temperature decreases owing to expansion in the exhaust manifold. The expansion process itself causes an extremely rapid cooling or 'adiabatic quenching'. The concentration of NO that was established at the peak combustion temperatures is 'frozen' by the expansion process because the rates of several of the reactions fall too rapidly for there to be time for the chemical equilibrium to adjust to the new temperature. These criteria are discussed in detail by Heywood [261]. Enhanced NO yields may come also from the super-equilibrium concentrations of O atoms and OH radicals, attained in the reaction zone and burned gas, which can accelerate the rate of the thermal NO formation.

'Prompt' NO formation [290], which is initiated by hydrocarbon radicals, such as [291]

$$CH + N_2 \rightarrow HCN + N \tag{14.41}$$

$$CH_2 + N_2 \rightarrow HCN + NH \tag{14.42}$$

can also occur. The N and NH concentrations may be balanced in the flame by the equilibrium

$$NH + OH \Leftrightarrow N + H_2O \tag{14.43}$$

The most probable route to NO from HCN is

$$HCN + O \rightarrow NCO + OH \tag{14.44}$$

followed by the conversion of NCO to NH and N (see also Fig. 14.4). Another supplementary route to NO is via the formation of N_2O

$$O + N_2 + M \rightarrow N_2O + M \tag{14.45}$$

followed by

$$N_2O + O \rightarrow NO + NO \tag{14.46}$$

14.3.3 Unburned hydrocarbons

There is little correspondence between the chemical composition of the unburned hydrocarbons measured in the exhaust and the original gasoline used in s.i. engines. Not only are they of lower molar mass, but also carbonyl compounds and peroxides are present in the engine exhaust. The term 'unburned hydrocarbons' usually implies the components, other than CO, that are not completely oxidised.

Potential sources of unburned hydrocarbons are the quench layer and the crevices between the piston and the cylinder wall (typically ~0.1 mm). As the flame propagates across the combustion chamber and approaches the wall, heat transfer and radical destruction processes quench the flame close to the wall. Nevertheless, appreciable reaction takes place in these hot gases, resulting in the formation of products by partial oxidation and pyrolysis of the fuel. In addition, a proportion of the total emissions of unburned hydrocarbons may come from the crevices around the piston rings. Lubricating oil on the cylinder walls, and oil stored in carbon deposits on the piston crown and elsewhere may also be sources. Engine design and manufacturing techniques, and the development of new lubricants have contributed significantly to the reduction of unburned hydrocarbons originating from these latter sources.

Unburned hydrocarbon concentrations in the exhaust decrease appreciably as the mixture is made more lean, but eventually there is a further increase in the unburned material in the exhaust (Fig. 14.1). Combustion in very lean mixtures is relatively weak and is prone to being quenched easily or even can fail to reach completion before the exhaust valve opens. These problems can be overcome by enhanced turbulence to maintain faster combustion processes, so that the efficient combustion range can be extended to air to fuel > 20:1 in lean-burn engines (section 13.2.5). Misfiring can give rise to a considerable increase in the unburned hydrocarbon emissions in the lean limit, the composition for which is governed by the prevailing pressure and temperature of the turbulent gases in a small combustion chamber.

In a diesel engine there is no well defined flame front as in an s.i. engine. The fuel–air mixture burns wherever the conditions of composition and temperature are suitable. Although the fuel is injected as a spray, the evaporation rate is normally fast enough to ensure homogeneous rather than heterogeneous droplet combustion except where the spray strikes the walls. Most diesel engines, if correctly operated and maintained, emit less carbon monoxide and unburned hydrocarbons than comparable s.i. engines.

14.3.4 PAH, smoke and particulates

The visible material in exhaust gases (smoke), that is associated with diesel engines, includes carbon particles and droplets of unburned fuel. Polynuclear aromatic hydrocarbons (PAH) are also present in diesel engine exhaust [292, 293]. They tend to accumulate on the surface of the soot, and are undesirable products since many of them are carcinogenic.

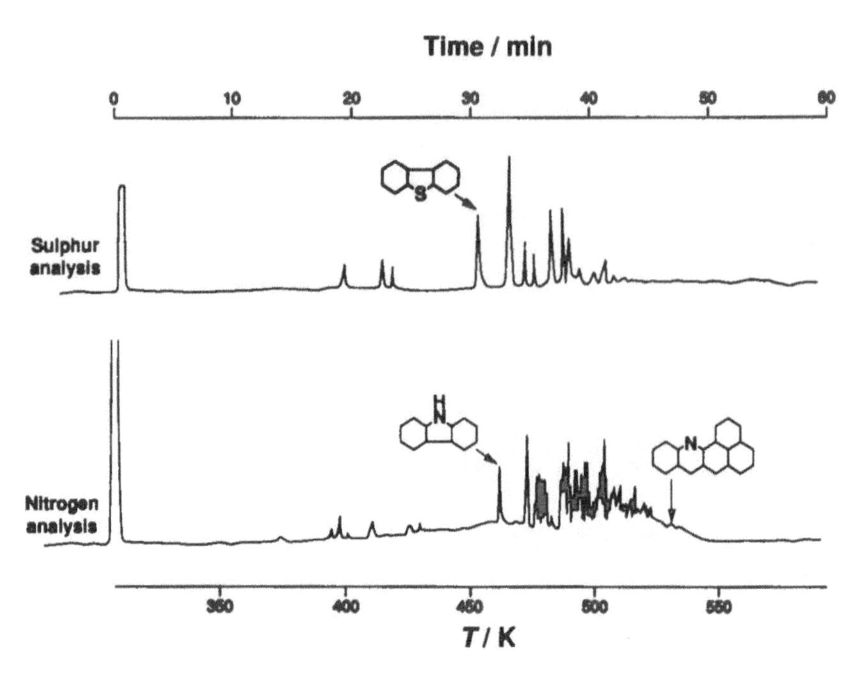

Figure 14.2 A gas chromatographic record of the PAC yields in diesel engine exhaust. These emissions are residuals from the fuel (after Williams *et al.* [292]).

They may be products of combustion (section 6.9), but they can also be residual fuel components. There is evidence that sulphur and nitrogen-containing polycyclic aromatic compounds can be emitted in diesel exhaust gases (Fig. 14.2 [292]). These are certainly not synthesised in the combustion process and must be carried over from the diesel fuel [292].

Although large soot particles are fairly resistant to oxidation, small species are removed by reaction with OH radicals (section 6.9.2). However, the concentration of OH cannot be maintained since the only source in the exhaust gases is by dissociation of H_2O, which has a very high activation energy. Smoke emissions can be suppressed quite effectively, even under fuel-rich conditions, by the use of compounds such as barium hydroxide ($Ba(OH)_2$) in the fuel. Its function is to enhance the OH radical concentration, probably through a sequence of free radical processes beginning with its decomposition

$$Ba(OH)_2 + M \rightarrow BaO + H_2O + M \tag{14.47}$$

followed by

$$BaO + H_2 \rightarrow BaOH + H \tag{14.48}$$

Lead antiknock additives are converted to solid lead oxide (PbO) in s.i. engines (section 13.2.3). Particles of PbO, and also sulphides and halides of lead, are detectable in the exhaust gas when leaded fuel is used. The nature of the lead-containing particles depends on several factors. In city streets there is already a background aerosol and the lead is found as coagulated, chain-aggregate particles, whereas on open roads, the lead is emitted as very small discrete particles about 0.02 μm in diameter. The lead concentrations in urban atmospheres may be about 3 $\mu g\, m^{-3}$, compared with a background of about 0.8 $\mu g\, m^{-3}$. Close to heavily used motorways atmospheric lead may be up to 15 $\mu g\, m^{-3}$, although these concentrations should decrease as the use of leaded gasoline diminishes [294].

14.4 Measurements of exhaust emissions

The performance and efficiency of motor vehicles are tested in research and development laboratories with a 'rolling road' facility. A relationship to road use is obtained from a prescribed driving cycle that simulates the demands put on a vehicle in typical urban or open road conditions. Fuel consumption figures quoted in advertising literature for new vehicles are usually obtained in this type of test. In many countries exhaust emissions from s.i. engines are measured in formal tests of motor vehicles in current road use. Typical results obtained at a UK testing station are shown in Table 14.3.

The detection methods used in fully equipped test facilities are infrared detection of CO, chemiluminescent detection of NO_x and flame ionisation or katharometer detection of unburned hydrocarbons. For CO measurement the infra-red analyser is tuned to a beam at the fundamental vibrational frequency of the CO bond. Absorption occurs by CO present in a test cell as the beam is passed through it, the fractional absorption being proportional to CO present. The chemiluminescent detection of NO_x is derived from the reaction

$$NO + O_3 \rightarrow NO_2^* + O_2 \tag{14.49}$$

in which NO_2 is formed in an electronically excited state. The intensity of the pale yellow light emitted from the chemiluminescent process

$$NO_2^* \rightarrow NO_2 + h\nu \tag{14.50}$$

Table 14.3 Exhaust gas analyses by SUN Gas Analyzer obtained during UK Ministry of Transport annual test with engine at idle (800 rpm) and at normal operating temperature

	1983 vehicle, 90 000 mi, 3.0 litre. Bosch K Jetronic fuel injection, with air injection to exhaust manifold	1993 vehicle, 17 000 mi, 1.8 litre. Bosch L Jetronic fuel injection, with catalytic converter in exhaust
CO (vol%)	0.92	0.00
CO_2 (vol%)	12.55	15.46
HC (ppm)	51	0
Air to fuel mass ratio	16.86	14.89

Data courtesy of Colin Pitt, Otley, W. Yorks

is measured by photomultiplier. The proportion of NO present is obtained from the calibration of the instrument operating with an excess of ozone, so that eqn (14.49) is under a pseudo first order dependence on the concentration of NO. Measurement of the total NO_x emission is more complicated because any higher oxides of nitrogen have to be reduced to NO by passing the effluent gas over heated copper.

The unburned hydrocarbons are measured by flame ionisation or katharometer detectors, each of which are used in gas chromatographic studies. The flame ionisation detector is described in Chapter 3. The katharometer is a temperature-sensitive device, and its temperature is controlled by the thermal conductivity of the gases flowing through it. Any change in thermal conductivity, as is brought about when organic species flow through the detection cell is monitored by a change in resistance of the device. No separation of the components is performed, so the signal represents the total organic content, but is reported as 'unburned hydrocarbons'. Samples for each of these tests are taken either from a bag in which the exhaust effluent has been collected or by continuous pumping to the analyser devices while the engine is running.

14.5 Suppression of pollutants from s.i. engines

Although increasingly stringent standards are being set, there have been considerable improvements in the performance of s.i. engines since the first concerns about air pollution emerged. The present position on emissions (g km^{-1}) from s.i. engines is reviewed in Table 14.4 [273]. Some averaged reference yields are shown for comparison with other sources in Table 14.5 [295], based on g MJ^{-1} fuel energy input.

Much of the success in the reduction of emissions from s.i. engines over the last two decades, has been achieved by the application of several technologies. One of these which successfully reduces nitric oxide is exhaust-gas recycle (EGR). This involves dilution of the fuel+air mixture by adding up to 20% exhaust gas at the inlet, which is about the maximum that can be tolerated under part throttle conditions to maintain combustion without misfire. This lowers the combustion temperatures and thereby reduces the nitric oxide to less than half its normal yield, although carbon monoxide and unburned hydrocarbon levels are adversely affected. EGR has been applied to the large-capacity engines

Table 14.4 Typical exhaust and other emissions from s.i. engines at 80 000 km (50 000 mi) (US figures [273])

Component	Pre-1970	Pre-1990	Post-1995
NO_x (g km^{-1})	2.5	0.6	0.25
CO (g km^{-1})	50	2.1	2.1
HC (g km^{-1})	>6	0.25	0.16 (NMHC)
Evaporative HC (g km^{-1})	>45	2.0	2.0 (revised test)

Table 14.5 Typical emissions from combustion systems g MJ^{-1}, fuel energy input (US figures, 1980 [295])

System	HC	CO	NO_x	SO_x
s.i. automobile (1980)	0.06	1.3	0.2	0.02
Turbojet aircraft	0.08	0.4	0.1	0.02
Coal-fired electricity generation	0.005	0.02	0.3	1.3
Gas-fired electricity generation	5×10^{-4}	0.01	0.2	2×10^{-4}
Domestic gas-fired heating	0.003	0.01	0.03	2×10^{-4}

commonly favoured in the USA. The technique bears some relationship to lean burn, although there is no excess of oxygen in EGR. About a decade ago, it was fashionable also to inject air into the hot exhaust gases just downstream from the exhaust outlet valve to promote oxidation of unburned HC and CO in the hot manifold.

The most widely used system fitted to the exhaust systems of production vehicles is the three-way catalytic converter [296]. This controls CO and HC by their oxidation to carbon dioxide and water and NO_x by its reduction to nitrogen. This method is capable of bringing about very high reductions in the proportions of pollutants leaving the engine (Table 14.3), but it is not without qualification, as noted below.

The catalytic surface comprises a mixture of the noble metals rhodium and platinum, which are dispersed as a very finely divided layer on an alumino-silicate monolith which is extruded to a honeycomb structure. The exhaust gases are exposed to an exceedingly large surface area as they pass through the unit at typical contact times of 100–400 ms. The action of the catalyst is to dissociatively adsorb NO_x on its surface, predominantly by the rhodium. The gaseous CO and HC are oxidised by the adsorbed O atoms when they come into contact with the surface. The adsorbed N atoms recombine to form N_2, which is then desorbed from the surface. The three-way activity of the catalyst is very effective, but it is achieved only within a very narrow range of engine operation close to the stoichiometric mixture (Fig. 14.3 [296]). Consequently, vehicles fitted with catalytic converters also require a zirconium detector which monitors the residual oxygen supply in the exhaust, and controls the inlet mixture so that the converter is always operated at the optimum condition. The overall cost of these systems is high, and becomes disproportionately so for cars at the lower priced end of the market.

The catalyst is most effective only when the temperature of the exhaust gas and the pipe is sufficiently high for 'ignition' to have taken place (Fig 10.6), typically at $T > 600$ K. The catalyst is ineffective at cold start-up. The heating up period, which may be only tens of seconds, is believed to contribute more than half of the emissions from vehicles used in urban environments. Methods for the minimisation of emissions during the cold-start period are being addressed at the present time. The catalyst

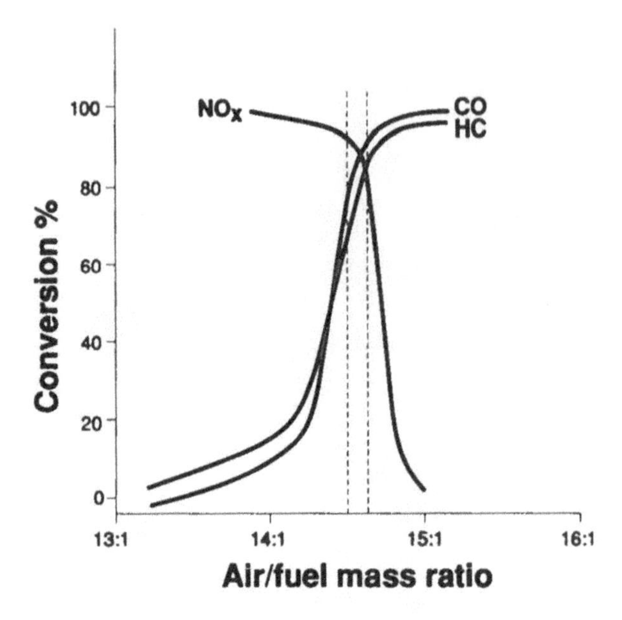

Figure 14.3 The performance of a three-way catalyst in the exhaust system of an s.i. engine as a function of air to fuel ratio (after Taylor [296]).

surface is poisoned by the presence of lead, which is an additional reason for the removal of lead from gasoline. The surface must also be kept free from 'coke' or tarry deposits, which has been part of the impetus for the reduction of oil consumption by modern s.i. engines.

Much has also been achieved recently in the reduction of emissions at source by improvements in s.i. engine design and in the management systems. In virtually all motor vehicles powered by s.i. engines there has been a shift from very simple carburettor and induction systems to much more sophisticated carburettors or to fuel injection with more efficient supply to each cylinder. Electronic ignition control has also been adopted in many cases. These have done much to reduce the variability in performance from cylinder to cylinder and to eliminate the marked variations in mixture compositions that used to occur with sudden variations of throttle opening.

The problems associated with increasing HC emissions and misfire of the engine in very lean burn conditions are being tackled through developments such as those discussed in section 13.2.5. The advantage of lean burn is a reduced fuel consumption and commensurately lower emission of CO_2. Although all modern engines are much more economical than their predecessors, the requirement for catalytic converters to operate at stoichiometric rather than fuel lean conditions is not so attractive (e.g. Table 14.3). Vehicles that will run at lean conditions and also have the benefit of a final catalytic 'clean up' are under development.

Unfortunately, these fall foul of current legislation in Europe: a vehicle fitted with a catalytic converter must, by law, operate at a stoichiometric air to fuel ratio.

14.6 Emissions from natural gas fired systems

The range of applications to which this applies includes electricity generation by conventional gas-fired boilers, stationary gas turbines and small-scale domestic or industrial heating. The main concerns are the release of NO_x and CO into the atmosphere and health hazards resulting from direct exposure within the home.

The methods used for NO_x and CO emission control are governed by the particular application, but may involve either primary or post-combustion methods. Since there is no fuel-bound nitrogen, NO_x formation is governed by the combustion temperature and the proportion of oxygen present. Methods that may be used in large-scale applications to reduce the temperature include flue gas recirculation, staged combustion, lean burn or water injection, but an attendant problem is that CO emissions may be increased in some cases.

Decreases of up to 70% NO_x emissions from boilers can be achieved if there is recycle of more than 20% flue gas in the burner. Staged combustion of large-scale natural gas utility boilers entails operation of the primary combustion zone under fuel rich conditions. Supplementary air is injected downstream in the second stage to give an overall lean burn. Reductions of 50% NO_x emissions relative to conventional firing can be achieved, which is comparable with the reduction obtained from staged fluidised-bed combustion (section 11.5.3). Staged combustion, as developed in aircraft applications, is also employed in stationary gas turbines which burn natural gas.

In general, these approaches are sufficient to reduce the NO_x emissions from large-scale natural gas fired appliances to within allowable limits. If they fail to do so then post-combustion removal has to be adopted. The principal methods used are *selective non-catalytic reduction* and *selective catalytic reduction* of NO_x. Selective non-catalytic reduction is brought about by the injection of a nitrogen-containing compounds, such as ammonia, into the combustion products. A similar procedure is adopted in selective catalytic reduction, but the injection is made in the presence of a catalytic surface, typically a metal oxide such as vanadium pentoxide. Both methods can be very effective within certain combustion gas temperature ranges. The catalytically assisted procedure is the more effective of the two, giving up to 90% removal of NO_x, and is operational at lower temperatures (500–700 K), compared with 1100–1400 K for non-catalysed reduction. The nitrogen-containing additive is believed to promote reaction of NO with NH_2 or NCO to form N_2 (Fig. 14.4 [291]).

Various options are open for limiting NO_x emissions from small appliances, such as domestic heating systems. Flame temperature reduc-

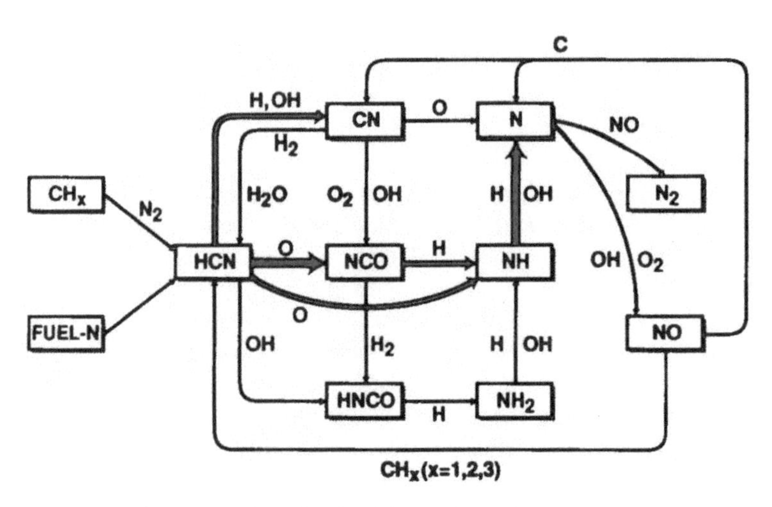

Figure 14.4 Reaction mechanisms involved in the formation of prompt NO and the conversion of fuel-bound nitrogen in flames (after Bowman [291]).

tion can be achieved by flame stabilisation at a porous ceramic surface from which there is considerable radiant heat loss. Typical emissions from a ceramic radiant burner at increasing excess air dilution are shown in Fig. 14.5 [291]. Catalytic combustors are also under development but have not yet found wide application in practical devices.

Burners that are in current use in most domestic applications are a derivative of the Bunsen burner. That is, a partially aerated flame is

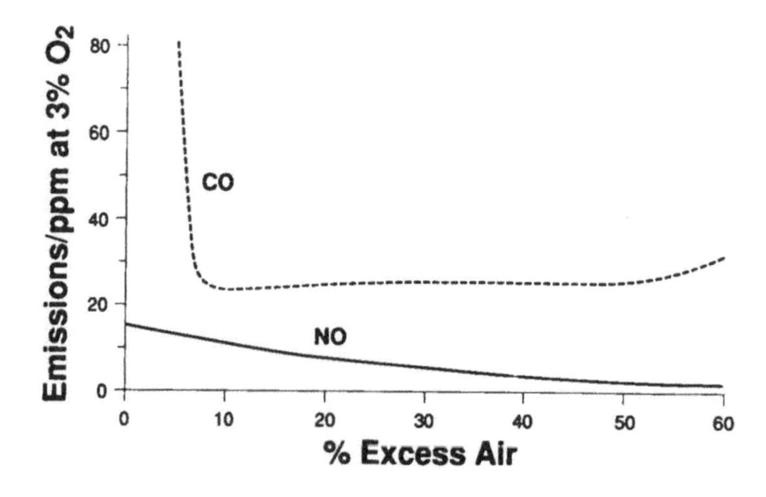

Figure 14.5 Representative pollutant emissions from a ceramic radiant burner operating on natural gas (after Bowman [291]).

formed by premixing primary air at the gas jet orifice, from which the non-luminous flame front develops. The secondary air to complete combustion is entrained above the burner by diffusion. Although not unduly serious, there is comparatively little control over pollutant emissions from this type of burner.

Legislation is being established in many parts of the world to satisfy 'clean air' requirements within the domestic and industrial environment, and to meet these regulations a shift towards fully premixed combustion is likely. The advantages lie in better control of fuel-lean operation and also compactness of the system, since shorter residence times are required for complete combustion. A potential problem of flame noise has to be addressed.

14.7 Emissions from liquid hydrocarbons and solid fuel combustion

Although the Zel'dovich thermal and prompt NO mechanisms contribute to NO_x emissions, the nitrogen that is chemically bound in fossil fuels is the main source of nitrogen oxides from fuel oil and coal combustion. Commonly, there is 1–2% nitrogen by mass in coal. The extent of conversion of the fuel-bound nitrogen during combustion depends mainly on fuel to air ratios and on the combustion temperature. The mechanism of oxidation is believed to be via HCN, CN, NCO and HNCO species to form NH_2, NH and N (Fig. 14.3) [291, 297]. Some reactions of these derivatives are common to those involved in prompt NO formation (e.g. eqns (14.41) and (14.42)). A fuel-rich primary combustion tends to favour a lower conversion of fuel nitrogen to NO as a result of a preferential conversion of CN and HCN to N_2 in reactions that involve NO itself. This may be regarded to be similar to the selective non-catalytic reduction of NO described above. Thus, staged combustion is a common method of minimising NO_x emission from combustion of fuels that have an appreciable nitrogen content.

Although gasoline and diesel fuels have a low sulphur content (~0.2% by mass) a typical coal contains 2% sulphur, while a heavy fuel oil may have double this amount. Very large amounts of coal and fuel oil are consumed in generating electricity and the sulphur emissions from power stations and similar industrial sources represent a considerable environmental hazard. The sulphur in the fuel is converted to sulphur dioxide

$$S + O_2 \rightarrow SO_2 \qquad (14.51)$$

the concentration of which in the flue gas from large furnaces and boilers may be as high as 2000 ppm. At flame temperatures in the presence of excess air, some sulphur trioxide is also formed

$$SO_2 + O + M \rightarrow SO_3 + M \qquad (14.52)$$

Even only a small amount of sulphur trioxide is undesirable since it leads to sulphuric acid, which causes severe corrosion. Although reduction of the excess air helps to control the SO_3 formation during combustion, the need to avoid soot formation dictates that the excess air level cannot be lowered sufficiently to eliminate SO_3 entirely.

While it is technically feasible to reduce the sulphur content of oils and even to some extent that of coal, a more economic alternative is the introduction of finely ground dolomite or limestone which removes the acid gases from the exhaust. This is particularly easy in fluidised-bed combustion where the powdered limestone can be added to the fuel feed and the calcium sulphate removed continuously with the ash. Other methods involve scrubbing the flue gas with the slurry of a suitable basic material, adsorption on carbon or alumina, or catalytic conversion to hydrogen sulphide which is then absorbed in ethanolamine solution.

The non-volatile oxidation products of the inorganic constituents of the fuel (e.g. Na, K, V compounds) are left as ash and can form corrosive deposits. With atomised fuel and pulverised fuel combustion, a large proportion of the ash is carried out of the combustion chamber with the exhaust gas stream and this fly ash has to be removed as far as possible, usually in electrostatic precipitators. Small amounts escape to the atmosphere, where the solid particles act as condensation nuclei and as surfaces at which sulphur dioxide may be converted to the trioxide. Good design and operation of large furnaces can reduce solid emissions and keep the carbon monoxide concentration in the flue gas below 0.01%.

14.8 Quantifying gaseous emissions

In order to compare the performance of different appliances which have different oxygen requirements and to ensure that emissions legislation is not being satisfied merely by dilution of flue gases, it is necessary to specify a reference oxygen concentration. The oxygen measured in a flue gas represents the dilution of the flue gas from the stoichiometric condition. The proportion of excess air (x/V) is given by

$$\frac{x}{V}=\frac{(O_2)_m}{20.9-(O_2)_m} \tag{14.53}$$

where x is the volume of excess air, V is the stoichiometric volume of the flue gas and $(O_2)_m$ is the percentage of oxygen in the flue gas. The dilution factor, D, at $(O_2)_m$ is

$$D=\frac{20.9}{20.9-(O_2)_m} \tag{14.54}$$

In order to specify pollutant concentration at a reference condition (c_{ref}), such as at 3% excess O_2 (Fig. 14.5), a conversion factor for the measured pollutant concentration (c_m) is required. This parameter relates to the respective dilution factors at the measured and reference conditions, and is given by [298]

$$c_{ref}=c_m\frac{(20.9-(O_2)_{ref})}{(20.9-(O_2)_m)} \tag{14.55}$$

Further reading

Bowman, C.T. (1976). *Prog. Energy Combust. Sci.*, **1**, 87.
Bowman, C.T. (1992). Control of combustion-generated nitrogen oxide emissions: Technology driven by regulation. In *Twenty-Fourth Symposium (International) on Combustion.* The Combustion Institute, Pittsburgh, PA, USA, p. 859.
Calvert, J.A. and Pitts, J.N. (1967). *Photochemistry and Reaction Kinetics.* Wiley, New York, USA.
Campbell, I.M. (1986). *Energy and the Atmosphere* (2nd edn). Wiley, New York, USA.
Finlayson-Pitts, B.J. and Pitts, J.N. (1986). *Atmospheric Chemistry.* Wiley Interscience, New York, USA.
Harrison, R.M. (ed.) (1992). *Understanding our Environment: An Introduction to Environmental Chemistry and Pollution.* Royal Society of Chemistry, Cambridge, UK.
Henein, N.A. (1976). *Prog. Energy Combust. Sci.*, **1**, 165.
Heywood, J.B. (1976). *Prog. Energy Combust. Sci.*, **1**, 135.
Heywood, J.B. (1988). *Internal Combustion Engine Fundamentals.* McGraw Hill, New York.
Jones, J.C. (1994). The behaviour of sulphur and nitrogen in combustion processes. In *Combustion Science, Principles and Practice.* Millennium Books, NSW, Australia, Chapters 8 and 9.
Miller, J.A. and Bowman, C.T. (1989). *Prog. Energy Combust. Sci.*, **15**, 287.
Pilling, M.J. and Seakins, P.W. (1995). *Reaction Kinetics.* Oxford University Press, Oxford, UK.
Prather, M. and Logan, J. (1994). Combustion's impact on the global atmosphere. In *Twenty-Fifth Symposium (International) on Combustion.* The Combustion Institute, Pittsburgh, PA, USA, in press.
Sawyer, R.F. (1981). The formation and destruction of pollutants in combustion processes: Clearing the air on the role of combustion research. In *Eighteenth Symposium (International) on Combustion.* The Combustion Institute, Pittsburgh, PA, USA, p.1.

Answers to problems with numerical solutions

Chapter 2

(1)

Fuel	$MJ\,kg^{-1}$		$MJ\,m^{-3}$
	Reactant	Fuel	
H_2	3.86	143	128
CH_4	2.93	55.6	39.7
C_2H_6	2.92	52.0	69.6
C_3H_8	2.91	50.5	99.1
C_2H_2	3.37	50.0	58.0
C_2H_4	3.07	50.4	63.0
$CH_3OH(g)$	3.08	23.9	34.1
$CH_3OH(l)$	2.92	22.6	1.9×10^4

(The maximum thermal energy is released when water is produced in the *liquid* state.)

(2)

	Estimates ($J\,mol^{-1}\,K^{-1}$)		C_p($J\,mol^{-1}\,K^{-1}$) from Table 2.2		
	C_v	C_p	298K	1500K	Mean
CO_2	37.4	45.7	37.14	58.36	47.8
H_2O	37.4	45.7	33.63	46.69	40.2
N_2	25.0	33.0	29.0	34.52	31.8

(3)

	Constant volume	Constant pressure
O_2	8115K	6687K
Air	2984K	2371K

(4) 8×10^{-17}, 8×10^{-6}, 2×10^{-3}, 5×10^{-2}.

(5) $v(mol\,dm^{-3}s^{-1}) = 13.2$; 718; 381; 1016; 0.40.

(6) $A = 10^{18.6}\,dm^{3/2}\,mol^{-1/2}\,s^{-1}$; $E = 358\,kJ\,mol^{-1}$.
C_2H_4 and H_2; CH_4 and C_4H_{10}
$[CH_3] = 9 \times 10^{-13}\,mol\,dm^{-3}$; $[H] = 1 \times 10^{12}\,mol\,dm^{-3}$;
$[C_2H_5] = 4 \times 10^{-10}\,mol\,dm^{-3}$; $7.5 \times 10^{-7}\,mol\,dm^{-3}\,s^{-1}$;
$6 \times 10^{-11}\,mol\,dm^{-3}\,s^{-1}$.

(7)

T(K)	k($mol^{-1}\,cm^3\,s^{-1}$)
750	9.294×10^{10}
1500	4.722×10^{12}
2250	2.296×10^{13}

from which $E/R = 5892\,K$, $A = 2.399 \times 10^{14}\,mol^{-1}\,cm^3\,s^{-1}$, and giving $k_{2250} = 1.749 \times 10^{13}\,mol^{-1}\,cm^3\,s^{-1}$, i.e. about 30% low.

(8) $D \propto$ (pressure)$^{-1}$; (temperature)$^{3/2}$; (molar mass)$^{-1/2}$;
$\kappa \propto$ (pressure)^0; (temperature)$^{1/2}$; (molar mass)$^{1/2}$.

(9) $466\,ms^{-1}$; $1.03 \times 10^{-7}\,m$; $1.60 \times 10^{-5}\,m\,s^{-2}$; $1.48 \times 10^{-2}\,W\,m^{-1}\,K^{-1}$; $Le = 0.714$ and γ^{-1}.

Chapter 3

(1)

	d_Q(mm)	d_T(mm)
H_2	0.84	1.29
CH_4	2.45	3.77
C_3H_8	2.41	3.72
C_2H_2	0.66	1.02

(2) 24.1 mm.

Chapter 4

(1) $1300\,mol\,m^{-3}\,s^{-1}$; 1 mm.

(2) Increased by a factor of 4.4.

Chapter 5

(1) An ideal gas ($\gamma = 5/3$) is compressed to 4.67 MPa and 1356 K.
Shock $T = 3671\,K$.

(2) H_2 3878 K, $2505\,m\,s^{-1}$, 2.23 MPa.
CH_4 2730 K, $1825\,m\,s^{-1}$, 1.58 MPa.

Chapter 6

(1) The final equation to be solved, expressed in terms of $p = p(O_2)/p^o$, is $6p^{3/2} + 3.3375p + 0.225p^{1/2} - 1 = 0$. The partial pressures are: $p(CO_2) = 44.0$ kPa, $p(CO) = 36.8$ kPa, $p(O_2) = 16.1$ kPa, $p(O) = 4.5$ kPa.

(2) $p(CO_2) = 26.27$ kPa, $p(CO) = 3.70$ kPa, $p(O_2) = 0.80$ kPa, $p(O) = 0.014$ kPa, $p(NO) = 0.335$ kPa.

Chapter 7

(1) $v_a/v_b = 6.84$ and 0.017.

(2) 623 K.

Chapter 8

(1) 14.89 K, 23.82 K.

(2) $\varepsilon = 0.029$.

(5) 0.534, 1.325 kPa.

(6) 1520 m, 1.92 m.

(7) 10.9 K.

(8) $\delta_c = 3.32$, $\Delta T_c = 19.4$ K.

Chapter 9

(2) 724 K, 469 Pa.

(4) $A = 1.54 \times 10^8$ m^3 mol^{-1} s^{-1}, $E = 67.3$ kJ mol^{-1}.

(5) 67.5 kJ mol^{-1}.

(6) 0.40.

Chapter 10

(2) (a) 720 K, (b) 346 K, (c) 50.6, (d) 2.78.

(3) $E = 54.64\,kJ\,mol^{-1}$, $A = 3.763 \times 10^6\,s^{-1}$, (a) 0.0334, (b) 312.5 s, (c) 0.912, (d) $0.44\,kmol\,m^{-3}$.

Chapter 11

(1) (a) 1513 K, (b) 1158 K, (c) 963 K.

(2) Relative rate (liquid/solid) = 4.9.

(3) 0.3 s.

Chapter 12

(1) 277 K.

References

1 Faraday, M. (1862). *The Chemical History of a Candle* (ed. W.R. Fielding, 1920). Dent, London, UK.

2 Dixon-Lewis, G. and Williams, D.J. (1977). *Comprehensive Chemical Kinetics* (Vol. 17) (eds C.H. Bamford and C.F.H. Tipper). Elsevier, Amsterdam, The Netherlands, p. 1.

3 Peters, N. and Rogg, B. (eds) (1993). *Reduced Kinetic Mechanisms for Application in Combustion Systems*. Springer-Verlag, Berlin, Germany.

4 Griffiths, J.F. (1995). *Prog. Energy Combust. Sci.*, **21**, 25.

5 Stull, D.R., Westrum, E.F. and Sinke, G.C. (1969). *The Chemical Thermodynamics of Organic Compounds*. John Wiley, New York, USA.

6 Stull, D.R. and Prophet, H. (1971). *JANAF Thermochemical Tables* (2nd edn). NSRDS-NBS 37. National Bureau of Standards, Washington, DC, USA.

7 Kee, R.J., Rupley, F.M. and Miller, J.A. (1992). *The Chemkin Thermodynamic Data Base*. Sandia Report, SANDIA NAT. LAB., Livermore, Ca.

8 Benson S.W. (1976). *Thermochemical Kinetics* (2nd edn). John Wiley, New York, USA.

9 Prothero, A. (1969). *Combust. Flame*, **13**, 399.

10 Gaydon, A.G. and Wolfhard, H.G. (1979). *Flames, Their Structure, Radiation and Temperature* (4th edn). Chapman and Hall, London, UK.

11 Van Zeggeren, F. and Storey, S.H. (1970). *The Computation of Chemical Equilibria*. Cambridge University Press, Cambridge, UK.

12 Kaufman, F. (1982). *Nineteenth Symposium (International) on Combustion*. The Combustion Institute, Pittsburgh, PA, USA, p. 1.

13 Robinson, P.J. and Holbrook, K.A. (1972). *Unimolecular Reactions*. John Wiley, New York, USA.

14 Baulch, D.L., Cobos, C.J., Cox, R.A., Frank, P., Hayman, G., Just, Th., Kerr, J.A., Murrells, T., Pilling, M.J., Troe, J., Walker, R.W. and Warnatz, J. (1994). *Combust. Flame*, **98**, 59.

15 Benson, S.W. (1960). *The Foundations of Chemical Kinetics*. McGraw-Hill, New York, USA.

16 Dixon-Lewis, G. and Islam, S.M. (1982). *Nineteenth Symposium (International) on Combustion*. The Combustion Institute, Pittsburgh, PA, USA, p. 245.

17 Andrews, G.E. and Bradley, D. (1972). *Combust. Flame*, **18**, 133.

18 Coward, H.F. and Jones, G.W. (1952). *Limits of Flammability of Gases and Vapors*. US Government Printing Office, Washington DC.

19 Lovachev, L.A., Babkin, V.S., Bunev, V.A., V'yan, A.V., Krivulin, V. and Baratov, A.N. (1973). *Combust. Flame*, **20**, 259.

20 Potter, A.E. (1960). *Prog. Energy Combust. Sci.*, **1**, 145.

21 Rozlovskii, A.I and Zakaznov, V.F. (1971). *Combust. Flame*, **17**, 215.

22 Smithells, A. and Ingle, H. (1892). *Trans. Chem. Soc.*, **61**, 204.

23 Lewis, B. and von Elbe G. (1987). *Combustion, Flames and Explosions of Gases* (3rd edn). Academic Press, Orlando, FL, USA.

24 Glassman, I. (1977). *Combustion*. Academic Press, New York, USA.

25 Hottel, H.C. and Hawthorne, W.R. (1949). *Third Symposium (International) on Combustion*. The Combustion Institute, Pittsburgh, PA, USA, p. 254.

26 Barr, J. (1954). *Fuel*, **33**, 51.

27 Smith, S.R. and Gordon, A.S. (1956). *J. Phys. Chem.*, **60**, 759, 1059; (1957) *J. Phys. Chem.*, **61**, 553.

28 Barnard, J.A. and Cullis, C.F. (1962). *Eighth Symposium (International) on Combustion*. The Combustion Institute, Pittsburgh, PA, USA, p. 481.
29 Biordi, J.C., Lazzar, C.P. and Papp, J.F. (1982). *Nineteenth Symposium (International) on Combustion*. The Combustion Institute, Pittsburgh, PA, USA, p. 1097.
30 Vandooren, J., Branch, M.C. and van Tiggelen, P.J. (1992). *Combust. Flame*, **90**, 247.
31 Zel'dovich, Ya.B., Barenblatt, G., Librovich, V.B. and Makhviladze, G.Y., (1985). *Mathematical Theory of Combustion and Explosion*. Consultants' Bureau, New York, USA.
32 Williams. F.A. (1985). *Combustion Theory* (2nd edn). Addison Wesley, New Jersey, USA.
33 Oppenheim, A.K. (1993). *Modern Developments in Energy, Combustion and Spectroscopy* (eds F.A. Williams, A.K. Oppenheim, D.B. Olfe and M. Lapp). Pergamon Press, Oxford, UK, p. 1.
34 Dixon-Lewis, G. (1990). *Twenty-Third Symposium (International) on Combustion*. The Combustion Institute, Pittsburgh, PA, USA, p. 305.
35 Tanford, C. and Pease, R.N. (1947). *J. Chem. Phys.*, **15**, 861.
36 Gray, P. and Scott, S.K. (1990). *Chemical Oscillations and Instabilities*. Clarendon Press, Oxford, UK.
37 Scott, S.K. and Showalter, K. (1992). *J. Phys. Chem.*, **96**, 8702.
38 Zel'dovich, Ya.B. and Frank-Kamenetskii, D.A. (1938). *Dokl. Akad. Nauk. SSSR*, **19**, 693.
39 Frank-Kamenetskii, D.A. (1969). *Diffusion and Heat Transfer in Chemical Kinetics* (2nd edn) (translated by J.P. Appleton). Plenum Press, New York, USA.
40 Spalding, D.B. (1957). *Combust. Flame*, **1**, 287, 296.
41 Strehlow, R.A. (1984). *Combustion Fundamentals*. McGraw Hill, New York, USA.
42 Buckmaster, J.D. and Ludford, G.S.S. (1982). *Theory of Laminar Flames*. Cambridge University Press, Cambridge, UK.
43 Williams, F.A. (1992). *Twenty-Fifth Symposium (International) on Combustion*. The Combustion Institute, Pittsburgh, PA, USA, p. 1.
44 Borghi R. (1988). *Prog. Energy Combust. Sci.*, **14**, 245.
45 Markstein, G.H. (1964). *Nonsteady Flame Propagation*. Pergamon Press, Oxford, UK.
46 Abdel-Gayed, R.G., Bradley, D. and Lawes, M. (1987). *Proc. Royal Soc. London*, **A414**, 389.
47 Karlovitz, B., Denniston, D.W., Knapschaeffer, D.H. and Wells, F.E. (1953). *Fourth Symposium (International) on Combustion*. The Combustion Institute, Pittsburgh, PA, USA, p. 613.
48 Law, C.K. (1988). *Twenty-Second Symposium (International) on Combustion*. The Combustion Institute, Pittsburgh, PA, USA, p. 1381.
49 Dowdy, R.D., Smith, D.B., Taylor, S.C. and Williams A. (1990). *Twenty-Third Symposium (International) on Combustion*. The Combustion Institute, Pittsburgh, PA, USA, p. 325.
50 Wu, C.K. and Law, C.K. (1984). *Twentieth Symposium (International) on Combustion*. The Combustion Institute, Pittsburgh, PA, USA, p. 1941.
51 Zhu, D.L., Egolfopoulos, F.N. and Law, C.K. (1988). *Twenty-Second Symposium (International) on Combustion*. The Combustion Institute, Pittsburgh, PA, USA, p. 1537.
52 Kee, R.J., Miller, J.A., Evans, G.H. and Dixon-Lewis, G. (1988). *Twenty-Second Symposium (International) on Combustion*. The Combustion Institute, Pittsburgh, PA, USA, p. 1479.
53 Tsuji, H. and Yamaoka, I. (1982). *Nineteenth Symposium (International) on Combustion*. The Combustion Institute, Pittsburgh, PA, USA, p. 1533.
54 Asato, K., Kawamura, T. and Ban, T. (1988). *Twenty-Second Symposium (International) on Combustion*. The Combustion Institute, Pittsburgh, PA, USA, p. 1509.
55 Palm-Leis, A. and Strehlow, R.A. (1969). *Combust. Flame*, **13**, 111.
56 Bradley, D.B. and Harper, C.M. (1994). *Twenty-Fifth Symposium (International) on Combustion*. The Combustion Institute, Pittsburgh, PA, USA, **99**, 562.
57 Klimov, A.M. (1963). *Zh. Prikl. Mekh. Tekh. Fiz.*, **3**, 49.
58 Williams, F.A. (1976). *Combust. Flame*, **26**, 269.
59 Bradley, D. (1992). *Twenty-Fourth Symposium (International) on Combustion*. The Combustion Institute, Pittsburgh, PA, USA, p. 247.
60 Peters, N. (1986). *Twenty-First Symposium (International) on Combustion*. The Combustion Institute, Pittsburgh, PA, USA, p. 1231.

61 Abdel-Gayed, R.G. and Bradley, D. (1981). *Phil. Trans. Royal Soc. London*, **A 301**, 1.
62 Bradley, D., Lau, A.K.C. and Lawes, M. (1992). *Phil. Trans. Royal Soc. London*, **A 338**, 359.
63 Abdel-Gayed, R.G. and Bradley, D. (1989). *Combust. Flame*, **76**, 213.
64 Jost, W. (1946). *Explosion and Combustion Processes*. McGraw Hill, New York, USA.
65 Burke, S.P. and Schumann, T.E.W. (1928). *Ind. Engng Chem.*, **20**, 998.
66 Roper, F.G. Smith, C. and Cunningham, A.C. (1977). *Combust. Flame*, **29**, 227.
67 Roper F.G. (1978). *Combust. Flame*, **31**, 251.
68 Bilger, R.W. (1988). *Twenty-Second Symposium (International) on Combustion*. The Combustion Institute, Pittsburgh, PA, USA, p. 475.
69 Penney, W.G. and Pike, H.H.M. (1950). *Rep. Prog. Phys.*, **13**, 46.
70 Davidson, D.F., Di Rosa, M.D., Chang, A.Y., Hanson, R.K. and Bowman, C.T. (1992). *Twenty-Fourth Symposium (International) on Combustion*. The Combustion Institute, Pittsburgh, PA, USA, p. 589.
71 Burcat, A., Lifshitz, A., Scheller, K. and Skinner, G.B. (1970). *Thirteenth Symposium (International) on Combustion*. The Combustion Institute, Pittsburgh, PA, USA, p. 745.
72 Ciezki and Adomeit (1993). *Combust. Flame*, **93**, 421.
73 Chapman, D.L. (1899). *Phil. Mag* **47**(5), 90.
74 Jouguet, E. (1905). *J. Mathematique*, 305.
75 Jouguet, E. (1906). *J. Mathematique*, 6.
76 Jouguet, E. (1917). *Mecanique des Explosifs*. Doin, Paris, France.
77 Zel'dovich, Y.B. (1940). *J. Exp. Theor. Phys. USSR*, **10**, 542.
78 Von Neumann, J. (1942). *OSRD Report*, No. 549.
79 Doring, W. (1943). *Ann. Physik*, **43**, 421.
80 Cher, M. and Kistiakowsky, G B. (1958). *J. Chem. Phys.*, **29**, 506.
81 Edwards, D.H., Williams, G T. and Breeze, J.C. (1959). *J. Fluid Mech.*, **6**, 497.
82 Edwards, D.H. (1969). *Twelfth Symposium (International) on Combustion*. The Combustion Institute, Pittsburgh, PA, USA, p. 819.
83 Bollinger, L.E. and Edse, R. (1961). *J. Am. Rocket Soc.*, **31**, 251, 588.
84 Shchelkin, K.I. and Troshin, Ya. K. (1965). *Gasdynamics of Combustion*. Mono Book Corp., Baltimore, MD, USA.
85 Strehlow, R.A. (1968). *Combust. Flame*, **12**, 90.
86 Oran, E.S., Young, T.R., Boris, J.P., Picone, J.M. and Edwards, D.H. (1982). *Nineteenth Symposium (International) on Combustion*. The Combustion Institute, Pittsburgh, PA, USA, p. 573.
87 Westbrook, C.K. and Dryer, F.L. (1984). *Prog. Energy Combust. Sci.*, **10**, 1.
88 Ritter, E.R. and Bonzell, J.W. (1991). *Int. J. Chem. Kin.*, **23**, 767.
89 Warnatz, J. (1984). *Twentieth Symposium (International) on Combustion*. The Combustion Institute, Pittsburgh, PA, USA, p. 845.
90 Baulch, D.L., Drysdale, D.D., Horne, D.G. and Lloyd, A.C. (1972). *Evaluated Rate Data for High Temperature Reactions* (Vol. 1). Butterworths. London, UK.
91 Frenklach, M., Wang, H. and Rabinowitz, M.J. (1992). *Prog. Energy Combust. Sci.*, **18**, 47.
92 Emdee, J.L., Brezinsky, K. and Glassman, I. (1992). *J. Phys. Chem.*, **96**, 2151.
93 Bradley, J.N. (1967). *Trans. Faraday Soc.*, **63**, 2945.
94 Maas, U. and Warnatz, J (1988). *Twenty-Second Symposium (International) on Combustion*. The Combustion Institute, Pittsburgh, PA, USA, p. 1695.
95 Warnatz, J. (1992). *Twenty-Fourth Symposium (International) on Combustion*. The Combustion Institute, Pittsburgh, PA, USA, p.553.
96 Polanyi, M. (1932). *Atomic Reactions*. Williams & Northgate, London, UK.
97 Laidler, K.J. (1955). *The Chemical Kinetics of Excited States*. Oxford University Press, Oxford, UK.
98 Gaydon A.G. (1974). *The Spectroscopy of Flames* (2nd edn). Chapman and Hall, London, UK.
99 Shuler, K.E. (1953). *J. Phys. Chem.*, **57**, 396.
100 Gaydon, A.G. and Wolfhard, H.G. (1951). *Proc. Royal Soc. London*, **A208**, 63.
101 Calcote, H.F. and King, I.R. (1955). *Fifth Symposium (International) on Combustion*. The Combustion Institute, Pittsburgh, PA, USA, p. 423.
102 Payne, K.G. and Weinberg, F.J. (1959). *Proc. Royal Soc. London*, **A250**, 316.
103 Palmer, H.B. and Cullis, C.F. (1965). *Chemistry and Physics of Carbon*, (Vol. 1) (ed. P.L. Walker). Marcel Dekker, New York, USA, p.200.

104 Homan, H.S. (1983). *Comb. Sci. Tech.*, **33**, 1.
105 Glassman, I. (1988). *Twenty-Second Symposium (International) on Combustion*. The Combustion Institute, Pittsburgh, PA, USA, p. 295.
106 Lam, F.W., Howard, J.B. and Longwell, J.P. (1988). *Twenty-Second Symposium (International) on Combustion*. The Combustion Institute, Pittsburgh, PA, USA, p. 323.
107 Williams, P., Bartle, K.D. and Andrews, G.E. (1986). *Fuel*, **65**, 1153.
108 Nelson, P.W. (1989). *Fuel*, **68**, 285.
109 Homann, K.H. (1984). *Twentieth Symposium (International) on Combustion*. The Combustion Institute, Pittsburgh, PA, USA, p. 857.
110 Cole, J.A., Bittner, J.D., Longwell, J.P. and Howard, J.B. (1984). *Combust. Flame*, **56**, 51.
111 Frenklach, M., Clary, D.W., Gardiner, W.C. and Stein, S.E. (1984). *Twentieth Symposium (International) on Combustion*. The Combustion Institute, Pittsburgh, PA, USA, p. 887.
112 Calcote, H.F. (1981). *Combust. Flame*, **42**, 215.
113 Howard, J.B. (1990). *Twenty-Third Symposium (International) on Combustion*. The Combustion Institute, Pittsburgh, PA, USA, p. 1107.
114 Kroto, H.W., Heath, J.R., O'Brien, S.C., Curl, R.F. and Smalley, R.E. (1985). *Nature*, **318**, 162.
115 Howard, J.B. (1992). *Twenty-Fourth Symposium (International) on Combustion*. The Combustion Institute, Pittsburgh, PA, USA, p. 933.
116 Murayama, M. and Uchida, K. (1992). *Combust. Flame*, **91**, 239.
117 Morgan, C.A., Pilling, M.J. and Tulloch, J.M. (1982). *J. Chem. Soc., Faraday Trans. II*, **78**, 1323.
118 Semenov, N.N. (1959). *Some Problems in Chemical Kinetics and Reactivity* (Vol. 2) (Translated by M. Boudart). Princeton University Press, New Jersey, USA, Chapter XII.
119 Fish, A. (1964). *Quarterly Rev. of The Chem. Soc.*, **XVIII**, 243.
120 Benson, S.W. (1965). *J. Am. Chem. Soc.*, **87**, 972.
121 Cartlidge, J. and Tipper, C.F.H. (1961). *Proc. Royal Soc. London*, **A261**, 388.
122 Fish, A., Haskell, W.W. and Read, I.A. (1969). *Proc. Royal Soc. London*, **A313**, 261.
123 Sahetchian, K.A., Blin, N., Rigny, R., Seydi, A. and Murat, M., (1991). *Combust. Flame*, **79**, 242.
124 Clague, A., Hughes, K.J. and Pilling, M., submitted to *J. Phys. Chem.*
125 Benson, S.W. (1981). *Prog. Energy Combust. Sci.*, **7**, 125.
126 Minetti, R., Ribaucour, M., Carlier, M., Fittschen, C. and Sochet, L.-R. (1994). *Combust. Flame*, **96**, 201.
127 Walker, R.W. (1975). *Specialist Periodical Report, Reaction Kinetics* (Vol. 1). The Chemical Society, London, UK, p. 161.
128 Walker, R.W. (1977). *Specialist Periodical Report, Gas Kinetics and Energy Transfer* (Vol. 2). The Chemical Society, London, UK, p. 296.
129 Hughes, K.J., Halford-Maw, P.A., Lightfoot, P.D., Turanyi, T. and Pilling, P.J. (1992). *Twenty-Fourth Symposium (International) on Combustion*. The Combustion Institute, Pittsburgh, PA, USA, p. 645.
130 Morley, C. (1988). *Twenty-Second Symposium (International) on Combustion*. The Combustion Institute, Pittsburgh, PA, USA, p. 911.
131 Nalbandyan, A.B., Oganessyan, E.A., Vardanyan, I.A. and Griffiths, J.F. (1975). *J. Chem. Soc. Faraday Trans. 1*, **71**, 1203.
132 Carlier, M., Corre, C., Minetti, R., Pawels, J.-F., Ribaucour, M. and Sochet, L.-R. (1990). *Twenty-Third Symposium (International) on Combustion*. The Combustion Institute, Pittsburgh, PA, USA, p. 1753.
133 Dagaut, P., Reuillon, M. and Cathonnet, M. (1994). *Comb. Sci. Tech.*, **95**, 233.
134 Pease, R.N. (1929). *J. Am. Chem. Soc.*, **51**, 1839.
135 Pease, R.N. (1938). *J. Am. Chem. Soc.*, **60**, 2244.
136 Wagner, A.F., Slagle, I.R., Sarzynski, D. and Gutman, D. (1990). *J. Phys. Chem.*, **94**, 1852; and (1991) **95**, 1014.
137 Griffiths, J.F. and Sykes, A.F. (1989). *Proc. Royal Soc. London*, **A422**, 289.
138 Fenter, F.F., Nozière, B., Caralp, F. and Lesclaux, R. (1994). *Int. J. Chem. Kinetics*, **26**, 171.
139 van't Hoff, J.H. (1896). *Studies in Chemical Dynamics* (Translated by T. Ewan). Williams and Northgate, London, UK.

140 Semenov, N.N. (1928). *Z. Physik.*, **48**, 571.
141 Taffenel, J. and Le Floch, M. (1913). *Comptes Rendus de l'Academie des Sciences de Paris*, **156**, 1544; **157**, 469.
142 Griffiths, J.F. and Scott, S.K. (1987). *Prog. Energy Combust. Sci.*, **13**, 161.
143 Gray, P. and Lee, P.R. (1968). *Oxidation and Combustion Reviews* (Vol 2) (ed. C.F.H. Tipper). Elsevier, Amsterdam, The Netherlands, p. 1.
144 Griffiths, J.F. and Phillips, C.H. (1990). *Combust. Flame*, **81**, 304.
145 Frank-Kamenetskii, D.A. (1939). *Z. Phys. Khim.*, **13**, 738.
146 Gray, B.F. and Wake, G.C. (1988). *Combust. Flame*, **71**, 101.
147 Bowes, P.C. (1984). *Self-Heating: Evaluating and Controlling the Hazard.* HMSO Books, London, UK.
148 Uppal, A., Ray, W.H. and Poore, A.B. (1974). *Chem. Engng. Sci.*, **29**, 967.
149 Boddington, T., Gray, P. and Wake, G.C. (1977). *Proc. Royal Soc. London*, **A357**, 403.
150 Tyler, B.J. and Wesley T.A.B. (1967). *Eleventh Symposium (International) on Combustion*. The Combustion Institute, Pittsburgh, PA, USA, p. 1115.
151 Boddington, T., Gray, P. and Harvey, D.I. (1971). *Phil. Trans. Royal Soc. London*, **A270**, 467.
152 Egeiban, O.M., Griffiths, J.F., Mullins, J.R. and Scott, S.K. (1982). *Nineteenth Symposium (International) on Combustion*. The Combustion Institute, Pittsburgh, PA, USA, p. 825.
153 Ashmore, P.G., Tyler, B.J. and Wesley, T.A.B. (1967). *Eleventh Symposium (International) on Combustion*. The Combustion Institute, Pittsburgh, PA, USA, p. 1133.
154 *Proceedings of the International Symposium on Runaway Reactions* (1989). AIChE, New York, USA.
155 Gray, P., Boddington, T. and Walker, I.K. (1980). *Proc. Royal Soc. London*, **A373**, 287.
156 Walker, I.K. (1967). *Fire Res. Abstr. Rev.*, **9**, 5.
157 Fine, D.H., Gray, P. and Mackinven, R. (1970). *Proc. Royal Soc. London*, **A316**, 255.
158 Griffiths, J.F. and Snee, T.J. (1989). *Combust. Flame*, **75**, 381.
159 Griffiths, J.F. and Phillips, C.H. (1989). *Royal Soc. Chem. Faraday Trans. I*, **85**, 3471.
160 Chinnick, K.J., Gibson, C., Griffiths, J.F. and Kordylewski, W. (1986). *Proc. Royal Soc. London*, **A405**, 117.
161 Gray, B.F. (1969). *Trans. Faraday Soc.*, **65**, 2133.
162 Semenov, N.N. (1935). *Chemical Kinetics and Chain Reactions.* Oxford University Press, Oxford, UK.
163 Hinshelwood, C.N. (1940). *The Kinetics of Chemical Change*. Oxford University Press, Oxford, UK.
164 Pease, R.N. (1942). *Equilibrium and Kinetics of Gas Reactions*. Princeton University Press, Princeton, USA.
165 Baldwin, R.R. and Walker, R.W. (1972). Branched-chain reactions: the hydrogen-oxygen reaction. In *Essays in Chemistry* (Vol. 3) (eds J.N. Bradley, R.D. Gillard and R.F. Hudson). Academic Press, London, UK.
166 Cheaney, D.E., Davies, D.A., Davis, A., Hoare, D.E., Protheroe, J. and Walsh, A.D. (1959). *Seventh Symposium (International) on Combustion*. The Combustion Institute, Pittsburgh, PA, USA, p. 183.
167 Sachyan, G.A., Mantashyan, A.A. and Nalbandyan, A.B. (1969). *Dokl. Akad. Nauk., SSSR*, **185**, 647.
168 Baulch, D.L., Griffiths, J.F., Pappin, A.J. and Sykes, A.F. (1988). *Royal Soc. Chem. Faraday Trans. I*, **84**, 1575.
169 Baulch, D.L., Griffiths, J.F. Kordylewski, W. and Richter, R. (1991). *Phil. Trans. Royal Soc.*, **A337**, 195.
170 Chinnick, K.J., Gibson, C. and Griffiths, J.F. (1986). *Proc. Royal Soc. London*, **A405**, 129.
171 Baldwin, R.R., Rossiter, B.N. and Walker, R.W. (1969). *Trans. Faraday Soc.*, **65**, 1044.
172 Barnard, J.A. and Platts, A.G. (1972). *Combustion Sci. Technol.*, **6**, 177.
173 Griffiths, J.F., Scott, S.K. and Vandamme, R. (1981). *J. Chem. Soc. Faraday Trans. I*, **77**, 2265.
174 Foo, K.K. and Yang, C.H. (1971). *Combust. Flame*, **17**, 223.
175 Kordylewski, W. and Scott, S.K. (1984). *Combust. Flame*, **57**, 127.
176 Baldwin, R.R. and Walker, R.W. (1981). *Eighteenth Symposium (International) on Combustion*. The Combustion Institute, Pittsburgh, PA, USA, p. 819.

177 Gray, B.F. (1970). *Trans. Faraday Soc.*, **66**, 1118.
178 Dickens, P.G., Dove, J.E. and Linnett, J.W. (1964). *Trans. Faraday Soc.*, **60**, 559.
179 Gordon, A.S. and Knipe, R.H. (1955). *J. Phys. Chem.*, **59**, 1160.
180 Gray, P., Griffiths, J.F. and Bond, J.R. (1978). *Seventeenth Symposium (International) on Combustion*. The Combustion Institute, Pittsburgh, PA, USA, p. 811.
181 Lewis, B., von Elbe, G. and Roth, W. (1955). *Fifth Symposium (International) on Combustion*. The Combustion Institute, Pittsburgh, PA, USA, p. 610.
182 Hadman, G., Thompson, H.W. and Hinshelwood, C.N. (1932). *Proc. Royal Soc. London*, **A137**, 87.
183 Bond, J.R., Gray, P. and Griffiths, J.F. (1982). *Proc. Royal Soc. London*, **A381**, 293.
184 Gray, P. and Scott, S.K. (1985). *Oscillations and Traveling Waves in Chemical Systems* (eds R.J. Field and M. Burger). Wiley, New York, USA, p. 493.
185 Gray, P., Griffiths, J.F. and Scott, S.K. (1985). *Proc, Royal Soc. London*, **A397**, 21.
186 Griffiths, J.F. and Sykes, A.F. (1989). *J. Chem. Soc. Faraday Trans. I*, **85**, 3059.
187 Alexandrov, E.N. and Azatyan, V.V. (1977). *Combust. Exp. Shock Waves*, **12**, 407.
188 Robinson, C. and Smith, D.B. (1984). *J. Hazard. Materials*, **8**, 199.
189 Gray, B.F., Gray, P. and Griffiths, J.F. (1971). *Thirteenth Symposium (International) on Combustion*. The Combustion Institute, Pittsburgh, PA, USA, p. 239.
190 Perkin, W.H. (1882). *J. Chem. Soc.*, **41**, 363.
191 Agnew, W.J. and Agnew, J.T. (1965). *Tenth Symposium (International) on Combustion*. The Combustion Institute, Pittsburgh, PA, USA, p. 123.
192 Bradley, J.N., Jones, G.A., Skirrow, G. and Tipper, C.F.H. (1966). *Combust. Flame*, **10**, 295.
193 Sheinson, R.S. and Williams, F.W. (1973). *Combustion Flame*, **21**, 221.
194 Gray, P., Griffiths, J.F., Hasko. S.M. and Lignola, P-G. (1981). *Proc. Royal Soc. London*, **A374**, 313.
195 Gray, P., Griffiths, J.F., Hasko, S.M. and Lignola, P.-G. (1981). *Combust. Flame*, **43**, 175.
196 Lignola, P.-G. and Reverchon, E. (1987). *Prog. Energy Combust. Sci.*, **13**, 75.
197 Lignola, P.G., Di Maio, F.P., Marzocchella, A., Mercogliano, R. and Reverchon, E. (1988). *Twenty-Second Symposium (International) on Combustion*. The Combustion Institute, Pittsburgh, PA, USA, p. 1625.
198 Kurzius. S.C. and Boudart, M. (1968). *Combust. Flame*, **12**, 477.
199 Zel'dovich, Ya. B. (1940). *Zh. Tekh.*, **XI**, 493.
200 Zel'dovich, Ya. B. and Zysin, Y.A. (1941). *Zh. Tekh.*, **XI**, 501.
201 Bilous, O. and Amundsen, N.R. (1955). *AIChEJ*, **1**, 513.
202 Longwell, J.P. and Weiss, M.A. (1955). *Ind. Engng Chem.*, **47**, 1634.
203 Uppal, A., Ray, W.H. and Poore, A.B. (1976). *Chem. Engng Sci.*, **31**, 203.
204 Minorsky, N. (1962). *Nonlinear Oscillations*. Krieger, New York, USA.
205 Andronov, A.A., Vitt, A.A. and Khaikin, S.E. (1966). *Theory of Oscillators*. Pergamon, Oxford, UK.
206 Doedel, E. (1986). *Auto-Continuation and Bifurcation Problems in Ordinary Differential Equations*. CALTECH, Pasadena, Ca.
207 Yang, C.H. and Gray, B.F. (1969). *J. Phys. Chem.*, **73**, 3395.
208 Yang, C.H. and Gray, B.F. (1969). *Trans. Faraday Soc.*, **65**, 1614.
209 Yang, C.H. (1969). *J. Phys. Chem.*, **73**, 3407.
210 Wang, X.J. and Mou, C.Y. (1985). *J. Chem. Phys.*, **83**, 4554.
211 Felton, P.G., Gray, P. and Shank, N. (1976). *Combust. Flame*, **27**, 363.
212 Di Maio, F.P., Lignola, P.G. and Talarico, P. (1993). *Combust. Sci. Tech.*, **91**, 119.
213 Fristrom, R.M. and Westenberg. A.A (1965). *Flame Structure*. McGraw Hill, New York, USA.
214 Long, V.D. (1964). *J. Inst. Fuel*, **37**, 522.
215 Williams, A. (1973). *Combust. Flame*, **21**, 1.
216 Kanury, A.M. (1975). *Introduction to Combustion Phenomena*. Gordon and Breach, New York, USA.
217 Williams, A. (1976). *The Combustion of Sprays of Liquid Fuels*. Paul Elek, London, UK.
218 Hayhurst, A.N. and Nedderman, R.M. (1987). *Chem. Engng Education*, **21**, 126.
219 Herzberg, M. (1973). *Combust. Flame*, **21**, 195.
220 De Ris, J. and Orloff, L. (1973). *Combust. Flame*, **18**, 381.
221 Orloff, L. and De Ris, J. (1982). *Nineteenth Symposium (International) on Combustion*. The Combustion Institute, Pittsburgh, PA, USA, p. 895.

222 Burgoyne, J.H. Roberts, A.F. and Quinton. P.G. (1968). *Proc. Royal Soc. Lond.*, **A308**, 39.

223 Burgoyne, J.H. and Roberts, A.F. (1968). *Proc. Royal Soc. London*, **A308**, 55, 69.

224 Roberts, A.F. and Quince. B.W. (1973). *Combust. Flame*, **20**, 245.

225 Suuberg, E.M., Peters, W.A. and Howard, J.B. (1979). *Seventeenth Symposium (International) on Combustion*. The Combustion Institute, Pittsburgh, PA, USA, p. 117.

226 Smith, I.W. (1982). *Nineteenth Symposium (International) on Combustion*. The Combustion Institute, Pittsburgh, PA, USA, p. 1045.

227 Mulcahy, M.F.R. (1978). *The Combustion of Carbon in Oxygen in The Metallurgical and Gaseous Fuel Industries*. The Chemical Society, London, UK, p. 175.

228 De Gray, A. (1922). *Rev. Metall.*, **19**, 645.

229 Thring, M.W. (1952). *Fuel*, **31**, 355.

230 Yates, J.G. (1983). *Fundamentals of Fluidized-Bed Chemical Processes*. Butterworths, London, UK.

231 Hoy, H.R. and Gill, D.W. (1987). *Principles of Combustion Engineering for Boilers* (ed. C.J. Lawn). Academic Press, London, UK, p. 521.

232 Davidson, J.F. and Harrison, D. (1963). *Fluidised Particles*. Cambridge University Press, Cambridge, UK.

233 La Nauze, R.D. (1986). *Advanced Combustion Methods* (ed. F.J. Weinberg). Academic Press, London, UK, p. 17.

234 Dry, R.J. and La Nauze, R.D. (1990). *Chem. Engng Prog.*, **86**, No 7, 31.

235 Lees, F.P. (1980). *Loss Prevention in the Process Industries* (Vols 1 and 2). Butterworths, London, UK.

236 *A Guide to Hazard and Operability Studies* (1977). Chemical Industries Association, London, UK.

237 Griffiths, J.F., Coppersthwaite, D.P., Phillips, C.H., Westbrook, C.K. and Pitz, W.J. (1990). *Twenty-Third Symposium (International) on Combustion*. The Combustion Institute, Pittsburgh, PA, USA, p. 1745.

238 Kono, M., Niu, K., Tsukamoto, T. and Ujiie, Y. (1988). *Twenty-Second Symposium (International) on Combustion*. The Combustion Institute, Pittsburgh, PA, USA, p. 643.

239 McQuaid, J. and Roberts, A.F. (1982). *I. Chem. E. 50th Jubilee Symposium*, **73**, B19.

240 Roberts, A.F. (1982). *I. Chem. E. Symp. Ser.*, **71**, 181.

241 Roberts, A.F. (1981). *Fire Safety J.*, **4**, 197.

242 Roberts, A.F. and Pritchard, D.K. (1982). *J. Occupational Accidents*, **3**, 231.

243 Parker, R.J. *Report of the Court of Enquiry on the Flixborough Disaster* (1975). HMSO, London, UK.

244 Bray, K.N.C. and Moss. J.B. (1981). *First Specialists Meeting (International) of the Combustion Institute*. The Combustion Institute (French Section), Bordeaux, France, p. 7.

245 Moore, S.R. and Weinberg, F.J. (1983). *Proc. Royal Soc. London*, **A385**. 373.

246 Bradley, D. and Mitcheson, A (1978). *Combust. Flame*, **32**, 237.

247 van Wingerden, C.J.M. and Zeeuwen, J.P. (1983). *I. Chem. E. Symposium Ser.*, **82**, 131.

248 Bond, J. (1991). *Sources of Ignition*. Butterworth Heinemann, Oxford, UK.

249 Zhang, D.K., Hills, P.C., Zheng, C., Wall. T.F. and Samson, P. (1992). *Twenty-Fourth Symposium (International) on Combustion*. The Combustion Institute, Pittsburgh, PA, USA, p. 1761.

250 Moussa, N.A., Toong, T.Y. and Garris, C.A. (1977). *Sixteenth Symposium (International) on Combustion*. The Combustion Institute, Pittsburgh, PA, USA, p. 1447.

251 Thomas, P.H., Bullen, M.L., Quintiere, J.G. and McCaffery (1980). *Combust. Flame*, **38**, 159.

252 Lockwood, F.C. amd Malasekara, W.M.G. (1988). *Twenty-Second Symposium (International) on Combustion*. The Combustion Institute, Pittsburgh, PA, USA, p. 1319.

253 Drysdale, D.D. (1985). *An Introduction to Fire Dynamics*. John Wiley, Chichester, UK, Chapter 9.

254 Cullis, C.F. and Hirschler, M.M. (1981). *The Combustion of Organic Polymers*. Clarendon Press, Oxford, UK.

255 Fenimore, C.P. and Martin, F.J. (1966). *Combust. Flame*, **10**, 133.

256 Babrauskas, V. (1984). *Fire and Materials*, **8**, 81.

257 Stuetz, D.E., Di Edwardo, A.H., Zitomer, F. and Barnes, B.P. (1975). *J. Polym. Sci. Polym. Chem. Edn.*, **13**, 585.

258 Burge, S.J. and Tipper, C.F.H. (1969). *Combust. Flame*, **13**, 495.
259 Sheinson, R.S., Eaton, H., Baldwin, S., Maranhides, A. and Smith (1994). *Twenty-Fourth Symposium (International) on Combustion*. The Combustion Institute, Pittsburgh, PA, USA (work in progress poster, Naval Research Labs, Washington, DC, USA).
260 Barnard, J.A. and Bradley, J.N. (1985). *Flame and Combustion* (2nd edn). Chapman and Hall, London, UK.
261 Heywood, J.B. (1988). *Internal Combustion Engine Fundamentals*. McGraw Hill, New York, USA.
262 Stone, R. (1992). *Introduction to Internal Combustion Engines* (2nd edn). Macmillan, London, UK.
263 American Petroleum Institute, New York (1956). *Research Project 45*, 18th Annual Report.
264 Lovell, W.G. (1948). *Ind. Engng Chem.*, **40**, 2388.
265 Hancock, G. (ed.) (1995). Oxidation kinetics and autoignition of hydrocarbons. In *Comprehensive Chemical Kinetics*. Elsevier, Amsterdam, The Netherlands.
266 Mills, G.A. and Ecklund, E.E. (1989). *CHEMTECH*, 549, 626.
267 Iborra, M., Izquierdo, J.F., Tejero, J. and Cunill, F. (1988). *CHEMTECH*, 120.
268 Brocard, J.C., Baronnet, F. and O'Neill, H.E. (1983). *Combust. Flame*, **53**, 25.
269 Downs, D., Griffiths. S.T. and Wheeler, R.W. (1961). *J. Inst. Petrol.*, **47**, 1.
270 Zimpel, C.F. and Graiff, L.B., (1967). *Eleventh Symposium (International) on Combustion*. The Combustion Institute, Pittsburgh, PA, USA, p. 1015.
271 Rao, V.K. and Prasad, C.R. (1972). *Combust. Flame*, **18**, 167.
272 Homer, J.B. and Hurle, I.R. (1972). *Proc. Royal Soc. London*, **A327**, 61.
273 Sawyer, R.F. (1992). *Twenty-Fourth Symposium (International) on Combustion*. The Combustion Institute, Pittsburgh, PA, USA, p. 1423.
274 Merdjani, S. and Sheppard, C.G.W. (1994). *SAE Technical paper*, 932640. Society of Automobile Engineers, Warrendale, PA, USA.
275 Hamidi, A.A. and Swithenbank, J. (1990). *Internal Combustion Engines: Science and Technology* (ed. J. Weaving). Elsevier, London, UK, p. 213.
276 Ham, R.W. and Smith, H.M. (1951). *Ind. Engng Chem.*, **43**, 2788.
277 Al-Rubaie, M.A.R., Griffiths, J.F. and Sheppard C.G.W. (1991). *SAE Technical Paper Series*, 912333, Society of Automobile Engineers, Warrendale, PA, USA.
278 Clothier, P.Q.E., Aguda, B.D., Moise, A. and Pritchard, H.O. (1993). *Chem. Soc. Rev.*, **22**, 101.
279 C. Read (ed.) (1994). *How Vehicle Pollution Affects our Health*. The Ashden Trust, London, UK.
280 Campbell, I.M. (1986). *Energy and the Atmosphere* (2nd edn). John Wiley, New York, USA.
281 Finlayson-Pitts, B.J. and Pitts, J.N. (1986). *Atmospheric Chemistry*. Wiley, New York, USA.
282 Kerr, J.A., Calvert, J.G. and Demerjian, K. (1972). *Chem. Br.*, **8**, 252.
283 Cox. R.A. (1979). *Phil. Trans. Royal Soc.*, **A290**, 543.
284 Houghton. J.T. (1979). *Phil. Trans. Royal Soc.*, **A290**, 515.
285 Prather, M. and Logan, J. (1994). *Twenty-Fifth Symposium (International) on Combustion*. The Combustion Institute, Pittsburgh, PA, USA (in press).
286 Newhall, H.K. (1969). *Twelfth Symposium (International) on Combustion*. The Combustion Institute, Pittsburgh, PA, USA, p. 603.
287 Starkman, E.S. (1969). *Twelfth Symposium (International) on Combustion*. The Combustion Institute, Pittsburgh, PA, USA, p. 598.
288 Suzuki, S. (1986). *SAE Technical paper*, 860408, Society of Automobile Engineers, Warrendale, PA, USA.
289 Heywood, J.B. (1976). *Prog. Energy Combust. Sci.*, **1**, 135.
290 Fenimore, C.P. (1971). *Thirteenth Symposium (International) on Combustion*. The Combustion Institute, Pittsburgh, PA, USA, p. 373.
291 Bowman, C.T. (1992). *Twenty-Fourth Symposium (International) on Combustion*. The Combustion Institute, Pittsburgh, PA, USA, p. 859.
292 Williams, P., Bartle, K.D. and Andrews, G.E. (1986). *Fuel*, **65**, 1153.
293 Nelson, P. (1989). *Fuel*, **68**, 285.
294 Chamberlain, A.C., Heard, M.J., Little, P. and Wiffen, R.D. (1979). *Phil. Trans. Royal Soc.*, **A290**, 577.

295 Sawyer, R.F. (1981). *Eighteenth Symposium (International) on Combustion*. The Combustion Institute, Pittsburgh, PA, USA, p. 859.
296 Taylor, K.C. (1990). *CHEMTECH*, 551.
297 Miller, J.A. and Bowman, C.T. (1989). *Prog. Energy Combust. Sci.*, **15**, 287.
298 HM Inspectorate of Pollution (1993). *Monitoring emissions of pollutants at source, Technical Guidance Notes (Monitoring) M2*. HMSO, London, UK.

Index

(Tabulated data are indicated in italics with *T* preceding the table number)

For Product Safety Concerns and Information please contact our EU representative GPSR@taylorandfrancis.com Taylor & Francis Verlag GmbH, Kaufingerstraße 24, 80331 München, Germany

T - #0066 - 090625 - C0 - 254/178/18 - PB - 9780751401998 - Gloss Lamination